From Quantum Cohomology to Integrable Systems

OXFORD GRADUATE TEXTS IN MATHEMATICS

Books in the series

1. Keith Hannabuss: *An introduction to quantum theory*
2. Reinhold Meise and Dietmar Vogt: *Introduction to functional analysis*
3. James G. Oxley: *Matroid theory*
4. N.J. Hitchin, G.B. Segal, and R.S. Ward: *Integrable systems: twistors, loop groups, and Riemann surfaces*
5. Wulf Rossmann: *Lie groups: An introduction through linear groups*
6. Qing Liu: *Algebraic geometry and arithmetic curves*
7. Martin R. Bridson and Simon M. Salamon (eds): *Invitations to geometry and topology*
8. Shumel Kantorovitz: *Introduction to modern analysis*
9. Terry Lawson: *Topology: A geometric approach*
10. Meinolf Geck: *An introduction to algebraic geometry and algebraic groups*
11. Alastair Fletcher and Vladimir Markovic: *Quasiconformal maps and Teichmüller theory*
12. Dominic Joyce: *Riemannian holonomy groups and calibrated geometry*
13. Fernando Villegas: *Experimental Number Theory*
14. Péter Medvegyev: *Stochastic Integration Theory*
15. Martin A. Guest: *From Quantum Cohomology to Integrable Systems*

From Quantum Cohomology to Integrable Systems

Martin A. Guest

Tokyo Metropolitan University

OXFORD
UNIVERSITY PRESS

OXFORD
UNIVERSITY PRESS

Great Clarendon Street, Oxford, OX2 6DP,
United Kingdom

Oxford University Press is a department of the University of Oxford.
It furthers the University's objective of excellence in research, scholarship, and education by publishing worldwide. Oxford is a registered trade mark of Oxford University Press in the UK and in certain other countries

First published 2008

Published in the United States of America by Oxford University Press
198 Madison Avenue, New York, NY 10016, United States of America

British Library Cataloguing in Publication Data
Data available

ISBN 978–0–19–856599–4

Preface

A glance at the current mathematics research literature reveals the torrent of ideas and results flowing from quantum cohomology, Frobenius manifolds, and mirror symmetry. These are concepts that were mostly unheard of 20 years ago. But the new ideas are not simply 'abstract nonsense'; they are deeply related to virtually all mainstream areas of mathematics and have already provided many new results in, and new connections between, those areas.

While I had always intended to write this book in the style of 'lecture notes', and it did indeed grow from several series of lectures which I have given, the end result is much less ambitious than I had initially hoped, and more philosophical. To some extent this is a consequence of lack of time and perseverance on my part, but it also reflects the novelty and vitality of the subject. Each bit of progress seems to lead in several new directions, all of which provide tempting diversions from the original task.

It is impossible for anyone to write a definitive book in these circumstances. I have settled for this introduction to the subject, emphasizing those aspects which are well established and unlikely to change much; it is just a starting point. On the other hand, this book is not a summary of research articles. It is more elementary and I have provided my own interpretation, which involved rethinking some aspects of the subject. However, the exposition relies very much on traditional mathematics, and traditional notation, and it is designed to be read by traditional mathematicians.

The first chapter of this book gives a very brief introduction to the ideas of algebraic topology for readers who are either not pure mathematicians or who have had little need for cohomology in their own work. The second chapter introduces quantum cohomology as a generalization of cohomology: in both cases, a certain product operation is defined in terms of intersection of cycles. For cohomology, these cycles are just subsets of a given manifold M; for quantum cohomology, they are subsets of a certain space of 'complex curves' in M. Instead of going into the details of the construction of this space, we just give an informal definition followed by some simple but important examples. The third chapter presents, in a similarly informal way, the quantum differential equations. These differential equations lie at the heart of the subject, and this book. They are a system of linear overdetermined partial differential equations. Although they can be defined for cohomology too, they have constant coefficients in that case, and so are of little interest. The non-triviality of the quantum differential equations sets quantum cohomology apart from cohomology, but, much more significantly, the quantum differential equations have a differential geometric interpretation, which leads to the link with the theory of integrable systems.

In Chapter 4, with three chapters of motivation behind us, we review the elementary theory of linear differential equations that will form the foundation of the rest of the book. This is mostly very straightforward, though it uses the language of D-modules, which is a somewhat non-standard topic. The quantum differential equations are most naturally described as a D-module, the quantum D-module, and we regard this language as extremely helpful. Nevertheless, the reader who prefers not to deal with D-modules may replace the word 'D-module' by 'flat connection' without being led too far astray.

Chapter 5 represents the true start of the book. Assuming the definition of quantum cohomology, we define the quantum D-module, and describe some of its key properties. From this point we have no further need of the topological or geometrical definition of quantum cohomology; we shall be concerned only with its associated D-module.

In Chapter 6 we approach the same object from a different direction. Starting from a certain type of D-module, we consider whether or not it has any of the properties of a quantum D-module. This leads to close links between quantum cohomology and integrable systems. To this end, we introduce a construction procedure, based on [62]. The main point of this construction is that it begins from easily recognizable data (a collection of scalar differential equations) and converts it to a D-module having many of the properties of a quantum D-module. In other words, since it is very difficult to recognize a fully fledged quantum D-module directly, we take an indirect approach. This has the advantage that our point of view can accommodate other integrable systems, which may only partially resemble quantum cohomology.

To put this in context, in Chapters 7 and 8 we review some of the famous (infinite-dimensional) integrable systems, concentrating on the KdV equation and the harmonic map equations, where D-modules and flat connections provide a natural framework. After a purely differential equation-theoretic discussion in Chapter 7, we review in Chapter 8 the infinite-dimensional Grassmannian, and how it can be used to produce important families of solutions to those differential equations.

The effectiveness of this method can be explained by the fact that the Grassmannian is a geometrical representation of the underlying D-module, and the geometry of finite-dimensional Grassmannians is a familiar source of intuition. In particular, Schubert cell decompositions arising from lower/upper triangular matrix group factorizations are a surprisingly useful tool. In the infinite-dimensional case these become Birkhoff loop group factorizations. Although this point of view is well known (it is simply a convenient way of handling the Riemann–Hilbert problem), it is not usually discussed in this way at the level of D-modules, where it also gives a decomposition into simpler components.

In the remaining two chapters we return to our main focus, quantum cohomology. First of all, in Chapter 9, we give the standard description of quantum cohomology as an integrable system; this says that the quantum cohomology of a manifold can be regarded as a solution of an integrable system known as the WDVV equations. This integrable system has some (though certainly not all) of the features of the KdV equation or the harmonic map equations. The full extent of these similarities is still to be investigated, but one aspect is already clear: the infinite-dimensional Grassmannian plays an indispensable role. This is because the Grassmannian point of view reveals mirror symmetry, in the sense that the quantum cohomology D-module is represented as an object which resembles a variation of Hodge structure. After a brief review of variations of Hodge structure at the beginning of Chapter 10, we explain this point of view, and show how most of the theory developed in this book contributes to it.

Let us try to summarize all this in a few brief sentences. The main purpose of this book is to explain how quantum cohomology is related to differential geometry and the theory of integrable systems. In concrete terms, the concept of D-module unifies several aspects of quantum cohomology, harmonic maps, and soliton equations like the KdV equation. It does this by providing natural conditions on families of flat connections and their 'extensions', from which these equations are derived. These conditions can be strong enough to determine the equations completely, despite their disparate geometric origins. Our goal is simply to explain this unified way of thinking.

A brief word about the notational conventions in this book is necessary. Whenever different areas of mathematics interact, well-established notation in one area can conflict with equally well-established notation in another. I have decided to tolerate such conflicts when the context can be relied upon to indicate the meaning, rather than make a desperate attempt to be systematic. For example, I allow the differential operators L, P of KdV theory to coexist with other L and P, such as the L in the next paragraph, and the parabolic subgroup P of a generalized flag manifold $G^{\mathbb{C}}/P$. I hope the reader will agree that an uncompromising attitude to such conflicts can do more harm than good.

Another kind of notational problem arises when the literature already contains several well-established names for the same object. I have tried to make sensible choices in such cases. Occasionally, however, I have introduced entirely new notation, in order to emphasize a new point of view. A minor example is the use of h (instead of $\hbar$) for the 'spectral parameter' of quantum cohomology; I use this to remind the reader that I generally have in mind 'abstract quantum cohomology' or even more general situations. A more troublesome matter is the choice of convention for the 'pullback of the Maurer–Cartan form'. I write this as

$$F^{-1}dF,$$

with F as 'general purpose' notation, but usually I use L instead of F when the map is holomorphic in the sense that $dL/d\bar{z} = 0$. Since it is overwhelmingly conventional to write matrix equations in the form $Y' = AY$ with Y a column vector, I introduce $H = F^t$ where H is interpreted as the fundamental solution matrix of such a system, so that $dHH^{-1} = A$. This avoids excessive use of transposes, at the cost of using both H and F.

There are several excellent books and survey articles on quantum cohomology, but these invariably take for granted a particular kind of background: physics, symplectic geometry, algebraic geometry, or singularity theory. Originally I intended to write an unbiased account, but, in fact, the book is heavily biased towards the 'quantum differential equations'. It is therefore not a substitute for other texts which cover quantum cohomology or Frobenius manifolds more systematically, and it should preferably be read in conjunction with such texts. Nevertheless, I have tried to write something which would be readable for people working in various fields.

With this in mind, it may be appropriate to say how I became involved in the subject ten years ago. I had been searching for applications of a result of Graeme Segal, on the approximation of spaces of continuous maps by spaces of rational maps. Despite the simplicity and plausibility of the statement (motivated by Morse theory), there were at least two unsatisfactory aspects: first, the lack of interesting applications; second, the restriction to manifolds which are generalized flag manifolds, toric varieties, or their mild generalizations. When I first heard about quantum cohomology, which evidently involves spaces of rational maps, I wondered whether Segal's theorem should be interpreted as evidence for the special nature of quantum cohomology of manifolds whose rational curves are 'sufficiently flexible'.

Around the same time I gave some lectures on elementary Morse theory. As a source of concrete examples, I paid special attention to Grassmannians and toric manifolds, where the distinguished collection of Morse functions provided by the torus action permits algorithmic calculations of the cohomology algebra as well as the Betti numbers. I was (like others before me) struck by the fact that manifolds with large[1] torus actions seem to acquire their cohomology from a rather small amount of data. At this point I ransacked the library looking for information about what I felt was surely part of a well known theory, but could not find anything. The case of cohomology seemed unlikely to be new (later on I realised that the theory of GKM manifolds addresses this question), and I hoped that quantum cohomology might present a more interesting test case.

With these two problems in the background, I was drawn to Alexander Givental's inspiring articles on 'homological geometry'. These demonstrated that the results of quantum cohomology computations are even more interesting than the fact that they can be carried out (a distinction not always clear in more mature research areas). Even better, the results indicated links with the theory of integrable systems and loop groups, about which I had just finished writing a book. I was hooked and began to read the literature. My activities intensified when I read about Givental's mysterious functions I and J, and I speculated idly (without knowing even the definitions of these functions) that they might be related by some well known integrable systems procedure like 'dressing'. I soon realized that the Birkhoff factorization is responsible for this, and that I had before my eyes an example of the 'DPW procedure' in the theory of harmonic maps, something that I was already very familiar with.

Subsequently, in measuring quantum cohomology against the harmonic map equations and equations of KdV type, I learned a lot about these more familiar integrable systems. For example, it was only after realising that the KdV equation arises through a simple 'D-module extension procedure' (section 4.4) that I began to understand the relation between the various standard approaches (section 8.5). I also became interested in previously shunned topics, such as the significance of coordinate changes in classical differential geometry. As a result, large chunks of the book are devoted to expositions of well known material, viewed retrospectively in the light of quantum cohomology.

[1] 'Large' means the existence of an algebraic torus orbit whose closure contains all critical points of the associated family of Morse functions. Generalized flag manifolds and toric manifolds both have this property.

Like many other people, I was motivated by the work of Edward Witten and Maxim Kontsevich which showed that the KdV equation appears in higher genus quantum cohomology theory, and I hoped to discuss what is behind this. However, it seems fair to say that the experts are not yet in full agreement regarding this point. In any case I found so much to say about the genus zero case that the Witten–Kontsevich theory has been squeezed out and left for another day; it needs an entire book of its own. Nevertheless, the presentation of the 'finite-dimensional' genus zero theory in this book may be helpful in understanding better the 'infinite-dimensional' higher genus theory. I am well aware of other major omissions but any of these topics would have led in very different directions, and I have made more effort to tell a coherent story than to be comprehensive.

Quantum cohomology is rapidly becoming a respectable area of mathematics, but it is still popularly regarded as somewhat obscure. I hope this book explains why the latter should not be so; that quantum cohomology is natural and related to ordinary geometry (though admittedly not in an ordinary way). As with any research topic which is related to several different areas, everyday language used by practitioners in one area may seem utterly mysterious to those in another. I have made an attempt to minimize such difficulties by avoiding specialized technical terms and considering simple cases wherever possible.

For beginners, a further word of warning may be in order. The research literature on quantum cohomology is full of brilliant ideas and promising new directions, but, perhaps unavoidably in such a fast-moving and competitive area bordering on theoretical physics, many authors are casually optimistic in their exposition. Jargon is rife, the distinction between 'is true' and 'should be true' is occasionally blurred, and, in moments of frustration, one might be forgiven for thinking that a paper has been written with the sole purpose of misleading the enemy. Do not be discouraged by this! Calculations are always ahead of theory, and in the quantum cohomology literature the results of spectacular calculations have often been published before a general result can be proved (or even stated), so the literature has acquired a certain messiness. There is a lot to be done, but, with the help of readers like you, it will all work out in the end.

Martin Guest

Tokyo, 2007

Acknowledgements

I am very grateful to Toshitake Kohno, Reiko Miyaoka, and Hiroaki Kanno who invited me to give short courses on quantum cohomology and integrable systems at Tokyo University, Kyushu University, and Nagoya University between 2004 and 2005. Preparing these lectures revealed huge gaps in my knowledge and stimulated me to try to write a book where everything would be explained properly.

As the book progressed, I received valuable advice, encouragement, and support, in many forms, from John Bolton, Josef Dorfmeister, Claus Hertling, Alan Huckleberry, Yoshinobu Kamishima, Ryoichi Kobayashi, Tosiaki Kori, Kee Lam, Yoshiaki Maeda, Seiki Nishikawa, Yoshihiro Ohnita, Mutsuo Oka, Richard Palais, John Parker, and Chuu-Lian Terng. My research during this period was generously supported by grants from the JSPS.

I would like to thank Amartuvshin Amarzaya, Hiroshi Iritani, Liviu Mare, Takashi Otofuji, Hironori Sakai, and Takashi Sakai, for discussions and collaboration.

I am grateful to John Bolton, Ramiro Carrillo, Josef Dorfmeister, Kee Lam, Thomas Reichelt, Wayne Rossman, and Hironori Sakai for pointing out errors in earlier versions of the manuscript.

Finally, I would like to express my deep gratitude to my colleagues at Tokyo Metropolitan University who, as well as providing a very stimulating environment, have gone out of their way to treat me professionally and personally with great kindness.

Contents

Preface v

Acknowledgements xi

Introduction xvii

1. Cohomology and quantum cohomology xvii
2. Differential equations and D-modules xx
3. Integrable systems xxii

1 The many faces of cohomology 1

1.1 Simplicial homology 2
1.2 Simplicial cohomology 3
1.3 Other versions of homology and cohomology 4
1.4 How to think about homology and cohomology 6
1.5 Notation 7
1.6 The symplectic volume function 10

2 Quantum cohomology 12

2.1 3-point Gromov–Witten invariants 12
2.2 The quantum product 16
2.3 Examples of the quantum cohomology algebra 19
2.4 Homological geometry 29

3 Quantum differential equations 33

3.1 The quantum differential equations 33
3.2 Examples of quantum differential equations 39
3.3 Intermission 43

4 Linear differential equations in general 46

4.1 Ordinary differential equations 46
4.2 Partial differential equations 53
4.3 Differential equations with spectral parameter 62
4.4 Flat connections from extensions of D-modules 67
4.5 Appendix: connections in differential geometry 71
4.6 Appendix: self-adjointness 89

5 The quantum D-module 100

5.1 The quantum D-module 100
5.2 The cyclic structure and the J-function 102
5.3 Other properties 106
5.4 Appendix: explicit formula for the J-function 112

6 Abstract quantum cohomology 116

6.1 The Birkhoff factorization 116
6.2 Quantization of an algebra 124
6.3 Digression on D^h-modules 125
6.4 Abstract quantum cohomology 130
6.5 Properties of abstract quantum cohomology 135
6.6 Computations for Fano type examples 138
6.7 Beyond Fano type examples 144
6.8 Towards integrable systems 152

7 Integrable systems 154

7.1 The KdV equation 155
7.2 The mKdV equation 160

7.3 Harmonic maps into Lie groups 164
7.4 Harmonic maps into symmetric spaces 171
7.5 Pluriharmonic maps (and quantum cohomology) 176
7.6 Summary: zero curvature equations 178

8 Solving integrable systems 182

8.1 The Grassmannian model 183
8.2 The fundamental construction 186
8.3 Solving the KdV equation: the Guiding Principle 191
8.4 Solving the KdV equation 197
8.5 Solving the KdV equation: summary 202
8.6 Solving the harmonic map equation 206
8.7 D-module aspects 218
8.8 Appendix: the Birkhoff and Iwasawa decompositions 219

9 Quantum cohomology as an integrable system 223

9.1 Large quantum cohomology 224
9.2 Frobenius manifolds 229
9.3 Homogeneity 236
9.4 Semisimple Frobenius manifolds 239

10 Integrable systems and quantum cohomology 243

10.1 Motivation: variations of Hodge structure (VHS) 244
10.2 Mirror symmetry: an example 255
10.3 *h*-version 265
10.4 Loop group version 270
10.5 Integrable systems of mirror symmetry type 276
10.6 Further developments 287

References 293

Index 303

Introduction

1. Cohomology and quantum cohomology

Although we shall end up in the largely uncharted territory of integrable systems, the natural starting point of this book is cohomology theory and its recent offshoot known as quantum cohomology theory. Algebraic topology in general, and cohomology theory in particular, was one of the truly new subjects of the 20th century. It is firmly based on geometric intuition, yet succeeds in making these intuitive notions very precise, to the extent that monodromy groups, Riemann surfaces, differential forms, and characteristic classes have become part of the everyday language of mathematicians and theoretical physicists. The foundations of the subject can be dry, because the framework has to be developed carefully and there are various choices to be made, in particular a choice of category of spaces and maps. However, over the past 50 years a working procedure has emerged whereby most mathematicians tend to think of homology in terms of submanifolds, and cohomology in terms of differential forms, and apply general principles without worrying too much about the category until it becomes absolutely necessary. This is natural precisely because it is close to the origins of the subject.

The same problems occur with quantum cohomology, although they are more acute because quantum cohomology is at an earlier stage of development. Quantum cohomology emerged from physics in the 1990's and quickly attracted the attention of mathematicians because of its spectacular predictions concerning enumeration problems in classical algebraic geometry. Even where such problems are not of immediate interest (such as in this book), no mathematician can fail to be impressed by the results, as they point to deeper connections beyond topology. Indeed, despite its definition and name, quantum cohomology does not behave like ordinary cohomology at all; it fails to be functorial in any naive sense, and it does not measure any obvious topological property. The mathematical foundations came only after great effort by researchers in algebraic geometry and symplectic geometry—and unfortunately these foundations are a significant barrier to anyone trying to learn the subject.

We present quantum cohomology in Chapter 2 as a natural generalization of ordinary cohomology, with some detailed examples, but without going into the technical foundations. The basic idea is that, while the cohomology of a manifold M involves studying the intersections of cycles in M itself, quantum cohomology involves the intersections of cycles in the space of 'complex curves' in M. This leads to the quantum product, which is a family of multiplications

$$\circ_t : H^*M \times H^*M \to H^*M$$

on the total cohomology group

$$H^*M = \bigoplus_{i=0}^{\dim M} H^iM$$

of M, generalizing the usual cup product. The parameter t will vary in H^2M for most of this book, but occasionally we shall allow it to vary in H^*M, or an even larger vector space. With respect to a basis $b_1, \dots, b_r$ of the complex vector space $H^2M = H^2(M;\mathbb{C})$, we write $t = \sum_{i=1}^r t_i b_i$, and we extend further to a basis $b_0, \dots, b_s$ of $H^*M = H^*(M;\mathbb{C})$ (usually choosing b_0 to be the identity element). Then, the quantum product is specified by the structure constants $c_{ijk} = c_{ijk}(t)$ of $\circ_t$:

$$b_i \circ_t b_j = \sum_{k=0}^{s} c_{ijk}\, b_k.$$

These structure constants are closely related to the 3-point Gromov–Witten invariants, which count those holomorphic maps $f : \mathbb{C}P^1 \to M$ (in each homotopy class) which 'hit' Poincaré dual cycles to the cohomology classes b_i, b_j, b_k. To an algebraic topologist it may seem inauspicious that we start immediately by choosing a basis, but it is convenient to do so because we shall soon be doing calculus and solving differential equations on the vector space H^2M, for which the local coordinates $t_1, \dots, t_r$ are undeniably useful. With this in mind let us introduce the partial derivatives

$$\partial_1 = \frac{\partial}{\partial t_1}, \quad \dots, \quad \partial_r = \frac{\partial}{\partial t_r}.$$

In terms of the new variables $q_i = e^{t_i}$, it turns out (for sufficiently nice manifolds M) that, if the subalgebra of H^*M which is generated multiplicatively by H^2M is written in the form

$$H^\sharp M \cong \mathbb{C}[b_1, \dots, b_r]/(R_1, \dots, R_u),$$

then the corresponding quantum subalgebra can be written in the form[2]

$$QH^\sharp M \cong \mathbb{C}[b_1,\dots,b_r,q_1,\dots,q_r]/(\mathcal{R}_1,\dots,\mathcal{R}_u),$$

where the relations $\mathcal{R}_i$ satisfy $\mathcal{R}_i|_{q=0} = R_i$. The relations are not unique, of course, but natural expressions for them often arise. For example, when $M = \mathbb{C}P^n$, the n-dimensional complex projective space, H^2M is one-dimensional and it generates the entire algebra H^*M, and one has

$$H^*\mathbb{C}P^n \cong \mathbb{C}[b]/(b^{n+1}), \quad QH^*\mathbb{C}P^n \cong \mathbb{C}[b,q]/(b^{n+1}-q).$$

The discussion so far applies when M is a homogeneous space such as $\mathbb{C}P^n$, and more generally when M is a Fano manifold. But quantum cohomology can be defined in much more general situations, and it soon becomes necessary to replace polynomials by more general functions of q, and to address convergence problems (in particular, around the crucial point $q = 0$). However, for the purpose of the current discussion we shall assume that we are in the Fano situation. To simplify the notation, let us assume also that $H^\sharp M = H^*M$ and $QH^\sharp M = QH^*M$.

One of the most intriguing aspects of quantum cohomology is its relation with differential equations. This leads to connections with the theory of integrable systems (Hamiltonian systems, soliton equations, etc.) and mirror symmetry (a duality between certain quantum field theories). The differential equations arise when one regards the underlying vector space of QH^*M as the space of polynomial functions $\Psi : H^2M \to H^*M$, or, more abstractly, as the space of polynomial sections of the trivial vector bundle

$$H^2M \times H^*M \to H^2M, \quad (t,x) \mapsto t.$$

This is because the bundle has a natural connection ∇, given by the quantum product in the following way. Since the bundle is trivial, we can write $\nabla = d+\omega$ where $\omega = \sum_{i=1}^r \omega_i dt_i$ is the local connection form. Then ω_i is defined to be the matrix of the linear transformation 'quantum multiplication by b_i'.

So far we have not used any property of the quantum product; any vector space with a family of products would give a connection in the same way. But for the quantum product it can be proved that the connection is flat, that is, its curvature $d\omega + \omega \wedge \omega$ is zero. This is equivalent to saying that the system of differential equations

$$\nabla_{\partial_i}\Psi = 0, \quad i = 1,\dots,r$$

[2] We use here the standard notation $\mathbb{C}[x_1,x_2,\dots]$ for the algebra of polynomials in the variables $x_1,x_2,\dots$, and $\mathbb{C}[x_1,x_2,\dots]/(F_1,F_2,\dots)$ denotes its quotient by the ideal generated by the polynomials $F_1,F_2,\dots$.

for flat (covariant constant) sections Ψ is consistent, in the sense that it has solutions other than $\Psi = 0$; indeed the solution space has dimension $s+1$, the dimension of H^*M. In fact a stronger result holds: for any (non-zero) complex number h, the connection $d \pm \frac{1}{\mathrm{h}}(\omega)$ is flat. This is equivalent to saying that both $d\omega$ and $\omega \wedge \omega$ are zero.

More simple-mindedly, without mentioning connections, one could introduce the quantum differential equations as the system of first-order linear partial differential equations

$$\mathrm{h}\partial_i \Psi = b_i \circ_t \Psi, \quad i = 1, \ldots, r$$

for maps $\Psi : H^2M \to H^*M$. This system has very special properties, arising from the properties of the quantum product, and as such it represents a very special mathematical object. An underlying theme of the book will be, *What kind of object is it? And how is it distinguished from other systems of partial differential equations?* We shall not answer these questions, but we hope at least to convince the reader that they are interesting and important.

It follows by general principles that there is a scalar system of p.d.e.

$$D_j y = 0, \quad j = 1, \ldots, u$$

which is equivalent to the above matrix system, and that the differential operators D_j produce the relations $\mathcal{R}_j$ by the following two-step procedure:

(1) first replace each occurrence of $\mathrm{h}\partial_i$ by b_i,
(2) then set any remaining occurrences of h equal to zero.

In the case of $\mathbb{C}P^n$, a suitable differential operator is $(\mathrm{h}\partial)^{n+1} - q$. In this case, the 'commutative object' $b^{n+1} - q$ and the 'non-commutative object' $(\mathrm{h}\partial)^{n+1} - q$ are completely equivalent. The differential operator would be obtained by reversing step one of the above procedure, that is, just replace b by $\mathrm{h}\partial$. But this is not typical; in general, the operators obtained from $\mathcal{R}_1, \ldots, \mathcal{R}_u$ by reversing step one do not recover the right answer. To explain this, it is convenient to introduce the concept of D-module.

2. Differential equations and D-modules

Let D denote the ring of all differential operators in the variables $q_1, \ldots, q_r$ with coefficients in some ring of functions $\mathcal{H}$. In the present context $\mathcal{H}$ would be the ring of polynomials in $q_1, \ldots, q_r$, but more usually $\mathcal{H}$ will denote the ring of functions which are holomorphic on some given open set. Let $D_1,\dots,D_u$ be differential operators, and let $(D_1,\dots,D_u)$ be the left ideal of D which they generate, that is, all differential operators of the form $X_1D_1+\cdots+X_rD_r$ where $X_1,\dots,X_r\in D$. Then the quotient

$$D/(D_1,\dots,D_u)$$

is a module over D, that is, a D-module. (It is not a ring, in general, because D is not commutative.) In particular, it is also a module over $\mathcal{H}$, and we shall be interested in free D-modules whose rank over $\mathcal{H}$ is finite.

In Chapter 4 we review the basic differential equations theory which will be needed in this book. A brief summary of some frequently used notation follows.

Let

$$T_1y=0,\ \dots,\ T_uy=0$$

be a system of linear differential equations for the scalar function $y(z_1,\dots,z_r)$, where $T_1,\dots,T_u$ are partial differential operators in $\partial_1=\partial/\partial z_1,\dots,\partial_r=\partial/\partial z_r$. (In the situation of the quantum differential equations we write t_i instead of z_i; for differential equations in general we use z_i.) We assume that the D-module $\mathcal{M}=D/(T_1,\dots,T_u)$ has finite rank $s+1$ over the ring of coefficient functions $\mathcal{H}$. On choosing a basis $[P_0],\dots,[P_s]$, we obtain a first-order $(s+1)\times(s+1)$ matrix system

$$\partial_1Y=A_1Y,\ \dots,\ \partial_rY=A_rY$$

where the vector function $Y(z_1,\dots,z_r)$ is defined by

$$Y=\begin{pmatrix}P_0Y\\ \vdots\\ P_sY\end{pmatrix}.$$

This generalizes the standard construction of a matrix system of o.d.e. from a scalar o.d.e., where one chooses $P_i=\partial^i$. The nature of the matrix functions $A_1,\dots,A_r$ depends on the differential operators $T_1,\dots,T_u$ and the basis $[P_0],\dots,[P_s]$.

To recover a scalar system from a matrix system, one starts with the D-module defined by the connection $d-A$, then chooses a cyclic element of the D-module defined by the dual connection $d+A^t$. The required scalar differential operators are generators of the ideal of operators which annihilate the cyclic element. It is clear that various choices are involved in passing between scalar and matrix systems. In Chapter 4 we explain these choices carefully, with numerous examples.

In Chapter 5 we put the quantum differential equations into this D-module framework. The ‘quantum D-module’ $\mathcal{M}$ which arises from quantum cohomology has the important property that its rank is $s + 1$, the same as the rank of the commutative ring QH^*M. For this reason, the D-module $\mathcal{M}_{\text{naive}}$ obtained from QH^*M by 'replacing b_i by $\hbar\partial_i$' will not in general be the quantum D-module: the rank of $\mathcal{M}_{\text{naive}}$ will in general be less than the rank of QH^*M.

For $QH^*\mathbb{C}P^n$ we have $\mathcal{M} = \mathcal{M}_{\text{naive}}$, but this is a very special case. Indeed, for any ordinary differential operator T of order n the D-module $D/(T)$ has rank n (this case also illustrates the fact that there are many D-modules with the correct rank, as any operator of the form $T + O(\hbar)$ will give the same result under the two step procedure described earlier).

This phenomenon leads one to ask more general questions, independent of quantum cohomology theory, concerning the 'matching' of non-commutative D-modules with commutative algebras. We take this point of view in Chapter 6.

3. Integrable systems

All this leads directly into the theory of integrable systems, because a free D-module of finite rank (over a field of functions) is essentially the same thing as a flat connection. For a D-module of specific type, the condition that its rank is n is equivalent to the condition that a specific connection is flat. This condition will be equivalent to a (usually non-linear) partial differential equation, and it is common practice to say that partial differential equations which can be written as zero curvature conditions are *integrable systems* or *integrable p.d.e.* We prefer the latter term, as 'integrable system' should have a more restricted meaning (though exactly what this meaning should be is still a matter of debate). A key example of this type is the KdV equation

$$u_t = uu_x + u_{xxx}.$$

This has a well known zero curvature representation. The D-module point of view (which seems much less well known) says that the zero curvature representation is equivalent to the condition that a certain D-module has rank 2—or that it 'matches' an appropriate commutative object.

The main theme of this book is that the quantum D-module, and more generally the idea of 'matching' a D-module with a commutative algebra, suggests a general scheme for constructing integrable systems. Optimistically, this could contribute to a more precise definition of the term 'integrable system'. Even more optimistically it could lead to a characterization of the quantum D-module of a manifold and a more efficient way of handling quantum cohomology, in the same way that de Rham cohomology has become a more efficient way of handling simplicial or singular cohomology (although it has to be said that the subject is still a long way from this point). To some extent, this justifies the lack of technical foundational material in the first three chapters of the book, as they may be regarded purely as motivation for the D-module approach which begins in Chapter 4.

The plan for the rest of the book is to examine three important examples (quantum cohomology, harmonic maps, and the KdV equation) from the D-module point of view. Although we have made a start on quantum cohomology in Chapters 5 and 6, its relation with integrable systems will not really be apparent until we have reviewed the more familiar cases of harmonic maps and the KdV equation in Chapters 7 and 8. Thus, we return to quantum cohomology in Chapters 9 and 10, by which time we are in a position to bring the various pieces together.

For practical purposes we use a geometrical manifestation of these D-modules, provided by the infinite-dimensional Grassmannian manifolds of Sato and Segal–Wilson. Our conceptual emphasis on D-modules is close in spirit to the work of the Sato school, while for computations we generally use the loop group methods of Segal–Wilson. The ubiquity of this Grassmannian as a computational tool is striking.

We devote a significant amount of space to a survey of the KdV equation and related integrable systems. This theory is very well known, and there are many references available, but none of these references were entirely suitable for our purposes. For some people, the Lax form of the KdV equation is all there is, and all results follow from that by computation; but this does not explain where the Lax equation comes from. At the opposite extreme, the Grassmannian model epitomizes the abstract point of view. The KdV equation is a kind of infinite-dimensional Plücker equation and its solutions correspond to points on an infinite-dimensional Grassmannian. This is very nice but in some sense 'too clever by half'. The gritty differential equation has disappeared completely and there is hardly anything left. Our exposition is a compromise between these two extremes. The abstract approach is undeniably accurate, as the KdV equation arises from very little input, basically just the positive integer 2. Our point of view is that the KdV equation is the simplest non-trivial extension of a general o.d.e. of order 2. However, this leads quickly to gritty formulae, and the compromise arises from how far one allows oneself to be led in this direction.

D-modules suggest another compromise, which we believe is helpful. On the one hand, the concept of Lax equation is too special; on the other hand, the concept of integrable p.d.e. (differential equation admitting a zero curvature representation) is too broad. It is possible to bridge this gap by generalizing the definition of Lax equation, or by considering only connection matrices of certain shapes. D-modules (together with choices of additional data) provide a natural way of doing this, as we shall eventually see.

Let us explain briefly how the Grassmannian $\mathrm{Gr}^{(s+1)}$ arises. It is a Grassmannian of $\frac{\infty}{2}$-dimensional linear subspaces of a Hilbert space $H^{(s+1)}$; $\frac{\infty}{2}$-dimensional means commensurate with a fixed linear subspace $H_+^{(s+1)}$ (this can be made precise). If $y_{(0)}, \dots, y_{(s)}$ is a basis of solutions of a scalar system as above, and $Y_{(0)}, \dots, Y_{(s)}$ is a basis of solutions of the matrix system, then a fundamental solution matrix H of the latter may be written in two ways:

$$H = \begin{pmatrix} | & & | \\ Y_{(0)} & \cdots & Y_{(s)} \\ | & & | \end{pmatrix} = \begin{pmatrix} - & P_0 J & - \\ & \vdots & \\ - & P_s J & - \end{pmatrix}.$$

Here $J = (y_{(0)}, \dots, y_{(s)})$, and $P_0, \dots, P_s$ are differential operators such that $[P_0], \dots, [P_s]$ is a basis of the D-module. We usually choose $P_0 = 1$, so that $P_0 J = J$. In using the notation J we follow Givental, whose 'J-function' plays an important role in quantum cohomology.

When the differential equations contain a spectral parameter λ (in the case of quantum cohomology, $\lambda = \mathrm{h}$), all these functions depend on λ, and we may regard H as a map taking values in the loop group $\Lambda \mathrm{GL}_{s+1}\mathbb{C}$. The map

$$W = FH_+^{(s+1)} = \begin{pmatrix} | & & | \\ P_0 J & \cdots & P_s J \\ | & & | \end{pmatrix} H_+^{(s+1)}, \quad \text{where } F = H^t$$

is a Grassmannian-valued map which contains the essential features of the original system. This is standard for the KdV equation, but we regard it as fundamental for any D-module, because the map

$$[X] \mapsto XJ$$

is an isomorphism between the original D-module and the space of sections of the pullback by W of the tautologous bundle on the Grassmannian. In other words, the original D-module is represented geometrically by W.

We shall discuss in detail three examples: the KdV equation, harmonic maps, and quantum cohomology. Although they have quite different origins and features, they are linked in various intriguing ways, and it is our contention that this is best understood from the point of view of D-modules and their Grassmannian representations. The Grassmannian for the KdV and harmonic map equations appears in Chapter 8, and for quantum cohomology in Chapter 10.

Before giving brief summaries of these three examples, we must explain the main point of the Grassmannian approach. Fundamentally, it rests on a simple geometrical idea, the Schubert cell decomposition. The finite-dimensional Grassmannian $\mathrm{Gr}_k(\mathbb{C}^n)$ of k-planes in n-space has an open dense 'big cell' consisting of all k-planes transverse to a fixed $(n-k)$-plane $\mathbb{C}^{n-k}$ (with smaller cells specified by how a k-plane intersects various other $\mathbb{C}^i$). These cells can be described as the orbits of a certain group Δ_- of invertible lower-triangular matrices, which acts naturally on $\mathrm{Gr}_k(\mathbb{C}^n)$ (as does the group $\mathrm{GL}_n\mathbb{C}$ of all invertible matrices). In particular, if we represent a k-plane as $X \cdot \mathbb{C}^k$ for some $X \in \mathrm{GL}_n\mathbb{C}$, this lies in the big cell if and only if X admits a factorization $X = X_-X_+$ where $X_- \in \Delta_-$, $X_+ \in \Delta_+$, and where Δ_+ is a certain 'complementary' group which fixes $\mathbb{C}^{n-k}$ (essentially, the upper-triangular matrices). The big cell provides canonical local coordinates for the Grassmannian, and if $z \mapsto X(z) \cdot \mathbb{C}^k$ is a map into the Grassmannian then $z \mapsto X_-(z) \cdot \mathbb{C}^k$ provides a canonical local representation of the map.

This picture carries over to the infinite-dimensional Grassmannian $\mathrm{Gr}^{(s+1)}$, which is a homogeneous space of the loop group $\Lambda\mathrm{GL}_{s+1}\mathbb{C}$. Our map $W = FH_+^{(s+1)}$ can be represented as

$$W = F_-H_+^{(s+1)}$$

where

$$F = F_-F_+$$

is a factorization generalizing the lower-triangular/upper-triangular factorization of finite-dimensional matrices. This factorization, known as the *Birkhoff factorization,* says that 'almost every' loop $\gamma \in \Lambda\mathrm{GL}_{s+1}\mathbb{C}$ may be factorized in the form

$$\gamma(\lambda) = \underbrace{(a_0 + \tfrac{1}{\lambda}a_1 + \tfrac{1}{\lambda^2}a_2 + \cdots)}_{\gamma_-(\lambda)}\underbrace{(b_0 + \lambda b_1 + \lambda^2 b_2 + \cdots)}_{\gamma_+(\lambda)}.$$

It is a well known but remarkable fact, and one which appears throughout the book, that the canonical representation of W is the key to solving the original integrable system. More precisely, if we denote by

$$F = \check{F}_+\check{F}_-$$

the analogous factorization with λ and λ^{-1} reversed, it is easy to see that F is determined uniquely by F_- and $\check{F}_+$ and it turns out (at least, in our three examples) that *the correspondence*

$$F \longleftrightarrow F_-, \check{F}_+$$

decomposes F into simpler ingredients. These ingredients contain the 'data' for construction of solutions to the integrable system. At a deeper level, the Birkhoff decomposition reflects a decomposition of the underlying D-module into simpler D-modules.

This idea is at the heart of several tricks and techniques in the theory of integrable systems, notably the method of inverse scattering for soliton equations and the Weierstrass representation and its generalizations in differential geometry. As a refinement of this method one could consider whether the map into the Grassmannian lies in a smaller cell (than the big cell), or whether it satisfies other natural conditions of this type, which corresponds to imposing additional conditions on the solution of the original equation.

The KdV and similar equations

The usual Lax form $L_t = [P, L]$ plays no role in our derivation of the KdV equation in Chapter 4; what appears naturally there is an equation of the form

$$L_t = [P, L] \mod L,$$

and many other integrable p.d.e. can be constructed in the same way. However, the fact that P can be chosen so that $L_t = [P, L]$ is a very special feature of the KdV equation, and it leads to very special formulae for solutions. We present this in Chapter 8 as an extreme case, which should be compared and contrasted with, for example, quantum cohomology.

The result can be stated very simply as follows: an important class of solutions is given by

$$F_- = \tilde{\gamma}, \quad \check{F}_+ = \tilde{E},$$

where $\tilde{\gamma}$ is a constant loop and $\tilde{E}$ is the exponential of a *linear* function:

$$\tilde{E} = \exp\ x\begin{pmatrix} 0 & \lambda \\ 0 & 0 \end{pmatrix} + t\begin{pmatrix} 0 & \lambda^2 \\ \lambda & 0 \end{pmatrix}$$

(the reason for the tildes is that, in Chapter 8, we shall actually use γ, E where $\tilde{\gamma} = \gamma^{t-1}$, $\tilde{E} = E^t$).

This is an extraordinarily simple and elegant formula for the solution of the original problem. Or rather, it would be, if the solution u of the KdV equation could be extracted easily. In Chapter 8 we shall explain how u can be obtained from W, but perhaps the most important thing to say here is that W reveals the underlying linearity of solutions to the KdV equation, which would not be apparent from an explicit formula for u. Since $\check{F}_+$ is always the same, solutions correspond to points $W = F_- H_+^{(2)} = \tilde{\gamma} H_+^{(2)}$ of the Grassmannian. These can be thought of as initial values for solutions of the evolution equation, and the actual solutions can be represented as 'linear flows' on the Grassmannian. This attractive geometrical picture is very well known and has stimulated much research.

The result itself can be derived by various methods, but the links between these methods are somewhat difficult to find in the literature. The D-module version of the KdV equation unifies these methods in a way that is conveniently summarized in the following diagram:

$$\begin{array}{ccc} D_{x,t} \otimes \mathcal{H}_\lambda^+ \,/\, (L-\lambda, \partial_t - P) & \xrightarrow{[X]\mapsto XJ} & \Gamma W \\ \downarrow & & \downarrow \\ \mathcal{J} = D_x K & \xrightarrow{XK\mapsto XKe^{x\mu+t\mu^3}} & \Gamma W^{\mathrm{scalar}} \end{array}$$

We shall explain in detail the notation used here in Chapter 8. Briefly, the top left hand corner is the D-module of rank 2 corresponding to a solution of the KdV equation (the spectral parameter is λ and the usual Lax form of the KdV equation is $L_t = [P, L]$), and the top arrow is the geometrical representation of this D-module mentioned above. The bottom left hand corner is the Sato D-module, and the bottom arrow gives the analogous geometrical representation of this, which is the 'scalar Grassmannian' used by Segal and Wilson.

Harmonic maps

Harmonic maps from a domain in $\mathbb{C}$ to a (compact or non-compact) symmetric space G/K can be defined as critical points of an 'energy' functional. The Euler-Lagrange equations are the harmonic map equations. More generally, one can consider pluriharmonic maps from a domain in $\mathbb{C}^r$ to G/K, although we shall focus on the case $r = 1$ here.

The harmonic map equations have a zero curvature formulation with spectral parameter, through which harmonic maps correspond to 'extended harmonic maps'. An extended harmonic map F takes values in a loop group, and there is a corresponding map W which takes values in an infinite-dimensional Grassmannian. Harmonic maps appear in a wide variety of situations and it is hopeless to expect a simple 'KdV-like' formula for

W. Nevertheless, the Birkhoff factorization simplifies the problem significantly, because F_- turns out to be a solution of an o.d.e. in the complex variable z, whereas F is a solution of a p.d.e. in the real variables x, y. Moreover, F_- is equivalent to $\check{F}_+$ here, so the holomorphic map F_- alone determines F. This is the origin of the Weierstrass representation of a minimal surface in terms of purely holomorphic data, which was generalized to the case of surfaces of constant mean curvature by Dorfmeister, Pedit, and Wu.

In general the relation between the harmonic map and its holomorphic data can be complicated, but there are two kinds of harmonic map where W describes all solutions efficiently: harmonic maps of finite type and harmonic maps of finite uniton number.

Harmonic maps of finite type are constructed in an analogous way to solutions of the KdV equation, and the formulae for W are similar. These are very special harmonic maps but they include important examples in differential geometry such as the Gauss maps of constant mean curvature tori.

Harmonic maps of finite uniton number are also very special, but they include all harmonic maps from two-spheres to compact Lie groups and symmetric spaces, in particular 'isotropic' harmonic maps and all harmonic maps obtained by the 'twistor construction'. In this case W is certainly not linear in any sense. However, the geometry of the Grassmannian plays an equally crucial role, as the meaning of 'finite uniton number' is that each solution is contained in a finite-dimensional Schubert subvariety, and the local coordinates given by the big cell of this subvariety permit a straightforward integration of the harmonic map equations in this case.

Quantum cohomology

The quantum D-module (or flat connection) which underlies the quantum cohomology algebra QH^*M can be regarded as a particular solution of an integrable system. However, in contrast to the KdV equation and the harmonic map equations, we have discussed the theory of quantum cohomology so far without mentioning any integrable system—and in fact there are various candidates, depending on one's point of view.

One candidate is given by the WDVV equations. The properties of quantum cohomology lead to the existence of a 'Gromov–Witten potential function'

$$\mathcal{F}^M : H^*M \to \mathbb{C}$$

which is related to the quantum product and the intersection form by

$$(b_i \circ_t b_j, b_k) = \partial_i \partial_j \partial_k \mathcal{F}(t).$$

(Here the variable t is in H^*M rather than H^2M, and the corresponding 'large' quantum product restricts to the earlier 'small' quantum product when $t \in H^2M$.) This observation can be reversed: for *any* function $\mathcal{F} : H^*M \to \mathbb{C}$, a corresponding product operation $*_t$ can be defined by

$$(b_i *_t b_j, b_k) = \partial_i \partial_j \partial_k \mathcal{F}(t).$$

It is obvious that $*_t$ is commutative and satisfies the 'Frobenius property'

$$(b_i *_t b_j, b_k) = (b_j, b_i *_t b_k).$$

In contrast, the associativity condition

$$(b_i *_t b_j) *_t b_k = b_i *_t (b_j *_t b_k)$$

is not obvious; it is a system of third-order non-linear partial differential equations for $\mathcal{F}$. These equations are the WDVV equations.

It turns out that the WDVV equations can be written as zero curvature equations, and the flat connection corresponding to the quantum cohomology of M gives a specific solution. General solutions to the WDVV equations correspond to Frobenius manifolds (the manifold being the domain of the function F, i.e. just a vector space in the above situation). In Chapter 9 we give some details of this theory of Frobenius manifolds.

Another point of view would be to regard the quantum cohomology of M as a particular example of a harmonic (or pluriharmonic) map. This is possible because the connection corresponding to the quantum cohomology of M is of exactly the same form as that corresponding to a harmonic map. As in the case of the WDVV equations, the quantum cohomology of each manifold M provides a solution of the harmonic map equations with very special properties. In Chapter 10 we discuss some of these properties. The most important is that the corresponding map

$$W = FH_+^{(s+1)}$$

has a deep and independent meaning: it is a variation of Hodge structures for a 'mirror partner' of M. Thus, this point of view reveals the phenomenon of mirror symmetry, and truly justifies the use of the infinite-dimensional Grassmannian in quantum cohomology. More generally, and again in analogy with the WDVV equations, it is natural to consider mathematical objects corresponding to other harmonic maps of this type (not necessarily the harmonic maps arising from quantum cohomology). This leads, for example, to the theory of tt^* structures, and encourages one to contemplate integrable systems under the umbrella of 'differential geometric mirror symmetry'.

1 The many faces of cohomology

This chapter is a brief essay on homology and cohomology, emphasizing those aspects which will reappear when we study quantum cohomology. The chapter can be skipped by the knowledgeable reader, although a brief glance at the notation in section 1.5 should be helpful, as we shall use it systematically throughout the book. In addition, section 1.6 mentions a less well known 'function-theoretic' point of view which contains a hint of our eventual approach to quantum cohomology.

The basic process of algebraic topology, and one of the great mathematical ideas of the 20th century, can be described as

$$\text{geometric object} \longrightarrow \text{algebraic object}$$

where, for example, the 'geometric object' could be a surface in three-dimensional space and the 'algebraic object' could be a number or a collection of numbers. More abstractly, the geometric object could be a manifold and the algebraic object a group.

One of the most famous examples is the Euler characteristic. If the geometric object M is a polyhedral surface constructed from v vertices, e edges and f faces, then its Euler characteristic $\chi(M)$ is the (positive or negative) integer $\chi(M) = v - e + f$. It can be proved that this number is the same even if the surface is subdivided in a different way (unlike, say, $v - 2e + 5f$, which does not have this property). Furthermore, it can be proved that this number is the same even if the surface is deformed continuously 'without tearing', for example, if a two-dimensional polyhedral cylinder is squashed into a one-dimensional polyhedral ring.

Another famous example is the fundamental group. To any topological space M one can associate a group denoted $\pi_1(M)$. It is a well known theorem that if two such topological spaces are homeomorphic (or homotopy equivalent), then their fundamental groups are isomorphic.

The rationale for studying such topological invariants is this: it is important in many parts of mathematics to distinguish between geometric objects (ideally, one attempts to classify all geometric objects of a certain type); on the other hand, algebraic objects are easier to classify than geometric objects; therefore, a topological invariant gives a way of converting a difficult geometric problem to an easier algebraic problem.

For example, in the case of surfaces, the sphere and the torus are distinguished by the Euler characteristic in the sense that $\chi = 2$ for the sphere but $\chi = 0$ for the torus. This example illustrates the main idea of 'geometric' algebraic topology very simply but it also illustrates its unavoidable awkwardness: any construction which is reasonably understandable seems to need a lot of additional (but ultimately irrelevant) data, such as the choice of subdivision into vertices, edges, and faces in the above example.

1.1 Simplicial homology

Let us illustrate the above remarks in the case of simplicial homology theory. This is a family of topological invariants which is directly related to the Euler characteristic described earlier; the point is that, underlying the number $\chi(M)$, there is a more sophisticated object, the homology group H_*M. The definition of the group H_*M involves three steps.

Step 1: We assume that the geometric object M is a particular kind of topological space called a simplicial complex of dimension m. The definition of simplicial complex can be found in books on algebraic topology; it formalizes the idea of dividing into vertices, edges, and faces. The basic unit is an i-dimensional simplex, or i-simplex (a vertex would be a 0-simplex, an edge would be a 1-simplex, and so on). Roughly speaking, a simplicial complex is an object formed when simplices of various dimensions are attached together in a certain specified way.

Step 2: For each i, the i-th chain groups C_iM are constructed. An element of C_iM is simply a linear combination $\sum_j n_jA_j$ where the n_j are integers and the A_j are i-simplices of M. This is just an abstract concept but one should have in mind the set-theoretic union, where A_j is counted n_j times (although this is not quite as simple as it seems, since n_j could be negative).

Step 3: The i-th homology groups H_iM are constructed. By definition, H_iM is the quotient group of (a) the subgroup of C_iM consisting of linear combinations with zero boundary (these are called 'cycles') by (b) the subgroup of C_iM consisting of elements which are boundaries of elements of $C_{i+1}M$ (these are called 'boundaries').

The rank of the group H_iM (called the i-th Betti number of M, and denoted by b_i) is, roughly speaking, the number of $(i+1)$-dimensional 'holes' in M.

A more accurate description of b_i is the number of (essentially different) i-dimensional cycles which are not boundaries of $(i+1)$-dimensional chains.

The total homology group H_*M is defined to be the group-theoretic direct sum

$$H_*M = H_0M \oplus H_1M \oplus \cdots \oplus H_mM$$

of the individual homology groups. It turns out that the homology groups H_iM are topological invariants in the same sense as the Euler characteristic (although the chain groups C_iM are not). Moreover, they are related to the Euler characteristic by the formula $\chi(M) = b_0 - b_1 + b_2 - \cdots$. In fact, there is another formula $\chi(M) = c_0 - c_1 + c_2 - \cdots$, where c_i is the rank of the group C_iM. From the definition of C_iM, c_i is simply the number of i-simplices in M, so this formula directly generalizes the earlier naive definition of the Euler characteristic.

To prove all these statements involves quite a lot of work, and establishing this 'simplicial homology theory' was an early triumph of algebraic topology. Nevertheless, the theory is rather intuitive, and (at least for simple examples) it is easy to work with. For example, the two-dimensional torus has Betti numbers given by $b_0 = 1$, $b_1 = 2$, $b_2 = 1$, and $b_i = 0$ for $i \geq 3$. This can be interpreted as follows. The torus has essentially one two-dimensional cycle, namely the torus itself, and two essentially different kinds of one-dimensional cycles, for example, $S^1 \times 1$ and $1 \times S^1$ if the torus is regarded as $S^1 \times S^1$. There is only one kind of zero-dimensional cycle, namely a point.

1.2 Simplicial cohomology

We have seen how the Euler characteristic can be enhanced to a more sophisticated invariant, the homology groups. There is no reason to stop here, as more refined algebraic objects give stronger invariants. The next step is to introduce a product structure, which corresponds geometrically to intersection of cycles. Roughly speaking, we obtain a map

$$H_{m-i}M \times H_{m-j}M \to H_{m-(i+j)}M, \quad (A, B) \mapsto A \cap B$$

where A, B are, respectively, cycles of dimension $m-i, m-j$. More precisely, A, B are equivalence classes of cycles, since homology groups are quotients of cycle groups, so it has to be verified that $A_1 \cap B_1$ is equivalent to $A_2 \cap B_2$ if A_1 is equivalent to A_2 and B_1 is equivalent to B_2.

However, there is a more obvious problem, since there is no guarantee that the intersection $A \cap B$ will have dimension $m-(i+j)$, and in fact it is easy to produce counterexamples. Resolving this problem leads to even more technical difficulties. Nevertheless, the intuition behind the above definition is absolutely correct, as the codimension of $A \cap B$ will be equal to the sum of the codimensions of A and B *providing* that A and B intersect 'transversally'. This is the reason for writing the dimensions of A, B as $m-i, m-j$; i, j are the codimensions of A, B and the codimension of the transversal intersection $A \cap B$ is $i+j$.

It is easier from the technical point of view to use 'dual' homology groups, or cohomology groups, denoted H^iM. Therefore, in the usual approach, one introduces a product

$$H^iM \times H^jM \to H^{i+j}M$$

in simplicial cohomology theory, and (after a considerable amount of work) one ends up with an algebraic invariant

$$H^*M = H^0M \oplus H^1M \oplus \cdots \oplus H^mM$$

which is a ring or an algebra (not just a group).

1.3 Other versions of homology and cohomology

Despite the technical difficulties of setting up the rigorous theory of simplicial cohomology, the ideas are sound, and other ways have been found to implement them. One of the standard approaches is singular (homology or) cohomology. Instead of constructing a topological space directly from simplices and then considering the structure of all i-simplices in that space, one starts with any topological space M and considers all possible continuous maps of a fixed i-simplex into M. Such a map is regarded as a singular i-simplex (the image may not look like an i-simplex at all; if the map is constant, then its image is just a single point). This situation is somewhat harder to think about, since the number of such singular simplices is uncountably infinite, but the basic constructions of the simplicial theory can still be carried out, and many of them become easier. For example, it is easy to show that a continuous map between topological spaces induces a homomorphism of chain groups, whereas in the simplicial theory a messy argument (the simplicial approximation theorem) is needed in order to convert the continuous map to an approximating map which takes simplices to simplices.

There are other, quite different approaches, for example, Cech cohomology and de Rham cohomology. These are convenient in their own ways, but less intuitive than the simplicial theory. In the case of de Rham cohomology, one starts with a differentiable manifold M, rather than a general topological space, and then defines a k-cochain to be a (smooth) differential form of degree k, that is, something which can be written using a local coordinate system $x_1, \ldots, x_m$ in the form

$$\sum_{i_1,\ldots,i_k} f_{i_1,\ldots,i_k}(x_1,\ldots,x_m)dx_{i_1} \wedge \cdots \wedge dx_{i_k}.$$

Instead of the boundary operator (which assigns to a simplicial chain its boundary chain), one has the exterior derivative operator. A differential form with exterior derivative zero is called a cocycle, and any differential form which is equal to the exterior derivative of another differential form is called a coboundary. The i-th de Rham cohomology group is defined to be the quotient of the group of i-cocycles by the group of i-coboundaries. It is denoted by exactly the same notation as before, H^iM.

To be more precise, we should write $H^i(M;\mathbb{R})$ to indicate that the 'scalars' are the real numbers. In the case of simplicial cohomology the scalars were the integers, so we should have written $H^i(M;\mathbb{Z})$ there (but $H^i(M;\mathbb{R})$ can be defined in simplicial cohomology theory in a similar way).

The reason for using the same notation is that the simplicial cohomology groups turn out to be isomorphic to the de Rham cohomology groups, at least if the manifold M is compact (any compact manifold turns out to be homeomorphic to a simplicial complex, so the simplicial cohomology groups of M make sense). This is a remarkable fact, and its proof is non-trivial. However, there is a simple and fundamental connection between the two theories, because a de Rham i-cocycle can be *integrated* over a simplicial i-chain, so a differential form can be regarded as a function on the space of cycles, and this is exactly what a cocycle should be. More precisely, there is a natural 'pairing'

$${}_{\text{de Rham}}H^i(M;\mathbb{R}) \times {}_{\text{simplicial}}H_i(M;\mathbb{R}) \to \mathbb{R}$$

given by integration. Over a cycle which is the boundary of another cycle, the integral of any differential form is zero, by Stokes' Theorem. Conversely, it is possible to show that any cycle over which the integrals of all differential forms are zero must be a boundary. Hence the pairing is non-degenerate, and it gives rise to an isomorphism between the vector space ${}_{\text{de Rham}}H^i(M;\mathbb{R})$ and the dual of the vector space ${}_{\text{simplicial}}H_i(M;\mathbb{R})$. The latter is isomorphic to ${}_{\text{simplicial}}H^i(M;\mathbb{R})$, so we obtain the desired isomorphism.

Even more remarkably, all other sensible definitions of homology and cohomology theory give the same answers. This indicates that the concept of (co)homology is more general than the original geometrical definition suggests, and it suggests an axiomatic approach to the subject. It can be shown that any (co)homology theory satisfying the famous Eilenberg–Steenrod axioms leads to isomorphic groups. Thus, in order to show that simplicial and de Rham cohomology (or any other of the standard theories) give the same answers it suffices to show that both theories satisfy the Eilenberg–Steenrod axioms.

1.4 How to think about homology and cohomology

Relying on the foundations of algebraic topology, most mathematicians nowadays think of homology and cohomology in a rather general way, without specifying the precise theory that they have in mind. This causes no difficulties so long as the spaces involved belong to certain well behaved types (such as compact manifolds, or simplicial complexes, or CW-complexes) to which the Eilenberg–Steenrod axioms apply. For example, one says that the first cohomology group of the torus is $\mathbb{Z}^2$ (or, with real coefficients, $\mathbb{R}^2$), without really thinking about simplices or differential forms. This flexibility is very helpful, because one has the freedom to revert to a specific theory when working on a specific problem.

For example, the product structure of de Rham cohomology theory is very easy to describe: it is given by the exterior product of differential forms. It is easier to work with the exterior product $\alpha \wedge \beta$ of differential forms on M than to work with the intersection $A \cap B$ of cycles in M, because one does not have to worry about specifying that the cycles are transversal. If M is a compact orientable manifold of dimension m, the Poincaré Duality Theorem says that

$$H^i M \cong H_{m-i} M,$$

a fact which is not at all clear from the point of view of simplicial theory. From the point of view of de Rham theory, the theorem says that the natural pairing

$$H^i M \times H^{m-i} M \to \mathbb{R}, \quad ([\alpha], [\beta]) \mapsto \int_M \alpha \wedge \beta$$

is non-degenerate, which is (also not obvious but) more plausible than the ingenious explanations of Poincaré duality which can be found in textbooks on simplicial theory.

Another example is the way that one thinks about cycles in a manifold M as submanifolds (or, at least, submanifolds with singular points). To be completely rigorous one should specify a theory such as simplicial theory, then regard the submanifold as approximated by small simplices, and one should also keep in mind that a homology class is an equivalence class of such objects. But it is convenient to think of the one-dimensional homology group $\mathbb{Z}$ of the cylinder $S^1 \times [0,1]$ as generated by the 1-cycle represented by the submanifold $S^1 \times \frac{1}{2}$, for example. Similarly, the circles $S^1 \times 1$ and $1 \times S^1$ represent generators of the one-dimensional homology group $\mathbb{Z}^2$ of the torus, and the product of these homology classes is represented by the intersection of these two circles, namely a single point, which in turn represents a generator of the zero-dimensional homology group $\mathbb{Z}$.

As a final example, the $2i$-th homology group of the $2n$-dimensional complex projective space $\mathbb{C}P^n$ is isomorphic to $\mathbb{Z}$ for $0 \le i \le n$. It is generated by the $2i$-cycle represented by the submanifold $\mathbb{C}P^i$ of $\mathbb{C}P^n$. In the case of complex algebraic varieties this way of thinking can be made into a rigorous theory (see the next section): every cohomology class can be represented by an algebraic cycle, and all basic calculations can be carried out directly with such cycles.

1.5 Notation

In this section we summarize the notation for cohomology theory which will be used throughout this book. In view of the previous remarks we shall talk about homology classes as though they are geometrical objects like submanifolds, and we shall use the language of differential forms whenever it is helpful in describing cohomology classes and their properties.

For most of the book we shall be concerned with complex manifolds (because quantum cohomology involves holomorphic maps), and with even-dimensional cohomology groups (because the product of such classes is commutative). Therefore, unless stated otherwise, the following assumption will be in force:

Assumption M is *a connected simply connected* compact Kähler manifold, of complex dimension n. The integral cohomology groups of M are even-dimensional and torsion-free, that is,

$$H^*(M;\mathbb{Z}) = \bigoplus_{i=0}^{n} H^{2i}(M;\mathbb{Z}), \quad H^{2i}(M;\mathbb{Z}) \cong \underbrace{\mathbb{Z} \oplus \cdots \oplus \mathbb{Z}}_{m_{2i}}$$

for some non-negative integers m_{2i}.

Let

$$\mathrm{PD} : H^i(M;\mathbb{Z}) \to H_{2n-i}(M;\mathbb{Z})$$

be the Poincaré duality isomorphism. Generally we shall use lower-case letters

$$a, b, c, \ldots \in H^*(M; \mathbb{Z})$$

for cohomology classes, and $|a|, |b|, |c|, \ldots$ for their degrees (dimensions). We shall use upper-case letters

$$A = \mathrm{PD}(a),\ B = \mathrm{PD}(b),\ C = \mathrm{PD}(c),\ \ \ldots\ \in H_*(M; \mathbb{Z})$$

for the Poincaré dual homology classes, and $|A|, |B|, |C|, \ldots$ for their degrees. As explained earlier, we shall often speak of $A, B, C, \ldots$ as though they are cycles (rather than equivalences classes of cycles), and we shall often think of them as submanifolds or subvarieties of M. Similarly, we shall regard $a, b, c, \ldots$ as differential forms on M. The fundamental homology class of the manifold M, that is, the homology class represented by the cycle M, is an element of $H_{2n}(M; \mathbb{Z})$, but its Poincaré dual cohomology class—the identity element of the cohomology algebra—will be denoted by $1 \in H^0(M; \mathbb{Z})$ (rather than m).

Let

$$\langle\ ,\ \rangle : H^i(M; \mathbb{Z}) \times H_i(M; \mathbb{Z}) \to \mathbb{Z}$$

denote the natural pairing, and also the extended pairing

$$\langle\ ,\ \rangle : H^*(M; \mathbb{Z}) \times H_*(M; \mathbb{Z}) \to \mathbb{Z}$$

where $\langle a, B\rangle = 0$ when $|a| \neq |B|$. In de Rham notation, $\langle a, B\rangle = \int_B a$. Since there is no torsion, these pairings are non-degenerate.

The *intersection pairing* is defined by

$$(\ ,\ ) : H^*(M; \mathbb{Z}) \times H^*(M; \mathbb{Z}) \to \mathbb{Z}, \quad (a, b) = \langle ab, M\rangle = \int_M a \wedge b.$$

We have $\langle ab, M\rangle = \langle a, B\rangle = \langle b, A\rangle$. It follows that the intersection pairing $(\ ,\ )$ is a non-degenerate symmetric bilinear form.

The product structure of quantum cohomology (quantum product) is a generalization of the product structure of cohomology (cup product). It will be convenient for future purposes to specify the cup product by giving its 'structure constants' with respect to a basis. Therefore, we choose generators as follows

$$H_*(M; \mathbb{Z}) = \bigoplus_{i=0}^{s} \mathbb{Z}A_i, \quad H^*(M; \mathbb{Z}) = \bigoplus_{i=0}^{s} \mathbb{Z}a_i$$

and we define dual cohomology classes $b_0, \ldots, b_s$ by $(a_i, b_j) = \delta_{ij}$. Then for any i, j we have

$$a_i a_j = \sum_k \mu_{ijk} a_k = \sum_k \lambda_{ijk} b_k$$

for some $\mu_{ijk}, \lambda_{ijk} \in \mathbb{Z}$. These structure constants are given by

$$\mu_{ijk} = \langle a_i a_j b_k, M \rangle, \quad \lambda_{ijk} = \langle a_i a_j a_k, M \rangle.$$

The μ's and the λ's are, of course, equivalent, but it seems more elegant to focus on

$$\lambda_{ijk} = \langle a_i a_j a_k, M \rangle = \int_M a_i \wedge a_j \wedge a_k = \sharp\, A_i \cap A_j \cap A_k.$$

Note that the intersection form itself can be specified in a similar way by the integers

$$(a_i, a_j) = \int_M a_i \wedge a_j = \sharp\, A_i \cap A_j.$$

To emphasize the geometrical point of view and the obvious symmetry it is natural to introduce the following notation:

Definition 1.1 *For cohomology classes a, b, c we define $\langle A|B|C \rangle_0 = \langle abc, M \rangle = \int_M a \wedge b \wedge c = \sharp\, A \cap B \cap C$.*

The reason for the suffix 0 is that the quantum product will be defined in terms of certain *Gromov–Witten invariants* $\langle A|B|C \rangle_D$, where D is not necessarily zero.

Let us reiterate that $\sharp\, A \cap B \cap C$, the number of points in the intersection $A \cap B \cap C$, has to be interpreted rather carefully: it is valid providing we use representative cycles A, B, C which intersect transversely, and providing we have a suitable notion of multiplicity. In the complex algebraic category, there is a simple criterion for this transversality condition: it holds automatically whenever there exist representative algebraic subvarieties A, B, C whose intersection is finite (or empty)—see the appendix of [44]. The most famous example where this method works is the case where M is the Grassmannian $\mathrm{Gr}_k(\mathbb{C}^n)$ of k-dimensional complex linear subspaces of $\mathbb{C}^n$ (or, more generally, a flag manifold). Here all the generators of the homology groups are representable by algebraic cycles (Schubert varieties), and for any three such generators a, b, c satisfying the condition $|a| + |b| + |c| = \dim M$ there exist representatives whose intersection is finite (or empty).

Example 1.2 The above result allows us to calculate the cohomology algebra $H^*\mathbb{C}P^n$ of complex projective space. We have $H^{2i}\mathbb{C}P^n \cong \mathbb{Z}x_i$

$(0 \leq i \leq n)$, where the Poincaré dual homology generator X_i (of dimension $2n - 2i$) can be represented by $\mathbb{P}(V)$, for any complex linear subspace $V \subseteq \mathbb{C}^{n+1}$ of codimension i. We write $x_0 = 1$.

First, we consider the intersection form. We have $(x_i, x_{n-j}) = \delta_{ij}$, as there exist linear subspaces V, W of $\mathbb{C}^{n+1}$ of codimensions $i, n-j$ such that $\mathbb{P}(V) \cap \mathbb{P}(W)$ is finite and non-empty if and only if $i = j$, and in this case the intersection consists of a single point (of multiplicity one).

Next, the product $x_i x_j$ must be of the form $\lambda_{ij} x_{i+j}$ when $i + j \leq n$ (and zero when $i + j > n$). Using the intersection form, we have $\lambda_{ij} = (x_i x_j, x_{n-(i+j)}) = \langle X_i | X_j | X_{n-(i+j)} \rangle_0$. To calculate this, we represent the three classes respectively by linear subspaces of $\mathbb{C}^{n+1}$ of codimensions $i, j, n-(i+j)$. If the subspaces are in general position, the codimension of the intersection is $i+j+n-(i+j) = n$, so this triple intersection is a line, and $X_i \cap X_j \cap X_{n-(i+j)}$ is a single point of $\mathbb{C}P^n$. Since we are taking intersections of linear subspaces, the multiplicity of this point is one. We conclude that $\lambda = 1$, so $x_i x_j = x_{i+j}$.

The cohomology algebra of $\mathbb{C}P^n$ is therefore isomorphic to $\mathbb{Z}[x_1]/(x_1^{n+1})$, the algebra of polynomials in x_1 divided by the ideal generated by x_1^{n+1}. □

1.6 The symplectic volume function

Although the origins of cohomology theory were discrete and combinatorial in nature, we have seen that differential forms provide an equally valid approach, and in this section we mention another 'function-theoretic' point of view. Although this is quite superficial in the case of cohomology, it contains the essence of an idea which will turn out to be very important for quantum cohomology.

First of all, the structure constants $\langle A|B|C \rangle_0$ can be generalized naturally to

$$\langle A_{i_1} | A_{i_2} | \dots | A_{i_j} \rangle_0 = \langle a_{i_1} a_{i_2} \dots a_{i_k}, M \rangle,$$

and the function

$$\left\langle e^{\sum_0^s t_i a_i}, M \right\rangle$$

is a formal generating function for all these numbers. (In fact, this is clearly a polynomial in $t_0, \dots, t_s$; only finitely many of the numbers can be non-zero.) The function 'packages' all the numbers in an efficient way, and they can be extracted from it by the simple device of taking derivatives. Conversely, the (constant coefficient) differential operators which annihilate this generating function constitute an ideal in the ring of all constant coefficient differential operators, and the quotient ring is isomorphic to the cohomology algebra.

If H^2M generates H^*M multiplicatively, the generating function $\langle e^{\sum_1^r t_i a_i}, M\rangle$ (in which we sum over two-dimensional classes) contains exactly the same information. For example, in the case of $\mathbb{C}P^n$, the function is just $t_1^n/n!$, and the ideal of constant coefficient differential operators annihilating this function is generated by $(d/dt_1)^{n+1}$.

So far, this is just formal algebra. However, an observation of [64] shows that the generating function can have an independent geometrical meaning (and can be calculated in an independent way). This applies only to the rather special situation where M is a Kähler manifold obtained by symplectic reduction, and it is based on the existence of a family of Kähler forms of the form $\omega_\xi = \omega + \sum_1^r \xi_i a_i$. Let the complex dimension of M be n, as usual. Then the 'symplectic volume function'

$$\frac{1}{n!}\int_M \underbrace{\omega_\xi \wedge \cdots \wedge \omega_\xi}_{n} = \int_M e^{\omega_\xi}$$

is essentially the same as the above generating function. For certain (even more special) manifolds such as toric varieties or flag manifolds, this symplectic volume of M can be identified with the Euclidean volume of a polytope associated to M, and calculated explicitly.

To summarize, the cohomology algebra of such a manifold M can be calculated as follows. First, compute the Euclidean volume of a certain polytope (as a function of ξ). Next, find the ideal of constant coefficient differential operators which annihilate this function. Then the cohomology algebra is isomorphic to the quotient of the ring of all constant coefficient differential operators by this ideal.

2 Quantum cohomology

The motivation for quantum cohomology comes from symplectic geometry and from physics (see [26]). We shall just proceed directly to a definition, in which quantum cohomology is simply ordinary cohomology but with a new 'quantum product' which extends the usual cup product. The structure constants of the quantum product are numbers denoted by $\langle A|B|C\rangle_D$, where D varies in $H_2(M;\mathbb{Z})$, and when $D = 0$ we obtain the structure constants of ordinary cohomology. These numbers are called 3-point Gromov–Witten invariants. They are invariants of the symplectic structure of M, although this aspect will not play a direct role in this book.

2.1 3-point Gromov–Witten invariants

Recall that the definition of $\langle A|B|C\rangle_0$ is

$$\langle A|B|C\rangle_0 = \langle abc, M\rangle = \int_M a\wedge b\wedge c = \sharp\, A\cap B\cap C.$$

The usual definition of $\langle A|B|C\rangle_D$ extends this, although it is more difficult because the space M has to be replaced by a certain moduli space (of J-holomorphic curves in the symplectic category, or stable curves in the algebraic category). The construction of this moduli space is complicated, and integration is problematical because the moduli space is not in general compact. Instead of this, we shall just use a naive definition of $\langle A|B|C\rangle_D$, which extends the geometrical formula $\langle A|B|C\rangle_0 = \sharp\, A\cap B\cap C$. This will be sufficient for our purposes, since we shall adopt an entirely different point of view later on.

Let M be a complex manifold. Let p, q, r be three distinct points in $\mathbb{C}P^1$. We define

$$\langle A|B|C\rangle_D = \sharp\ \mathrm{Hol}_D^{A,p} \cap \mathrm{Hol}_D^{B,q} \cap \mathrm{Hol}_D^{C,r}$$

where

$$\mathrm{Hol}_D^{A,p} = \{\text{holomorphic maps } f : \mathbb{C}P^1 \to M \mid f(p) \in A \text{ and } [f] = D\}$$

and $\mathrm{Hol}_D^{B,q}, \mathrm{Hol}_D^{C,r}$ are defined in a similar way. The notation $[f]$ denotes the homotopy class of f, which is an element of $\pi_2(M) \cong H_2(M;\mathbb{Z})$. As in the case $D = 0$, the definition employs specific cycles A, B, C, although it turns out that the value of $\langle A|B|C\rangle_D$ depends only on their homology classes. The choice of p, q, r is also irrelevant, as any three points can be moved to any other three points by a linear fractional transformation. Holomorphic maps $\mathbb{C}P^1 \to M$ are sometimes called 'rational curves'.

Evidently this definition inherits the defects of the definition of $\langle A|B|C\rangle_0$, to which it reduces when $D = 0$. We shall assume that these defects may be remedied in a similar way, at least for 'good' manifolds M. It is certainly plausible that, under mild restrictions, $\mathrm{Hol}_D^{M,p}$ is a complex manifold. Moreover, it should have complex dimension $n + \langle c_1(TM), D\rangle$, because the hypothetical tangent space at $f \in \mathrm{Hol}_D^{M,p}$ may be identified with the space of holomorphic sections of the bundle f^*TM, and by the Riemann–Roch theorem, the complex dimension of this vector space is $n + \langle c_1(TM), D\rangle$. (This argument is correct if M is convex in the sense that $H^1(\mathbb{C}P^1, f^*TM) = 0$ for all holomorphic $f : \mathbb{C}P^1 \to M$.) It is also plausible that $\mathrm{Hol}_D^{A,p}$ is a complex submanifold of $\mathrm{Hol}_D^{M,p}$, whose complex codimension in $\mathrm{Hol}_D^{M,p}$ is equal to the complex codimension of A in M. When A, B, C are tranversal, the intersection

$$\mathrm{Hol}_D^{A,p} \cap \mathrm{Hol}_D^{B,q} \cap \mathrm{Hol}_D^{C,r}$$

should be zero-dimensional when

$$\mathrm{codim}_{\mathbb{R}} \mathrm{Hol}_D^{A,p} + \mathrm{codim}_{\mathbb{R}} \mathrm{Hol}_D^{B,q} + \mathrm{codim}_{\mathbb{R}} \mathrm{Hol}_D^{C,r} = \dim_{\mathbb{R}} \mathrm{Hol}_D^{M,p},$$

that is, when $|a| + |b| + |c| = 2n + 2\langle c_1(TM), D\rangle$. Thus, when $\langle A|B|C\rangle_D$ is non-zero, we expect this numerical condition to hold.

If M is a Fano manifold, in the sense that the cohomology class $c_1(TM)$ can be represented by a Kähler 2-form, then $\langle c_1(TM), D\rangle > 0$ for each homotopy class $D \in \pi_2(M)$ which contains a holomorphic map $\mathbb{C}P^1 \to M$. Thus, for Fano manifolds, only a finite number of $\langle A|B|C\rangle_D$ can be non-zero (when A, B, C belong to a given basis).

As explained in [26] (for example) it turns out that $\langle A|B|C\rangle_D$ can be defined rigorously under very general conditions. The definition[1]

[1] A more common notation for $\langle A|B|C\rangle_D$ is $\langle a, b, c\rangle_{0,3,D}$ where 0 indicates that we consider curves of genus zero (and 3 is redundant). We prefer A, B, C as all our explicit calculations will use homology classes directly.

has the form

$$\langle A|B|C\rangle_D = \int_{[\overline{M}(D)]^{\text{virt}}} ev_1^* a \wedge ev_2^* b \wedge ev_3^* c$$

where $M(D)$ is a certain moduli space of 'curves', $\overline{M}(D)$ is a compactification of $M(D)$, obtained by adding suitable 'boundary components', and $[\overline{M}(D)]^{\text{virt}}$ denotes the 'virtual fundamental class' over which integration is carried out. The evaluation map $ev_i : \overline{M}(D) \to M$ assigns to a curve its value at a given i-th basepoint ($i = 1, 2, 3$). As suggested by the above discussion, $\langle A|B|C\rangle_D$ has the following properties:

(1) $\langle A|B|C\rangle_D \in \mathbb{Z}$ is defined for $A, B, C \in H_*(M; \mathbb{Z})$, $D \in H_2(M; \mathbb{Z})$.
(2) $\langle A|B|C\rangle_D$ is $\mathbb{Z}$-linear and symmetric in A, B, C.
(3) $\langle A|B|C\rangle_D \neq 0 \Rightarrow |a| + |b| + |c| = 2n + 2\langle c_1(TM), D\rangle$. (This 'numerical condition' is useful when performing concrete calculations.)

Any compact homogeneous Kähler manifold (i.e. generalized flag manifold) is convex and Fano. It can be shown that, in this case, the naive geometrical definition of $\langle A|B|C\rangle_D$ coincides with the rigorous definition. Moreover, the (generalized) Schubert classes provide a natural homology basis. Thanks to the Kleiman Transversality Theorem all Gromov–Witten invariants involving this basis may be computed as in the case $D = 0$ by using Schubert cycles.

To understand the definition better, let us consider the case where $\pi_2(M) \cong \mathbb{Z}$. For a map $f : \mathbb{C}P^1 \to M$, we call $[f] = D \in \mathbb{Z}$ the 'degree' of f. Assume that $\langle c_1(TM), 1\rangle = N > 0$ and that the degree of any holomorphic map is non-negative. Let us examine briefly the cases $D = 0, 1, 2$.

For $D = 0$, we have $\text{Hol}_D^{A,p} = A$, so the definition of $\langle A|B|C\rangle_D$ reduces to the definition of $\langle A|B|C\rangle_0$, as it should. We are simply counting the points of the triple intersection $A \cap B \cap C$ (Fig. 2.1).

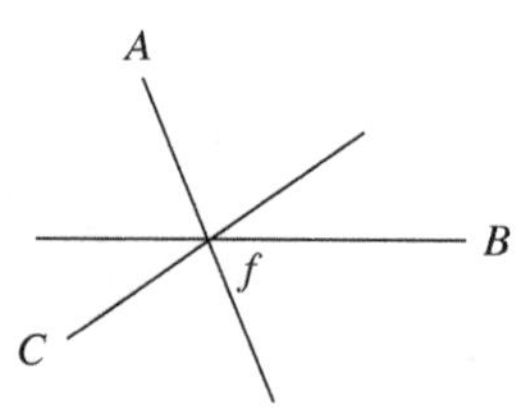

Figure 2.1. $\langle A|B|C\rangle_0$.

For $D = 1$, $\langle A|B|C\rangle_1$ counts the holomorphic maps f of degree 1 such that $f(p) \in A, f(q) \in B, f(r) \in C$ (Fig. 2.2). In this case, $|a|+|b|+|c| = 2n+2N$, so $|a| + |b| + |c| > 2n$, and the triple intersection $A \cap B \cap C$ is in general empty.

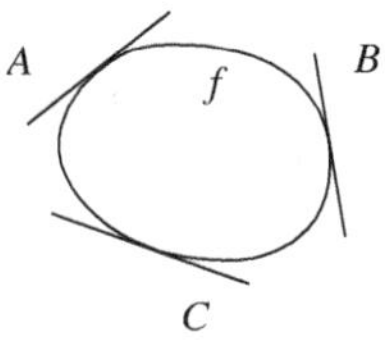

Figure 2.2. $\langle A|B|C\rangle_1$.

For $D = 2$, $\langle A|B|C\rangle_2$ counts the holomorphic maps f of degree 2 such that $f(p) \in A$, $f(q) \in B$, $f(r) \in C$ (Fig. 2.3). If $\langle A|B|C\rangle_2$ is non-zero, then both $\langle A|B|C\rangle_0$ and $\langle A|B|C\rangle_1$ are zero.

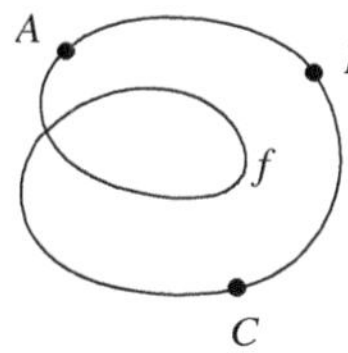

Figure 2.3. $\langle A|B|C\rangle_2$.

Example 2.1 Let us compute $\langle A|B|C\rangle_D$ for the manifold $M = \mathbb{C}P^n$. In this case $N = n+1$, so a necessary condition for $\langle A|B|C\rangle_D \neq 0$ is $|a|+|b|+|c| = 2n + 2D(n + 1)$. Since $0 \leq |a|, |b|, |c| \leq 2n$, it follows immediately that $D = 0$ and $D = 1$ are the only relevant values, that is, that $\langle A|B|C\rangle_D = 0$ for $D \neq 0, 1$.

For $D = 0$ we already know that

$$\langle X_i|X_j|X_k\rangle_0 = \begin{cases} 1 & \text{if } i + j + k = n \\ 0 & \text{otherwise.} \end{cases}$$

As we saw in Example 1.2, the calculation reduces to consideration of the intersection of three linear subspaces in $\mathbb{C}^{n+1}$. For $D = 1$ we claim that

$$\langle X_i|X_j|X_k\rangle_1 = \begin{cases} 1 & \text{if } i + j + k = 2n + 1 \\ 0 & \text{otherwise.} \end{cases}$$

To prove this, one shows that, when $i + j + k = 2n + 1$, there exist complex linear subspaces E^i, E^j, E^k of $\mathbb{C}^{n+1}$, of codimensions i, j, k, with the following property: there exist unique complex lines L', L'', L''' such that $L' \subseteq E^i$, $L'' \subseteq E^j$, $L''' \subseteq E^k$ and such that L', L'', L''' span a subspace E of dimension 2. The holomorphic map of degree 1 defined by the inclusion $\mathbb{P}(E) \subseteq \mathbb{C}P^{n+1}$ is then the unique point of the triple intersection $\mathrm{Hol}_1^{\mathbb{P}(E^i),p} \cap \mathrm{Hol}_1^{\mathbb{P}(E^j),q} \cap \mathrm{Hol}_1^{\mathbb{P}(E^k),r}$, and we obtain $\langle X_i|X_j|X_k\rangle_1 = 1$. While elementary, this kind of messy calculation can be avoided by appealing to general properties. The basic point is that it is not necessary to compute all the Gromov–Witten invariants one by one, as they are not independent. There are relations between them, and it turns out that a small subset of Gromov–Witten invariants generates the rest. The relations exist precisely because Gromov–Witten invariants are structure constants of the quantum product, as we shall see in the next section. □

To end this section we remark that the i-point Gromov–Witten invariants

$$\langle X_1|X_2|\dots|X_i\rangle_D = \sharp\ \mathrm{Hol}_D^{X_1,p_1} \cap \mathrm{Hol}_D^{X_2,p_2} \cap \dots \cap \mathrm{Hol}_D^{X_i,p_i}$$

can be defined for any $i \geq 3$. However, unless $i = 3$, this does not coincide with the rigorous definition

$$\langle X_1|X_2|\dots|X_i\rangle_D = \int_{[\overline{M}(D)]^{\mathrm{virt}}} ev_1^* x_1 \wedge ev_2^* x_2 \wedge \dots \wedge ev_i^* x_i\,.$$

The first definition turns out to be of minor interest (as we shall see in the next section), so in most of the quantum cohomology literature (and the rest of this book) the second definition is used when $i \geq 4$. These i-point Gromov–Witten invariants (with $i \geq 3$) are called *primary* Gromov–Witten invariants. There are various other kinds, some of which we will mention later.

2.2 The quantum product

To define $a \circ_t b$ for $a, b \in H^*M$ and $t \in H^2M$, it suffices to define $\langle a \circ_t b, C\rangle$ for all $C \in H_*M$.

Definition 2.2 *Assume that M is a Fano manifold. Then the quantum product $a \circ_t b$ of two cohomology classes $a, b \in H^*M$ is defined by*

$$\langle a \circ_t b, C\rangle = \sum_{D \in H_2(M;\mathbb{Z})} \langle A|B|C\rangle_D\, e^{\langle t,D\rangle}.$$

The Fano condition ensures that the sum is finite. Observe that as '$t \to -\infty$' the right hand side converges to $\langle A|B|C\rangle_0$; hence $a \circ_t b$ converges to the cup product ab. In this sense, the quantum product is a deformation of the cup product.

The main result concerning the quantum product is:

Theorem 2.3 *For each $t \in H^2M$, $\circ_t$ is a commutative, associative product operation on H^*M.*

The most difficult part of this theorem is the associativity; the other properties are obvious (intuitively, at least). Details can be found in [26].

Later on we shall use generalizations of the above definition, for example, where the Fano assumption is weakened or where the parameter t is allowed to vary in a vector space larger than H^2M. For the moment, we just mention a simple but convenient variant, which is a product operation $\circ$ on

$$H^*M \otimes \Lambda$$

where Λ is the group algebra $\mathbb{Z}[H_2(M;\mathbb{Z})]$. Formally, an element of Λ is a finite sum $\sum_X \lambda_X q^X$, where $\lambda_X \in \mathbb{Z}$, $X \in H_2(M;\mathbb{Z})$, and where the symbols q^D are multiplied in the obvious way, that is, $q^D q^E = q^{D+E}$.

Definition 2.4 *Let M be a Fano manifold. Then*

$$a \circ b = \sum_{D \in H_2(M;\mathbb{Z})} (a \circ b)_D \, q^D,$$

where $(a \circ b)_D$ is defined by $\langle (a \circ b)_D, C\rangle = \langle A|B|C\rangle_D$ for all $C \in H_*M$. The definition extends in a Λ-linear fashion to $H^*M \otimes \Lambda$. The algebra $(H^*M \otimes \Lambda, \circ)$ is called the quantum cohomology algebra, and we shall denote it by QH^*M.

By the Fano assumption, the sum on the right hand side of the definition of $a \circ b$ is finite.

It is convenient to introduce a grading on QH^*M by

$$|aq^D| = |a| + 2\langle c_1(TM), D\rangle.$$

This extends the usual grading on H^*M. We claim that the quantum product preserves the grading in the sense that

$$|a \circ b| = |a| + |b|.$$

For this we need to show that $|(a \circ b)_D| + 2\langle c_1(TM), D\rangle = |a| + |b|$ whenever $(a \circ b)_D \neq 0$. If the latter condition holds, then there exists some C with $\langle (a \circ b)_D, C\rangle = \langle A|B|C\rangle_D \neq 0$. In this situation we must have $(a \circ b)_D = |C|$ and also $|a| + |b| + |c| = 2n + 2\langle c_1(TM), D\rangle$, which implies the desired result.

For explicit calculations we shall choose bases as in Chapter 1. It will be convenient to use bases for cohomology and homology which are dual in the sense that

$$\langle b_i, A_j\rangle = \int_{A_j} b_i = \delta_{ij}, \quad 0 \le i,j \le s.$$

Since the second (co)homology group plays a prominent role in quantum cohomology, we shall declare that

$$\dim H^2M = r$$

throughout the rest of the book, and we make the following addition to the notation of section 1.5:

Assumption For the bases $A_0,\dots,A_s$ of H_*M, and $b_0,\dots,b_s$ of H^*M, as in section 1.5, we have

$$H_0M = \mathbb{Z}A_0 = \mathbb{Z}\,\{\text{point}\}, \quad H^0M = \mathbb{Z}b_0 = \mathbb{Z}1$$

and

$$H_2M = \bigoplus_{i=1}^{r} \mathbb{Z}A_i, \quad H^2M = \bigoplus_{i=1}^{r} \mathbb{Z}b_i.$$

A general element of H^2M will be written $t = \sum_1^r t_ib_i \in H^2M$, and a general element of H_2M will be written $D = \sum_{i=1}^r D_iA_i$. If we introduce $q_i = q^{A_i}$, then $q^D = q_1^{D_1}\dots q_r^{D_r}$.

For Fano manifolds the products $\circ_t$ and $\circ$ are equivalent, and there is no reason to prefer one or the other. For non-Fano manifolds, similar definitions can be made, but it will be necessary to consider whether the infinite series in the definition are convergent, and this is a non-trivial problem. To avoid a direct assault, various strategies can be used. For example, using $\circ$ rather than $\circ_t$, one can introduce the Novikov ring of series in q (see the discussion in section 5.2.1 of [95]) or one can simply consider all series as formal series. To avoid such complications we shall focus on the Fano case in our exposition, pointing out generalizations where necessary. In practical terms, the relation between $\circ$ and $\circ_t$ is that the latter is obtained from the former by 'replacing q^D by $e^{\langle t,D\rangle}$', or, after choosing a basis as above, by 'putting $q_i = e^{t_i}$'.

Having defined the quantum product, we can explain why the naively defined i-point Gromov–Witten invariants

$$\langle X_1|X_2|\dots|X_i\rangle_D = \sharp\ \mathrm{Hol}_D^{X_1,p_1} \cap \mathrm{Hol}_D^{X_2,p_2} \cap \dots \cap \mathrm{Hol}_D^{X_i,p_i}$$

are not interesting for $i \geq 4$: one has

$$\langle X_1|X_2|\ldots|X_i\rangle_D = \langle (x_1 \circ x_2 \circ \cdots \circ x_{i-1})_D, X_i\rangle,$$

so these i-point invariants for $i > 3$ are easily expressible in terms of the 3-point invariants. (In certain situations, the rigorously defined i-point invariants can *also* be expressed in terms of the 3-point invariants, though this is a much more subtle matter, which we shall return to in Chapter 9.)

2.3 Examples of the quantum cohomology algebra

We shall compute some standard examples in this section (more detailed versions of some of these computations can be found in [63]). In each case the procedure is as follows:

(i) Specify the cohomology algebra.
(ii) Choose a basis $A_0, \ldots, A_s$ of H_*M. We shall always do this in such a way that a homology classes $D = \sum_1^r D_i A_i \in H_2M$ is representable by a holomorphic map $f : \mathbb{C}P^1 \to M$ only if $D_1, \ldots, D_r \geq 0$.
(iii) Identify the dual basis $b_0, \ldots, b_s$ of H^*M such that $\langle b_i, A_j\rangle = \delta_{ij}$.
(iv) Compute $|q_i| = 2\langle c_1(TM), A_i\rangle$. For this we need to know the first Chern class $c_1(TM)$ in terms of $b_1, \ldots, b_r$.
(v) Compute the Gromov–Witten invariants and the quantum product.

Example 2.5 Using the notation from Example 2.1 for $M = \mathbb{C}P^n$, let us choose $A_i = X_{n-i}$, hence $b_i = x_i$. Instead of $q_1 = q^{A_1}$ we shall just write q.

We shall calculate the quantum products $x_i \circ x_j$. We use $\circ$ rather than $\circ_t$ because the grading tells us immediately that the form of the answer is

$$x_i \circ x_j = (x_i \circ x_j)_0 + (x_i \circ x_j)_1 q$$

(as $|q^2| > |x_i| + |x_j|$, no higher powers of q can occur). The values of $\langle X_i|X_j|X_k\rangle_D$ from Example 2.1 allow us to calculate $(x_i \circ x_j)_0$ and $(x_i \circ x_j)_1$:

$$\langle (x_i \circ x_j)_0, X_k\rangle = \langle X_i|X_j|X_k\rangle_0 = \begin{cases} 1 & \text{if } i+j+k = n \\ 0 & \text{otherwise} \end{cases}$$

$$\langle (x_i \circ x_j)_1, X_k\rangle = \langle X_i|X_j|X_k\rangle_1 = \begin{cases} 1 & \text{if } i+j+k = 2n+1 \\ 0 & \text{otherwise.} \end{cases}$$

It follows that

$$x_i \circ x_j = \begin{cases} x_{i+j} & \text{if } 0 \le i+j \le n \\ x_{i+j-(n+1)}q & \text{if } n+1 \le i+j \le 2n. \end{cases}$$

In particular $QH^*\mathbb{C}P^n \cong \mathbb{Z}[x_1, q]/(x_1^{n+1} - q)$.

Before leaving this computation, let us mention two special cases. First, if $i+j \le n$, the term involving q cannot occur, because $|q| > 2n \ge |x_i| + |x_j|$, so we must have $x_i \circ x_j = x_i x_j = x_{i+j}$ (just the cup product). Next, consider the simplest case where $i+j = n+1$, namely $x_1 \circ x_n$. Since $|(x_1 \circ x_n)_0| \le 2n$, this term must be zero, so $x_1 \circ x_n = \lambda q$ where $\lambda = \langle X_1|X_n|X_n\rangle_1$. Now, $\langle X_1|X_n|X_n\rangle_1$ is the number of linear maps $\mathbb{C}P^1 \to \mathbb{C}P^n$ which 'hit' generic representatives of X_1 (a hyperplane), X_n (represented by a point not on the hyperplane), X_n (represented by another point not on the hyperplane) at three prescribed points of $\mathbb{C}P^1$. There is precisely one such map, namely, the (complex, projective) line through the two points (which automatically hits the hyperplane in one point), so $\lambda = 1$ and $x_1 \circ x_n = q$. From these two cases, it turns out all other quantum products $x_i \circ x_j$ may be deduced, for example, $x_2 \circ x_n = x_1^2 \circ x_n = (x_1 \circ x_1) \circ x_n = x_1 \circ (x_1 \circ x_n) = x_1 q$. Thus, essentially the only Gromov–Witten invariant which has to be computed 'by hand' is $\langle X_1|X_n|X_n\rangle_1$, and this is 1 because[2] there exists one line through two distinct points. □

Example 2.6 Let

$$F_3 = F_{1,2}(\mathbb{C}^3) = \{(L, V) \in \mathrm{Gr}_1(\mathbb{C}^3) \times \mathrm{Gr}_2(\mathbb{C}^3) \mid L \subseteq V\}$$
$$\cong \mathrm{U}_3/\mathrm{U}_1 \times \mathrm{U}_1 \times \mathrm{U}_1$$

be the (full) flag manifold of the unitary group U_3. The 'Borel description' of $H^*(F_3; \mathbb{Z})$ is

$$H^*(F_3; \mathbb{Z}) \cong \mathbb{Z}[x_1, x_2, x_3]/(\sigma_1, \sigma_2, \sigma_3)$$

where $\sigma_1, \sigma_2, \sigma_3$ are the elementary symmetric functions of x_1, x_2, x_3. Geometrically, $x_i = -c_1(\mathcal{L}_i)$, where $\mathcal{L}_1, \mathcal{L}_2, \mathcal{L}_3$ are the complex line bundles on F_3 whose fibres over (L, V) are L, $L^\perp \cap V$, $V^\perp$ respectively. On the other hand, the 'Schubert description' of $H_*(F_3; \mathbb{Z})$ involves specific cycles, the Schubert varieties, which generalize the subvarieties $\mathbb{C}P^i$ of $\mathbb{C}P^n$. With respect to a fixed reference flag $E_1 \subseteq E_2 \subseteq \mathbb{C}^3$ there are six Schubert varieties. To compute the intersection of two Schubert homology classes

[2] Those who are prone to exaggeration can say that Euclid knew the quantum cohomology of projective space.

(hence the products of the Poincaré dual cohomology classes) one chooses two reference flags so that the set-theoretic intersection of the two Schubert varieties is either empty or another Schubert variety.

Excluding the trivial cases 'F_3' and 'a point', we list the Schubert varieties below.

$$
\begin{array}{llllll}
(1) & \{L \subseteq E_2\} & = & \mathrm{PD}(a) & = & A \in H_4(F_3;\mathbb{Z}), \\
(2) & \{E_1 \subseteq V\} & = & \mathrm{PD}(b) & = & B \in H_4(F_3;\mathbb{Z}), \\
(3) & \{L = E_1\} & = & \mathrm{PD}(a^2) & = & A^2 \in H_2(F_3;\mathbb{Z}), \\
(4) & \{V = E_2\} & = & \mathrm{PD}(b^2) & = & B^2 \in H_2(F_3;\mathbb{Z}).
\end{array}
$$

Here, $\{L \subseteq E_2\}$ is an abbreviation for $\{(L, V) \in F_3 \mid L \subseteq E_2\}$, and A^2 is an abbreviation for $PD(a^2)$, etc. With this notation, the Schubert variety 'a point' would be $\{L = E_1, V = E_2\} = PD(a^2b) = A^2B$ or $PD(ab^2) = AB^2$.

Let us choose the basis $A_0, \ldots, A_5$ as follows:

$H_0(F_3;\mathbb{Z})$	$H_2(F_3;\mathbb{Z})$	$H_4(F_3;\mathbb{Z})$	$H_6(F_3;\mathbb{Z})$
$A_0 = \text{point}$	$A_1 = B^2$	$A_3 = B$	$A_5 = F_3$
	$A_2 = A^2$	$A_4 = A$	

The dual basis $b_0, \ldots, b_5$ is

$H^0(F_3;\mathbb{Z})$	$H^2(F_3;\mathbb{Z})$	$H^4(F_3;\mathbb{Z})$	$H^6(F_3;\mathbb{Z})$
$b_0 = 1$	$b_1 = a$	$b_3 = a^2$	$b_5 = a^2b = ab^2$
	$b_2 = b$	$b_4 = b^2$	

The remaining cup products are determined by $ab = a^2 + b^2$ (which implies $a^3 = 0, b^3 = 0$); this equation can also be verified by considering Schubert varieties. The relation with the Borel description is that

$$a = x_1, \quad b = x_1 + x_2.$$

Geometrically, $a = -c_1(\mathcal{L})$ and $b = -c_1(\mathcal{V})$, where $\mathcal{L}$, $\mathcal{V}$ are the holomorphic bundles whose fibres over (L, V) are L, V respectively.

In particular, we have $H_2(F_3;\mathbb{Z}) = \mathbb{Z}A_1 \oplus \mathbb{Z}A_2 = \mathbb{Z}B^2 \oplus \mathbb{Z}A^2$. If $[f] = D_1A_1 + D_2A_2$ then it follows that

$$D_1 = f^*a = -c_1 f^*\mathcal{L}, \quad D_2 = f^*b = -c_1 f^*\mathcal{V}.$$

It is easily verified that there exists a holomorphic map f such that $[f] = D_1A_1 + D_2A_2 = (D_1, D_2)$ if and only if either (a) $D_2 \geq D_1 \geq 0$, or (b) $D_2 = 0, D_1 \geq 0$.

It is known that $c_1(TF_3) = 2a+2b$. Therefore, for $q_1 = q^{A_1}$ and $q_2 = q^{A_2}$ we have

$$|q_1| = 2\langle 2a + 2b, B^2 \rangle = 4, \quad |q_2| = 2\langle 2a + 2b, A^2 \rangle = 4.$$

To calculate the quantum product $\circ$, we begin by investigating the numerical condition for $\langle X|Y|Z\rangle_D \neq 0$. We have $\langle c_1(TF_3), D\rangle = 2D_1 + 2D_2$, so the numerical condition is

$$|x| + |y| + |z| = 6 + 4D_1 + 4D_2.$$

Since $\mathrm{Hol}^{F_3,p}_{D_1,D_2}$ is empty when either of D_1 or D_2 is negative, we have

$$x \circ y = \sum_{D_1,D_2 \geq 0} (x \circ y)_{D_1,D_2}\, q_1^{D_1} q_2^{D_2}.$$

Since the degree $|(x \circ y)_{D_1,D_2}|$ is given by $|x| + |y| - 4D_1 - 4D_2$, and this must be 0, 2, 4, or 6, the relevant values of (D_1, D_2) are severely restricted. We give one example:

Proposition 2.7 $a \circ a = a^2 + q_1$.

Proof We have $a \circ a = (a \circ a)_{0,0} + (a \circ a)_{1,0}q_1 + (a \circ a)_{0,1}q_2$. Now, $(a \circ a)_{0,0}$ is necessarily a^2, so it remains to calculate the degree 0 cohomology classes $\lambda = (a \circ a)_{1,0}$ and $\mu = (a \circ a)_{0,1}$.

By definition, $\lambda = \langle A|A|Z\rangle_{1,0}$, where Z is a point. Any holomorphic map of degree $(1, 0)$ is of the form $\mathbb{P}(H) \to F_3$, $L \mapsto (L, H)$, where H is a fixed two-dimensional subspace of $\mathbb{C}^3$. Therefore, we must count the number of such maps which hit three fixed subvarieties of the form

(1) $\{L \subseteq E_2'\}$ (representing A),
(2) $\{L \subseteq E_2''\}$ (also representing A),
(3) (E_1, E_2) (a point).

The logic behind this argument is somewhat convoluted so we reiterate it. We are attempting to find a line E_1 and planes H, E_2', E_2'', E_2 such that the holomorphic map determined by H hits each of the three subvarieties defined by (1) E_2', (2) E_2'', (3) E_1, E_2. With E_1, E_2', E_2'', E_2 fixed, we count the number of such H. If this number is zero or finite, then it must be equal to λ. If it is infinite, then we can conclude nothing—except that this naive method of computing λ is inadequate. In fact, thanks to the homogeneity of the flag manifold, the naive method always works. We shall come across an example where it fails a little later, in the case of the Hirzebruch surface.

Returning to the calculation, let us hope for the best and choose E_2, E_2', E_2'' in general position, that is, such that the intersection of any two of them is a line, and the intersection of all three is the origin. Let us choose E_1 to be any line in E_2. Then there is at most one H with the required property, namely $H = E_2$, since the image of $\mathbb{P}(H)$ has to contain the point (E_1, E_2). The line $\mathbb{P}(H)$ intersects the subvariety $\{L \subseteq E_2'\}$ at the point $(E_2 \cap E_2', E_2)$, and the subvariety $\{L \subseteq E_2''\}$ at the point $(E_2 \cap E_2'', E_2)$, so there is precisely one such H. We conclude that $\lambda = 1$. A similar calculation gives $\mu = \langle A|A| Z\rangle_{0,1} = 0$. □

All quantum products of additive generators of H^*F_3 may be computed in a similar[3] way. The results are collected in the table below:

	1	a	b	a^2	b^2	$a^2b = ab^2$
1	1	a	b	a^2	b^2	a^2b
a	a	a^2+q_1	ab	bq_1	ab^2	$b^2q_1+q_1q_2$
b	b	ba	b^2+q_2	ba^2	aq_2	$a^2q_2+q_1q_2$
a^2	a^2	bq_1	a^2b	b^2q_1	q_1q_2	aq_1q_2
b^2	b^2	b^2a	aq_2	q_1q_2	a^2q_2	bq_1q_2
a^2b	a^2b	$b^2q_1+q_1q_2$	$a^2q_2+q_1q_2$	aq_1q_2	bq_1q_2	abq_1q_2

As in the case of $\mathbb{C}P^n$, it is not really necessary to calculate all quantum products in this table from Gromov–Witten invariants. Once the products of degree at most six have been computed, the rest follow from associativity and commutativity.

The relations defining the quantum cohomology algebra are ‘quantum modifications’ of the Borel relations

$$x_1 + x_2 + x_3, \quad x_1x_2 + x_2x_3 + x_3x_1, \quad x_1x_2x_3.$$

This follows from Theorem 2.2 of [114], where it is proved that any ‘quantum’ polynomial is equal to the same ‘classical’ polynomial plus classical polynomials of lower degree, and vice versa. From the table we obtain

$$x_1 \circ x_2 + x_2 \circ x_3 + x_3 \circ x_1 = -a \circ a - b \circ b + a \circ b = -q_1 - q_2$$

[3] As a therapeutic exercise, the author verified all products in [63].

and

$$x_1 \circ x_2 \circ x_3 = -a \circ b \circ b + a \circ a \circ b = bq_1 - aq_2.$$

The quantum cohomology algebra QH^*F_3 is therefore the quotient of the polynomial algebra $\mathbb{Z}[x_1, x_2, x_3, q_1, q_2]$ by the ideal generated by

$$x_1 + x_2 + x_3, \quad x_1x_2 + x_2x_3 + x_3x_1 + q_1 + q_2, \quad x_1x_2x_3 + x_3q_1 + x_1q_2,$$

or, more efficiently,

$$\frac{\mathbb{Z}[a, b, q_1, q_2]}{(a^2 + b^2 - ab - q_1 - q_2, ab^2 - a^2b - aq_2 + bq_1)}.$$

This concludes our calculation of the quantum cohomology of F_3. □

Example 2.8 The Hirzebruch surface $\Sigma_k = \mathbb{P}(\mathcal{O}(0) \oplus \mathcal{O}(-k))$, where $\mathcal{O}(i)$ denotes the holomorphic line bundle on $\mathbb{C}P^1$ with first Chern class i, may be described explicitly as a smooth projective algebraic variety as follows (see [69]):

$$\Sigma_k = \left\{([z_0; z_1; z_2], [w_1; w_2]) \in \mathbb{C}P^2 \times \mathbb{C}P^1 \mid z_1 w_1^k = z_2 w_2^k\right\}.$$

It suffices to consider the case $k \geq 0$ as $\Sigma_k \cong \Sigma_{-k}$. The subvarieties

$$\begin{aligned} X_1 &= \{z_2 = w_1 = 0\} \\ X_2 &= \{z_1 = z_2 = 0\} \\ X_3 &= \{z_1 = w_2 = 0\} \\ X_4 &= \{z_0 = 0\} \end{aligned}$$

are all isomorphic to $\mathbb{C}P^1$, and represent two-dimensional homology classes. If Σ_k is regarded as '$\mathcal{O}(-k) \cup \infty$-section', then X_1 and X_3 are fibres, X_2 is the 0-section, and X_4 is the ∞-section. We illustrate this schematically in Fig. 2.4. □

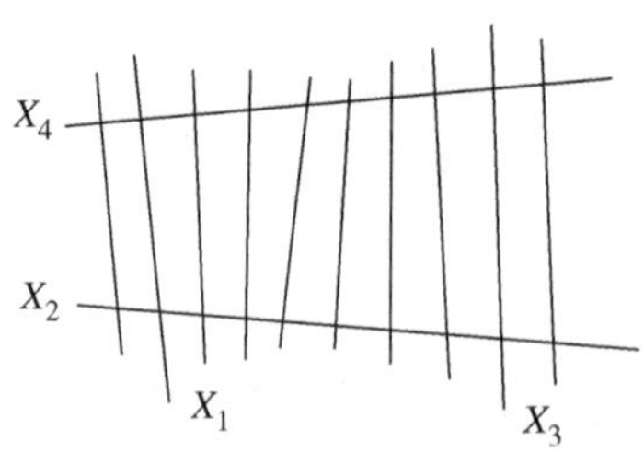

Figure 2.4. *The Hirzebruch surface.*

It can be shown that the relations between the Poincaré dual cohomology classes x_1, x_2, x_3, x_4 are

(1) $x_1 = x_3$, $x_4 = x_2 + kx_1$
(2) $x_1x_3 = x_2x_4 = 0$, $x_1x_2 = x_1x_4 = x_2x_3 = x_3x_4 = z$
(3) $x_1^2 (= x_3^2) = 0$, $x_2^2 = -kz$, $x_4^2 = kz$

where z is a generator of $H^4(\Sigma_k; \mathbb{Z}) \cong \mathbb{Z}$.

The cohomology algebra of Σ_k is

$$\begin{aligned} H^*(\Sigma_k; \mathbb{Z}) &\cong \mathbb{Z}[x_1, x_2, x_3, x_4]/(x_1 - x_3, x_4 - x_2 - kx_1, x_1x_3, x_2x_4) \\ &\cong \mathbb{Z}[x_1, x_4]/(x_1^2, x_4^2 - kz). \end{aligned}$$

Let H_1, H_2 be the restrictions to Σ_k of the tautologous line bundles on $\mathbb{C}P^1$, $\mathbb{C}P^2$. It may be verified that $c_1(H_1) = -x_1$ and $c_1(H_2) = -x_4$.

We shall choose the following homology basis:

$$A_0 = \text{point}, \quad A_1 = X_2, \quad A_2 = \text{fibre}, \quad A_3 = \Sigma_k.$$

The dual cohomology basis is

$$b_0 = 1, \quad b_1 = x_1, \quad b_2 = x_4, \quad b_3 = x_1x_4 = z.$$

Thanks to the explicit embedding in $\mathbb{C}P^2 \times \mathbb{C}P^1$, we have an explicit description of holomorphic maps $f : \mathbb{C}P^1 \to \Sigma_k$ in terms of polynomials. Namely,

$$f = ([p_4; p_2p_3^k; p_2p_1^k], [p_1; p_3])$$

where p_1, p_2, p_3, p_4 are arbitrary complex polynomials such that p_1, p_3 have no common factor, and p_2, p_4 have no common factor. (The notation is chosen so that $p_i \equiv 0$ if and only if $f(\mathbb{C}P^1) \subseteq X_i$.) If $[f] = D_1A_1 + D_2A_2 = (D_1, D_2)$, we have

$$\begin{aligned} D_1 &= -c_1(f^*H_1) = \max\{\deg p_1, \deg p_3\}, \\ D_2 &= -c_1(f^*H_2) = \max\left\{\deg p_4, \deg p_2p_3^k, \deg p_2p_1^k\right\}. \end{aligned}$$

It follows that there exists a holomorphic map f such that $[f] = D_1A_1 + D_2A_2$ if and only if either

(a) $D_2 \geq kD_1 \geq 0$, or
(b) $D_2 = 0$, $D_1 > 0$.

It is known that $c_1(T\Sigma_k) = 2x_4 - (k-2)x_1$. Therefore, for $q_1 = q^{A_1} = q^{X_2}$ and $q_2 = q^{A_2} = q^{X_1}$ we have

$$|q_1| = 2\langle 2x_4 - (k-2)x_1, X_2\rangle = 2(2-k),$$
$$|q_2| = 2\langle 2x_4 - (k-2)x_1, X_1\rangle = 4.$$

Here we have the first intimation of trouble: Σ_k is Fano only for $k = 0, 1$. Moreover, it is convex only for $k = 0$, which is the trivial case $\mathbb{C}P^1 \times \mathbb{C}P^1$ (whose quantum cohomology is $QH^*\mathbb{C}P^1 \otimes QH^*\mathbb{C}P^1$). Therefore, there is *not a single interesting case* where we can be confident that the naive geometrical method can be used to compute Gromov–Witten invariants of Σ_k. In fact the situation is even worse, as (unlike in the flag manifold example) we cannot be sure that the above cycles X_i are a sufficiently flexible collection—and indeed they are not, as, even for ordinary cohomology, we cannot compute x_2^2 because we have identified only one cycle of type X_2.

However, purely by good luck, we will be able to compute enough Gromov–Witten invariants in the case $k = 1$ to arrive at the correct answer. *From now on, therefore, we assume that* $k = 1$. We have $|q_1| = 2$ and $|q_2| = 4$.

Before starting, we note that (from the polynomial representation):

(1) the holomorphic maps of degree $(1, 0)$ are given by the 0-section X_2,
(2) the holomorphic maps of degree $(0, 1)$ are given by the fibres.

Proposition 2.9 $x_1 \circ x_1 = x_2 q_1$.

Proof Let us write $x_1 \circ x_1 = x_1^2 + \alpha q_1 + \beta q_2 + \gamma q_1^2$ where $\alpha \in H^2(\Sigma_1; \mathbb{Z})$ and $\beta, \gamma \in H^0(\Sigma_1; \mathbb{Z})$.

To calculate α, we must calculate $\langle \alpha, Y\rangle = \langle X_1|X_1|Y\rangle_{1,0}$, for two independent homology classes $Y \in H_2(\Sigma_1; \mathbb{Z})$. The product $\langle X_1|X_1|X_1\rangle_{1,0}$ is equal to 1, as we may choose three distinct fibres representing X_1, and then the holomorphic map represented by the 0-section X_2 intersects these fibres in three distinct points. The product $\langle X_1|X_1|X_4\rangle_{1,0}$ is equal to zero, because $X_2 \cap X_4 = \emptyset$. We conclude that $\langle \alpha, X_1\rangle = 1$ and $\langle \alpha, X_4\rangle = 0$, which forces $\alpha = x_2$.

Next, we have $\beta = \langle X_1|X_1|Z\rangle_{0,1}$, where Z is the generator of $H_0(\Sigma_1; \mathbb{Z})$. Let us choose X_1, X_3, and any point of Σ_1 as representatives of the three homology classes. Then each holomorphic map of degree $(0, 1)$ (i.e. each fibre) fails to intersect all three representatives. Hence $\beta = 0$. Finally, $\gamma = \langle X_1|X_1|Z\rangle_{2,0}$ and this is zero because a holomorphic map of degree (2,0) must be a map $\mathbb{C}P^1 \to X_2$ of degree two. □

Proposition 2.10 $x_1 \circ x_4 = x_1 x_4$.

Proof Let us write $x_1 \circ x_4 = x_1 x_4 + \alpha q_1 + \beta q_2 + \gamma q_1^2$. As in the previous proposition, the fact that $X_2 \cap X_4 = \emptyset$ implies $\alpha = 0$. We have $\beta = \langle X_1 | X_4 | Z \rangle_{0,1}$. By taking Z as any point in the complement of $X_1 \cup X_4$, we see that $\beta = 0$. Similarly, $\gamma = 0$. □

Proposition 2.11 $x_2 \circ x_4 = q_2$.

Proof Let us write $x_2 \circ x_4 = x_2 x_4 + \alpha q_1 + \beta q_2 + \gamma q_1^2$. As $X_2 \cap X_4 = \emptyset$, we obtain $\alpha = 0$. We have $\beta = \langle X_2 | X_4 | Z \rangle_{0,1}$, and this is equal to 1, as there is a unique fibre which intersects X_2, X_4 and a point in the complement of $X_2 \cup X_4$. Similarly, $\gamma = 0$. □

The three remaining quantum products are not amenable to direct computation in our present framework, but we can obtain them indirectly from the relation $x_4 = x_2 + x_1$:

$$x_4 \circ x_4 = z + q_2$$

$$x_1 \circ x_2 = z - x_2 q_1$$

$$x_2 \circ x_2 = -z + q_2 + x_2 q_1.$$

The quantum cohomology algebra is therefore

$$QH^* \Sigma_1 \cong \frac{\mathbb{Z}[x_1, x_4, q_1, q_2]}{(x_1^2 - (x_4 - x_1) q_1, x_4^2 - z - q_2)}.$$

Although we have not completed the calculation rigorously (even by the low standards of our naive definition of quantum cohomology), it can be proved that the above result is correct.

Example 2.12 Let $M = M_N^k$ be a non-singular complex hypersurface of degree k in $\mathbb{C}P^{N-1}$. For fixed k and N, all such hypersurfaces have the same cohomology algebra.

The Lefschetz Theorems show that $H_i M_N^k \cong H_i \mathbb{C}P^{N-1}$ for $0 \le i \le 2N-4$ except possibly for the middle dimension $i = N-2$, and that the subalgebra $H^\sharp M_N^k$ generated by $H^2 M_N^k$ has additive generators over $\mathbb{C}$ represented by cycles of the form $M_N^k \cap \mathbb{C}P^j$. To avoid odd-dimensional cohomology, and make use of the above cycles, it is natural to study the subalgebra $QH^\sharp M_N^k$ of $QH^* M_N^k$ generated by $H^2 M_N^k$.

Let us write $b = b_1$ for the 'hyperplane class', that is, the cohomology class Poincaré dual to $M_N^k \cap \mathbb{C}P^{N-2}$. This is an additive generator[4]

[4] Except for the case $M_4^2 \cong \mathbb{C}P^1 \times \mathbb{C}P^1$; here we interpret $H^\sharp M_4^2$ as the subalgebra generated by b.

of $H^2(M_N^k;\mathbb{Z}) \cong \mathbb{Z}$, and we have $c_1(TM_N^k) = (N-k)b$. It follows that M_N^k is Fano if and only if $1 \le k \le N-1$. The classes $1, b, \ldots, b^{N-2}$ are an additive basis (over $\mathbb{C}$) for $H^\sharp M_N^k$. The intersection form is given by $(b^i, b^j) = k\delta_{i+j,N-2}$ (as $(b^i, b^{N-i-2}) = (b^0, b^{N-2}) = \sharp M_N^k \cap \mathbb{C}P^1 = k$).

The quantum product for Fano hypersurfaces M_N^k was studied in [14], [24], [49], [76], and [106]; see also [26]. As the general formulae are complicated, let us just give a simple typical result here, namely for $k=3, N=5$. All quantum products in this case follow from

$$\begin{aligned} b \circ 1 &= b \\ b \circ b &= b^2 + 6q \\ b \circ b^2 &= b^3 + 15qb \\ b \circ b^3 &= 6qb^2 + 36q^2. \end{aligned}$$

In particular, $b \circ b \circ b = b^3 + 21qb$ and $b \circ b \circ b \circ b = 27qb \circ b$. Taking coefficients in $\mathbb{C}$, we deduce that

$$H^\sharp M_5^3 \cong \frac{\mathbb{C}[b]}{(b^4)}, \qquad QH^\sharp M_5^3 \cong \frac{\mathbb{C}[b,q]}{(b^4 - 27qb^2)}.$$

For $1 \le k \le N-2$ in general we have $H^\sharp M_N^k \cong \mathbb{C}[b]/(b^{N-1})$ and $QH^\sharp M_N^k \cong \mathbb{C}[b,q]/(b^{N-1} - k^k q b^{k-1})$. However, for $k = N-1$ the quantum cohomology algebra is $\mathbb{C}[b,q]/((b+k!q)^k - k^k q(b+k!q)^{k-1})$. □

We end the section with a few general comments. As in the case of projective space, we managed to compute all $\langle A|B|C\rangle_D$ in the above examples from a few simple cases, relying on the quantum product structure to generate the rest. These numbers (for general A, B, C and D) are extremely difficult to compute directly. The quantum product structure is therefore very powerful, and in turn reflects deep properties of the moduli space and its compactification.

Although the quantum cohomology algebra QH^*M has been the most visible ambassador for quantum cohomology theory so far, it should be emphasized that QH^*M contains less information than the set of all 3-point Gromov–Witten invariants. We have $QH^*M \cong H^*M \otimes \mathbb{C}[q]$ as $\mathbb{C}[q]$-modules, but not as algebras; the map which 'evaluates quantum products in terms of cup products' gives an explicit identification. We shall say more about this in section 6.3.

It will be useful to keep in mind the above examples throughout the book. Although many more examples have been computed (see [26] for references), they are mainly of the above types, that is, homogeneous spaces and toric manifolds (or complete intersections therein). A glance at the list suggests already that quantum cohomology is not functorial in any obvious sense, in contrast to ordinary cohomology. The reason for this lack of functoriality is that the quantum cohomology algebra QH^*M is only the tip of the iceberg; underlying it there is a deeper structure, and exploring this deeper structure will be our main task from now on.

2.4 Homological geometry

Significant first steps towards uncovering the deeper structure of quantum cohomology were taken by A. Givental in [47] and [48]. He observed that the additional structure on the cohomology vector space provided by quantum cohomology has a differential geometric or symplectic aspect, and he introduced the term 'homological geometry' to describe it. In this section we shall give a very brief description, primarily as motivation. In addition to the articles just mentioned, we recommend [55] as well as [6] and [7].

Let M be a complex manifold whose quantum cohomology is defined (as in sections 2.2 and 2.3). The construction of this section requires an additional assumption, that $H^*(M;\mathbb{Z})$ is generated as an algebra by the subgroup $H^2(M;\mathbb{Z})$. When this condition does not hold, one may replace $H^*(M;\mathbb{Z})$ by the subalgebra $H^\sharp(M;\mathbb{Z})$ generated by $H^2(M;\mathbb{Z})$ and work with that.

As in sections 2.2 and 2.3 we choose bases

$$H_*M = \bigoplus_{i=0}^{s} \mathbb{Z}A_i, \quad H^*M = \bigoplus_{i=0}^{s} \mathbb{Z}b_i$$

such that $\langle b_i, A_j \rangle = \delta_{ij}$. In anticipation of a symplectic interpretation, we denote a general element of $H_2(M;\mathbb{C})$ by $\sum_{i=1}^r p_i A_i$ (modifying our earlier notation $\sum_{i=1}^r D_i A_i \in H_2(M;\mathbb{Z})$). A general element of $H^2(M;\mathbb{C})$ will be denoted by $\sum_{i=1}^r t_i b_i$ as usual.

Thus, we regard p_i, t_j as the standard coordinate functions:

$$p_i = \langle b_i, \ \rangle : H_2(M;\mathbb{C}) \cong \mathbb{C}^r \to \mathbb{C}$$

$$t_j = \langle \ , A_j \rangle : H^2(M;\mathbb{C}) \cong \mathbb{C}^r \to \mathbb{C}.$$

Similarly, we regard $q_j = e^{t_j}$ as a map

$$q_j = e^{\langle \ , A_j \rangle} : H^2(M;\mathbb{C}) \cong \mathbb{C}^r \to \mathbb{C}^* = \mathbb{C} - \{0\}.$$

The map $\exp : \mathbb{C} \to \mathbb{C}^*$ induces an identification $\mathbb{C}/2\pi\sqrt{-1}\mathbb{Z} \cong \mathbb{C}^*$. Using this, we have an identification

$$B = H^2(M; \mathbb{C}/2\pi\sqrt{-1}\mathbb{Z}) \cong H^2(M; \mathbb{C}^*) \cong (\mathbb{C}^*)^r.$$

Thus, there is an induced map $q_j = e^{\langle\ ,A_j\rangle} : B \to \mathbb{C}^* = \mathbb{C} - \{0\}$.

By the above assumption, $H^*(M; \mathbb{C})$ has the form

$$H^*(M; \mathbb{C}) \cong \mathbb{C}[p_1, \ldots, p_r]/(R_1, \ldots, R_u)$$

where $(R_1, \ldots, R_u)$ denotes the ideal generated by some relations $R_1, \ldots, R_u$. It follows (Theorem 2.2 of [114]) that

$$QH^*M \cong \mathbb{C}[p_1, \ldots, p_r, q_1, \ldots, q_r]/(\mathcal{R}_1, \ldots, \mathcal{R}_u)$$

for some relations $\mathcal{R}_1, \ldots, \mathcal{R}_u$ which are 'quantum versions' of $R_1, \ldots, R_u$.

Homological geometry begins with the manifold T^*B, the cotangent bundle of the complex algebraic torus B. As B is a group, we have canonical isomorphisms

$$TB \cong B \times H^2(M; \mathbb{C}), \quad T^*B \cong B \times H^2(M; \mathbb{C})^*.$$

Via this identification, the natural (complex) symplectic form of T^*B is $d\lambda$, where

$$\lambda = \sum_{i=1}^{r} \frac{dq_i}{q_i} \wedge p_i.$$

With respect to this symplectic structure, the Poisson bracket of two functions $f, g : T^*B \to \mathbb{C}$ is given by

$$\{f, g\} = \sum_{i=1}^{r} q_i \left(\frac{\partial f}{\partial p_i} \frac{\partial g}{\partial q_i} - \frac{\partial g}{\partial p_i} \frac{\partial f}{\partial q_i} \right).$$

Let us consider next the algebraic variety V_M defined by

$$V_M = \{(q_1, \ldots, q_r, p_1, \ldots, p_r) \in T^*B \mid \mathcal{R}_1 = \cdots = \mathcal{R}_u = 0\}.$$

It is shown in [55] and [7] that, under certain conditions, V_M is a Lagrangian subvariety of T^*B. This means that V_M is maximal isotropic with respect to the symplectic form $d\lambda$, that is, $d\lambda$ is identically zero on V_M, and V_M is maximal with respect to this property. The underlying reason for this property is that V_M is the characteristic variety of the 'quantum D-module', a topic that we shall begin to study in the next chapter. For the moment we just remark that the Lagrangian property is related to the ideal $(\mathcal{R}_1, \ldots, \mathcal{R}_u)$ being closed under Poisson bracket.

Let us identify the variety V_M and verify the closure property for the examples of the previous section.

Example 2.13 For $M = \mathbb{C}P^n$ (see Example 2.5) we have

$$V_M = \{(q,p) \in \mathbb{C}^* \times \mathbb{C} \mid p^{n+1} = q\}.$$

It is one-dimensional and therefore automatically Lagrangian. The case of M_N^k (Example 2.12) is similar. □

Example 2.14 For the flag manifold $M = F_3$ (see Example 2.6), V_M is the subvariety of

$$T^*B = \{(q_1,q_2,p_1,p_2) \in (\mathbb{C}^*)^2 \times \mathbb{C}^2\}$$

defined by the equations $\mathcal{R}_1 = 0, \mathcal{R}_2 = 0$ where

$$\mathcal{R}_1 = p_1^2 + p_2^2 - p_1p_2 - q_1 - q_2$$

$$\mathcal{R}_2 = p_1^2p_2 - p_1p_2^2 - p_2q_1 + p_1q_2.$$

A simple computation gives $\{\mathcal{R}_1, \mathcal{R}_2\} = 0$, so the closure property certainly holds in this case. □

Example 2.15 For $M = \Sigma_1 = \mathbb{P}(\mathcal{O}(0) \oplus \mathcal{O}(-1))$, V_M is the subvariety of

$$T^*B = \{(q_1,q_2,p_1,p_2) \in (\mathbb{C}^*)^2 \times \mathbb{C}^2\}$$

defined by the equations $\mathcal{R}_1 = 0, \mathcal{R}_2 = 0$ where

$$\mathcal{R}_1 = p_1^2 - (p_2 - p_1)q_1$$

$$\mathcal{R}_2 = p_2^2 - p_1p_2 - q_2.$$

This time we obtain $\{\mathcal{R}_1, \mathcal{R}_2\} = q_1\mathcal{R}_2$, so the closure property holds. □

In the case $M = F_3$, the stronger than expected property $\{\mathcal{R}_1, \mathcal{R}_2\} = 0$ suggests that this example is special. In fact the same property (Poisson brackets of the natural relations of QH^*M are zero) holds for for $M = F_n$. This fact has the intriguing interpretation that V_M is the phase space for a completely integrable Hamiltonian system in the sense of classical mechanics. Each relation $\mathcal{R}_i$ can be regarded as a Hamiltonian function on the symplectic manifold T^*B, and the integral curves of the Hamiltonian vector field corresponding to $\mathcal{R}_i$ are the solution curves of an ordinary differential equation[5] on T^*B. Each such equation is said to be completely integrable

[5] In classical mechanics the symplectic manifold and differential equations are real, but all relevant concepts have natural extensions to the complex case, which is what we are using here.

because T^*B has dimension $2(n-1)$ and there are $n-1$ Poisson-commuting 'conserved quantities' (or 'integrals of motion'), namely $\mathcal{R}_1, \ldots, \mathcal{R}_{n-1}$. Remarkably, the completely integrable Hamiltonian system arising in this way from QH^*F_n was already well known; it is the so-called Toda lattice, describing the motion of a system of n masses connected by springs with non-linear interaction. Although the case $M = F_n$ is clearly very special, the connection with the Toda lattice came as a surprise when it was discovered by A. Givental and B. Kim (see [55]), and was an early stimulus for research on quantum cohomology and integrable systems.

3 Quantum differential equations

In section 1.6 we mentioned the curious fact that the cohomology of certain symplectic manifolds can be computed by finding the differential operators which annihilate the symplectic volume function. According to A. Givental (see [47] and [48]), the symplectic volume function for the Floer homology of the free loop space of a symplectic manifold should play the same role for quantum cohomology. It should be a formal generating function for Gromov–Witten invariants, and the differential operators annihilating this function should produce the quantum cohomology algebra. Without going into the justification for this, we shall simply study an equivalent system of differential equations, which are known as the quantum differential equations.

3.1 The quantum differential equations

The quantum differential equations are the equations

$$\mathrm{h}\partial_i \Psi = b_i \circ_t \Psi, \quad i = 1, \dots, r$$

for $\Psi : H^2(M;\mathbb{C}) \to H^*(M;\mathbb{C})$, where[1] h is a complex parameter.

We use $\circ_t$ rather than $\circ$ and we use cohomology with complex coefficients, because both these conventions are convenient from the point of view of differential equations. So far we have generally used integer coefficients, but from now on we use complex coefficients, and abbreviate $H^*(M;\mathbb{C})$ to H^*M.

A deeper understanding of quantum cohomology can be obtained by studying these equations, and that is the first reason for doing so. Perhaps

[1] In the quantum cohomology literature the parameter is usually called $\hbar$; in the integrable systems literature it is usually called λ. We adopt h as a compromise since we shall be considering quantum cohomology-like integrable systems.

surprisingly, it turns out that geometrical/topological aspects of quantum cohomology are reflected in algebraic/analytic properties of the quantum differential equations. For example, the coefficients of power series solutions of the quantum differential equations turn out to be related to Gromov–Witten invariants. Singularities of solutions can be traced back to properties of the manifold M. The very existence of solutions (i.e. the compatibility of the equations, in the partial differential equation case) is related to the associativity of the quantum product.

With hindsight the 'surprising' quickly becomes the 'obvious', and the case of the quantum differential equations is no exception. One could say that Gromov–Witten invariants satisfy certain combinatorial identities, and that such identities may be rewritten as differential equations satisfied by appropriate generating functions for Gromov–Witten invariants; and therefore it is not too surprising that quantum cohomology is governed by such differential equations. Alternatively, from the point of view of quantum field theory, it is certainly not surprising that differential equations appear. Indeed, from that point of view, what is most surprising is that topological ideas (of any kind) should play a role in physics.

We shall not pursue these lines of thought, because we prefer a different theory: that quantum cohomology, like its ordinary predecessor, is a rather more universal concept than its original definition suggests. This idea has been advanced particularly by B. Dubrovin and his collaborators. Moreover, there is already quite a lot of evidence that the quantum differential equations may be *characterized* (amongst general systems of differential equations) in certain situations, thereby initiating a process of separating quantum cohomology from the daunting technicalities of its original geometrical definition. All these aspects point in the direction of the theory of integrable systems. This connection with integrable systems is perhaps the strongest justification for introducing and studying the quantum differential equations.

We begin by looking at the quantum differential equations in two somewhat different ways, one algebraic and one differential geometric.

First version: matrix of structure constants

The quantum differential equations can be written as a matrix system

$$\mathrm{h}\partial_i \begin{pmatrix} \psi_0 \\ \vdots \\ \psi_s \end{pmatrix} = C_i \begin{pmatrix} \psi_0 \\ \vdots \\ \psi_s \end{pmatrix}$$

where we write $\Psi = \sum_0^s \psi_j b_j$ and consider b_j to be the column vector with 1 in position j and zeros elsewhere ($0 \le j \le s$). The matrix C_i is the matrix of 'quantum multiplication by b_i' with respect to the basis $b_0, \dots, b_s$, where $1 \le i \le r$. That is,

$$b_i \circ_t b_j = \sum_0^s (C_i)_{kj} b_k,$$

and the coefficients $(C_i)_{kj}$ of the differential equations are the 'structure constants' of the quantum cohomology algebra.

Second version: a flat connection

This is more interesting, and is the origin of the link to integrable systems. Let us rewrite the system as

$$\left(\partial_i - \frac{1}{\mathrm{h}} C_i\right) \begin{pmatrix} \psi_0 \\ \vdots \\ \psi_s \end{pmatrix} = 0.$$

These equations say that the covariant derivative of Ψ is zero, with respect to the connection $\nabla = d - \frac{1}{\mathrm{h}} \sum_1^r C_i dt_i$.

Let us recall briefly some terminology from the theory of connections (more details can be found in appendix 4.5). We regard $\Psi : H^2M \to H^*M$ as a section of the trivial vector bundle

$$H^2M \times H^*M \to H^2M, \quad (t,x) \mapsto t.$$

A connection (or covariant derivative operator) on this vector bundle is a differential operator ∇ on sections of the bundle. More precisely, for each tangent vector field X on the base manifold $H^2M \cong \mathbb{C}^r$, we have a first-order differential operator ∇_X, and $\nabla_X \Psi$ may be interpreted as the derivative of Ψ in the direction of X. For example, in the case of the 'trivial' connection $\nabla = d$, $\nabla_X \Psi$ is, by definition, the usual directional derivative of Ψ in the direction of X, sometimes denoted $X\Psi$. Any other connection may be written in the form $\nabla = d + \omega$, where ω is a matrix of 1-forms (matrix-valued 1-form). Thus, our connection $\nabla = d - \frac{1}{\mathrm{h}} \sum_1^r C_i dt_i$ corresponds to the particular matrix of 1-forms $-\frac{1}{\mathrm{h}} \sum_1^r C_i dt_i$.

For $r = 1$ the connection is not particularly helpful, but for $r \ge 2$ it is fundamental, because the condition for local existence of solutions—near a regular point—is precisely that the curvature of the connection is zero. (For $r = 1$, the curvature is always zero, and, as is well known, an ordinary differential equation always admits local solutions near a regular point.)

We shall see shortly that the properties of quantum cohomology guarantee that the curvature of the above connection is zero. This is the starting point of the entire theory. The connection is often referred to as the Dubrovin connection, or the Givental connection.

The condition for local existence of solutions (often called the 'compatibility condition' or 'consistency condition') is an elementary result, based on Frobenius' Theorem, but it is so important that we shall review the statement before we proceed any further.

Basic Lemma Let $\nabla = d - \beta$ be a connection in the trivial vector bundle $\mathbb{C}^a \times \mathbb{C}^b \to \mathbb{C}^a$, with $\beta = \sum_1^a \beta_i dt_i$. Each β_i is a $b \times b$ matrix function of $t \in \mathbb{C}^a$ on a neighbourhood N of the origin. Then, the dimension of the space of maps $\Psi : N \to \mathbb{C}^b$ which satisfy the system $(\partial_i - \beta_i)\Psi = 0$, $1 \le i \le a$ is either 0 or b. The dimension is b if and only if any of the following equivalent conditions holds:

(1) $d - \beta$ is a flat connection, that is, has zero curvature.
(2) $d\beta - \beta \wedge \beta = 0$.
(3) $\beta = dGG^{-1}$ for some map $G : N \to \mathrm{GL}_b\mathbb{C}$.

In (3), the map G is simply a 'fundamental solution matrix' for the linear system, that is,

$$G = \begin{pmatrix} | & & | \\ \Psi_{(1)} & \cdots & \Psi_{(b)} \\ | & & | \end{pmatrix}$$

where $\Psi_{(1)}, \ldots, \Psi_{(b)}$ are any basis of solutions. It is unique up to multiplication on the right by a constant invertible $b \times b$ matrix, that is, by an element of $\mathrm{GL}_b\mathbb{C}$. A more detailed discussion can be found in appendix 4.5.

Theorem 3.1 *The quantum differential equations are consistent, that is, $d - \frac{1}{\mathrm{h}}\mathrm{C}$ has zero curvature, for any non-zero value of the parameter* h. *Thus, $d\mathrm{C} = \mathrm{C} \wedge \mathrm{C} = 0$.*

Proof By condition (2) above, the zero curvature condition says that $d(\frac{1}{\mathrm{h}}\mathrm{C}) - (\frac{1}{\mathrm{h}}\mathrm{C}) \wedge (\frac{1}{\mathrm{h}}\mathrm{C}) = 0$ for all h. Comparing coefficients of powers of h, we see that this condition is equivalent to $d\mathrm{C} = \mathrm{C} \wedge \mathrm{C} = 0$. Now, the fact that $\mathrm{C} \wedge \mathrm{C} = 0$ is an immediate consequence of the commutativity and associativity of the quantum product. The fact that $d\mathrm{C} = 0$, that is, $\partial_i C_j = \partial_j C_i$, follows from a property of Gromov–Witten invariants known as the divisor equation. □

As in the Basic Lemma, we write

$$\tfrac{1}{\mathrm{h}}\mathrm{C} = dGG^{-1}$$

but now G is an $(s+1) \times (s+1)$ matrix-valued function of $t \in H^2M$ and $\mathrm{h} \in \mathbb{C}^* = \mathbb{C} - \{0\}$. It is holomorphic in any simply connected neighbourhood of any point where the matrices C_i are holomorphic functions; we shall not make any assumptions about such neighbourhoods yet, nor about a specific choice of G (which is unique only up to multiplication on the right by an $(s+1) \times (s+1)$ matrix-valued function of h).

Next we discuss a well known feature of first-order linear differential equations for vector functions, which will be very important for quantum cohomology: such systems may be rewritten as higher-order systems of differential equations for *scalar* functions. In the next chapter we shall review this theory. For the rest of this chapter, however, we shall just proceed in an experimental fashion, as motivation for the general theory.

Let us assume that the quantum differential equations for $\Psi = (\psi_0, \ldots, \psi_s)$ are equivalent to a system of p.d.e.

$$D_1\psi = 0, \ \ldots, \ D_u\psi = 0$$

for a scalar function ψ, the equivalence being given through expressions of the form

$$\psi_i = \text{ differential polynomial in } \psi.$$

The o.d.e. case is familiar: passage from an n-th-order equation for ψ to a system of n first-order equations for $\psi_0, \ldots, \psi_s$ may be achieved by writing $\psi_i = \partial^i\psi$. But for p.d.e., it is not immediately obvious how, or whether, such a procedure can be carried out (and the same can be said of the converse procedure of passing from a vector system to a scalar system—even in the o.d.e. case).

However, if they exist, the higher-order operators D_i have a very interesting interpretation, in terms of *relations* of the quantum cohomology algebra QH^*M. If $\mathcal{R}_i$ is a polynomial such that $\mathcal{R}_i(b_1, \ldots, b_r, q_1, \ldots, q_r) = 0$ in QH^*M, then the quantum differential equations suggest that $\mathcal{R}_i(\mathrm{h}\partial_1, \ldots, \mathrm{h}\partial_r, q_1, \ldots, q_r)\psi$ might also be zero—but this turns out not to be correct.

If $\mathcal{R}_i(\mathrm{h}\partial_1, \ldots, \mathrm{h}\partial_r, q_1, \ldots, q_r)\psi = 0$ then in fact $\mathcal{R}_i(b_1, \ldots, b_r, q_1, \ldots, q_r) = 0$. But, in the opposite direction, the 'differential relation' D_i corresponding to $\mathcal{R}_i$ contains additional terms involving h, in general. The following observation of [55] elucidates this point:

Theorem 3.2 *Let $P(X_0, \ldots, X_{2r})$ be a polynomial in $2r+1$ variables, written so that, in each monomial term, X_i precedes X_j if $1 \le i \le r$ and $r+1 \le j \le 2r$. If*

$$P(\mathrm{h}, q_1, \ldots, q_r, \mathrm{h}\partial_1, \ldots, \mathrm{h}\partial_r)(\Psi_{(v)}, 1) = 0,$$

for $v = 0, \ldots, s$, then the relation $P(0, q_1, \ldots, q_r, b_1, \ldots, b_r) = 0$ holds in the quantum cohomology algebra QH^*M.

Proof For any $f : H^2M \to H^*M$,

$$\begin{aligned} \mathrm{h}\partial_i(\Psi_{(v)}, f) &= (\mathrm{h}\partial_i\Psi_{(v)}, f) + (\Psi_{(v)}, \mathrm{h}\partial_i f) \\ &= (b_i \circ_t \Psi_{(v)}, f) + (\Psi_{(v)}, \mathrm{h}\partial_i f) \\ &= (\Psi_{(v)}, b_i \circ_t f) + (\Psi_{(v)}, \mathrm{h}\partial_i f) \\ &= (\Psi_{(v)}, b_i \circ_t f + \mathrm{h}\partial_i f) \\ &= (\Psi_{(v)}, B_i f), \end{aligned}$$

where $B_i = b_i \circ_t + \mathrm{h}\partial_i$. In the third line we have used the formula $(a \circ_t b, c) = (b, a \circ_t c)$, a consequence of the symmetry of the Gromov–Witten invariants. It follows that

$$\mathrm{h}\partial_{i_j} \ldots \mathrm{h}\partial_{i_1}(\Psi_{(v)}, f) = (\Psi_{(v)}, B_{i_j} \ldots B_{i_1} f)$$

and similarly for any polynomial in $\mathrm{h}\partial_1, \ldots, \mathrm{h}\partial_r$.

Let us take f to be the constant function 1 in these formulae. We have

$$\begin{aligned} B_i 1 &= b_i \\ B_j B_i 1 &= b_j \circ_t b_i \\ B_k B_j B_i 1 &= b_k \circ_t b_j \circ_t b_i + \mathrm{h}\partial_k(b_j \circ_t b_i) \end{aligned}$$

and more generally $B_{i_j} \ldots B_{i_1} 1 = b_{i_j} \circ_t \cdots \circ_t b_{i_1} + O(\mathrm{h})$. Hence

$$\begin{aligned} &P(\mathrm{h}, q_1, \ldots, q_r, \mathrm{h}\partial_1, \ldots, \mathrm{h}\partial_r)(\Psi_{(v)}, 1) \\ &= (\Psi_{(v)}, P(0, q_1, \ldots, q_r, b_1, \ldots, b_r)) + O(\mathrm{h}). \end{aligned}$$

By hypothesis the left hand side is zero for all v. Setting $\mathrm{h} = 0$ and using the fact that $\Psi_{(0)}(t), \ldots, \Psi_{(s)}(t)$ are a basis of H^*M, we obtain the required result $P(0, q_1, \ldots, q_r, b_1, \ldots, b_r) = 0$. □

Remarks on notation Evidently

$$d - \tfrac{1}{\mathrm{h}}\mathrm{C} \text{ is flat for all h} \iff d + \tfrac{1}{\mathrm{h}}\mathrm{C} \text{ is flat for all h}$$

since both are equivalent to $d\mathrm{C} = \mathrm{C} \wedge \mathrm{C} = 0$. This shows that the quantum differential equations may equally well be introduced as '$\mathrm{h}\partial_i\Psi + b_i \circ_t \Psi = 0$'. Since the dual connection $d + \beta^t$ of a flat connection $d - \beta$ is also flat, there are in fact four flat connections in quantum cohomology:

$$d - \tfrac{1}{\mathrm{h}}\mathrm{C}, \quad d + \tfrac{1}{\mathrm{h}}\mathrm{C}, \quad d - \tfrac{1}{\mathrm{h}}\mathrm{C}^t, \quad d + \tfrac{1}{\mathrm{h}}\mathrm{C}^t.$$

To clarify the relations between them, we mention a simple modification of the Basic Lemma, for connections of the form $\nabla = d + \alpha$, where $\alpha =$

$\sum_1^a \alpha_i dt_i$. In this version the following conditions are equivalent:

(1) $d + \alpha$ is a flat connection.
(2) $d\alpha + \alpha \wedge \alpha = 0$.
(3) $\alpha = F^{-1}dF$ for some map $F : N \to \mathrm{GL}_b\mathbb{C}$.

Of course, this connection can be obtained from $d - \beta$ simply by putting $\beta = -\alpha$, so the equivalence of (1) and (2) is obvious. To deduce (3), we use the fact that the dual connection $d - \alpha^t$ is also flat, so the Basic Lemma (with $\beta = \alpha^t$, now) gives a map F such that $\alpha^t = d(F^t)(F^t)^{-1}$. We have

$$F^t = \begin{pmatrix} | & & | \\ \Phi_{(1)} & \cdots & \Phi_{(a)} \\ | & & | \end{pmatrix}$$

where $\Phi_{(1)}, \ldots, \Phi_{(a)}$ are any basis of solutions of $\partial_i \Phi = \alpha_i^t \Phi$. It should be noted that the relation between F and G is non-trivial: G is the fundamental solution matrix of $(\partial_i - \frac{1}{\hbar}C_i)\Psi = 0$, whereas F^t is the fundamental solution matrix of $(\partial_i - \frac{1}{\hbar}C_i^t)\Phi = 0$.

3.2 Examples of quantum differential equations

To gain some experience, we shall convert some concrete examples of quantum differential equations to higher-order scalar equations. For more detailed versions of some of these calculations, see [63].

Example 3.3 We begin with $\mathbb{C}P^n$. With respect to the usual additive basis $1, x_1, \ldots, x_n$ of $H^*\mathbb{C}P^n$, the quantum differential equations are the system

$$\hbar\partial \begin{pmatrix} \psi_0 \\ \vdots \\ \vdots \\ \psi_n \end{pmatrix} = \begin{pmatrix} 0 & & & q \\ 1 & \ddots & & \\ & \ddots & \ddots & \\ & & 1 & 0 \end{pmatrix} \begin{pmatrix} \psi_0 \\ \vdots \\ \vdots \\ \psi_n \end{pmatrix}$$

for the vector function $\Psi(q) = \sum_0^n \psi_j(q) x_j$, with $q = e^t$. Let us write $\psi = \psi_n$. Then, we obtain

$$\psi_{n-1} = \hbar\partial\psi, \ \psi_{n-2} = (\hbar\partial)^2\psi, \ \ldots, \ \psi_0 = (\hbar\partial)^n\psi$$

(which express $\psi_0, \ldots, \psi_{n-1}$ in terms of ψ) together with

$$(\hbar\partial)^{n+1}\psi = q\psi.$$

Conversely, if we start with the o.d.e. $(\text{h}\partial)^{n+1}\psi = q\psi$, and define $\psi_0, \ldots, \psi_{n-1}$ in terms of $\psi = \psi_n$, we obtain the original quantum differential equations. □

Example 3.4 Next we consider the flag manifold F_3. With $t = t_1 a + t_2 b$, we have $C = C_1 dt_1 + C_2 dt_2$ where C_1, C_2 are the matrices of the quantum multiplication operators $a\circ_t$, $b\circ_t$ on $H^* F_3$. From the multiplication table given earlier we have

$$C_1 = \left(\begin{array}{c|cc|cc|c} 0 & q_1 & 0 & 0 & 0 & q_1 q_2 \\ \hline 1 & 0 & 0 & 0 & 0 & 0 \\ 0 & 0 & 0 & q_1 & 0 & 0 \\ \hline 0 & 1 & 1 & 0 & 0 & 0 \\ 0 & 0 & 1 & 0 & 0 & q_1 \\ \hline 0 & 0 & 0 & 0 & 1 & 0 \end{array}\right), \quad C_2 = \left(\begin{array}{c|cc|cc|c} 0 & 0 & q_2 & 0 & 0 & q_1 q_2 \\ \hline 0 & 0 & 0 & 0 & q_2 & 0 \\ 1 & 0 & 0 & 0 & 0 & 0 \\ \hline 0 & 1 & 0 & 0 & 0 & q_2 \\ 0 & 1 & 1 & 0 & 0 & 0 \\ \hline 0 & 0 & 0 & 1 & 0 & 0 \end{array}\right).$$

The quantum differential equations are

$$\text{h}\partial_1 \Psi = C_1 \Psi, \quad \text{h}\partial_2 \Psi = C_2 \Psi$$

namely

$$\begin{aligned} \text{h}\partial_1 \psi_0 &= q_1 \psi_1 + q_1 q_2 \psi_5 & \qquad \text{h}\partial_2 \psi_0 &= q_2 \psi_2 + q_1 q_2 \psi_5 \\ \text{h}\partial_1 \psi_1 &= \psi_0 & \text{h}\partial_2 \psi_1 &= q_2 \psi_4 \\ \text{h}\partial_1 \psi_2 &= q_1 \psi_3 & \text{h}\partial_2 \psi_2 &= \psi_0 \\ \text{h}\partial_1 \psi_3 &= \psi_1 + \psi_2 & \text{h}\partial_2 \psi_3 &= \psi_1 + q_2 \psi_5 \\ \text{h}\partial_1 \psi_4 &= \psi_2 + q_1 \psi_5 & \text{h}\partial_2 \psi_4 &= \psi_1 + \psi_2 \\ \text{h}\partial_1 \psi_5 &= \psi_4 & \text{h}\partial_2 \psi_5 &= \psi_3. \end{aligned}$$

Let us choose $\psi = \psi_5$. Then, five of the above twelve equations may be used to express $\psi_0, \psi_1, \psi_2, \psi_3, \psi_4$ in terms of ψ, as follows:

$$\begin{aligned} \psi_0 &= \text{h}^3 \partial_1^2 \partial_2 \psi - q_1 \text{h}\partial_2 \psi \\ \psi_1 &= \text{h}^2 \partial_1 \partial_2 \psi - \text{h}^2 \partial_1^2 \psi + q_1 \psi \\ \psi_2 &= \text{h}^2 \partial_1^2 \psi - q_1 \psi \\ \psi_3 &= \text{h}\partial_2 \psi \\ \psi_4 &= \text{h}\partial_1 \psi. \end{aligned}$$

The remaining seven equations reduce to the following system of five equations for ψ:

$$\begin{aligned}(\mathrm{h}^2\partial_1^2 + \mathrm{h}^2\partial_2^2 - \mathrm{h}^2\partial_1\partial_2 - q_1 - q_2)\psi &= 0\\ (\mathrm{h}^3\partial_1\partial_2^2 - \mathrm{h}^3\partial_1^2\partial_2 - q_2\mathrm{h}\partial_1 + q_1\mathrm{h}\partial_2)\psi &= 0\\ (\mathrm{h}^3\partial_1^3 - q_1\mathrm{h}\partial_1 - q_1\mathrm{h}\partial_2)\psi &= q_1\mathrm{h}\psi\\ (\mathrm{h}^4\partial_1^3\partial_2 - 2q_1\mathrm{h}^2\partial_1\partial_2 + q_1\mathrm{h}^2\partial_1^2 - q_1^2 - q_1q_2)\psi &= q_1\mathrm{h}^2\partial_2\psi\\ (\mathrm{h}^4\partial_1^2\partial_2^2 - q_1\mathrm{h}^2\partial_2^2 - q_2\mathrm{h}^2\partial_1^2)\psi &= 0.\end{aligned}$$

The last three equations follow from the first two. We conclude that the original system is equivalent to the scalar system $D_1\psi = 0, D_2\psi = 0$, where

$$\begin{aligned}D_1 &= \mathrm{h}^2\partial_1^2 + \mathrm{h}^2\partial_2^2 - \mathrm{h}^2\partial_1\partial_2 - q_1 - q_2\\ D_2 &= \mathrm{h}^3\partial_1\partial_2^2 - \mathrm{h}^3\partial_1^2\partial_2 - q_2\mathrm{h}\partial_1 + q_1\mathrm{h}\partial_2,\end{aligned}$$

the correspondence being given by

$$\Psi = \begin{pmatrix} (\mathrm{h}^3\partial_1^2\partial_2 - q_1\mathrm{h}\partial_2)\psi \\ (\mathrm{h}^2\partial_1\partial_2 - \mathrm{h}^2\partial_1^2 + q_1)\psi \\ (\mathrm{h}^2\partial_1^2 - q_1)\psi \\ (\mathrm{h}\partial_2)\psi \\ (\mathrm{h}\partial_1)\psi \\ \psi \end{pmatrix}.$$ □

Example 3.5 Let M be the Hirzebruch surface $\Sigma_1 = \mathbb{P}(\mathcal{O}(0) \oplus \mathcal{O}(-1))$. With $t = t_1x_1 + t_2x_4$, we have $C = C_1dt_1 + C_2dt_2$ where C_1, C_2 are the matrices of the quantum multiplication operators $x_1\circ_t, x_4\circ_t$ on $H^*\Sigma_1$. From our computations of the quantum products,

$$C_1 = \left(\begin{array}{c|cc|c} 0 & 0 & 0 & q_1q_2 \\ \hline 1 & -q_1 & 0 & 0 \\ 0 & q_1 & 0 & 0 \\ \hline 0 & 0 & 1 & 0 \end{array}\right), \quad C_2 = \left(\begin{array}{c|cc|c} 0 & 0 & q_2 & q_1q_2 \\ \hline 0 & 0 & 0 & q_2 \\ 1 & 0 & 0 & 0 \\ \hline 0 & 1 & 1 & 0 \end{array}\right).$$

The quantum differential equations are

$$\begin{aligned} \mathrm{h}\partial_1\psi_0 &= q_1q_2\psi_3 & \qquad \mathrm{h}\partial_2\psi_0 &= q_2\psi_2 + q_1q_2\psi_3\\ \mathrm{h}\partial_1\psi_1 &= \psi_0 - q_1\psi_1 & \mathrm{h}\partial_2\psi_1 &= q_2\psi_3\\ \mathrm{h}\partial_1\psi_2 &= q_1\psi_1 & \mathrm{h}\partial_2\psi_2 &= \psi_0\\ \mathrm{h}\partial_1\psi_3 &= \psi_2 & \mathrm{h}\partial_2\psi_3 &= \psi_1 + \psi_2.\end{aligned}$$

Let us choose $\psi = \psi_3$. Then, three of the above eight equations may be used to express ψ_0, ψ_1, ψ_2 in terms of ψ, as follows:

$$\begin{aligned}\psi_0 &= \mathrm{h}^2 \partial_1 \partial_2 \psi \\ \psi_1 &= \mathrm{h}\partial_2 \psi - \mathrm{h}\partial_1 \psi \\ \psi_2 &= \mathrm{h}\partial_1 \psi.\end{aligned}$$

The remaining five equations reduce to the following system of four equations for ψ:

$$\begin{aligned}(\mathrm{h}^2\partial_2^2 - \mathrm{h}^2\partial_1\partial_2 - q_2)\psi &= 0 \\ (\mathrm{h}^2\partial_1^2 - q_1\mathrm{h}\partial_2 + q_1\mathrm{h}\partial_1)\psi &= 0 \\ (\mathrm{h}^3\partial_1^2\partial_2 - q_1q_2)\psi &= 0 \\ (\mathrm{h}^3\partial_1\partial_2^2 - q_2\mathrm{h}\partial_1 - q_1q_2)\psi &= 0.\end{aligned}$$

The last two equations follow from the first two. Hence, the original system is equivalent to the system $D_1\psi = 0, D_2\psi = 0$, where

$$\begin{aligned}D_1 &= \mathrm{h}^2\partial_2^2 - \mathrm{h}^2\partial_1\partial_2 - q_2 \\ D_2 &= \mathrm{h}^2\partial_1^2 - q_1\mathrm{h}\partial_2 + q_1\mathrm{h}\partial_1,\end{aligned}$$

the correspondence being given by

$$\Psi = \begin{pmatrix} (\mathrm{h}^2\partial_1\partial_2)\psi \\ (\mathrm{h}\partial_2 - \mathrm{h}\partial_1)\psi \\ (\mathrm{h}\partial_1)\psi \\ \psi \end{pmatrix}.$$ □

Example 3.6 Finally, let us look at the degree three hypersurface M_5^3 in $\mathbb{C}P^4$, whose quantum products were given in Example 2.12.

We have $\mathrm{h}\partial\Psi = C\Psi$ where

$$C = \begin{pmatrix} & 6q & & 36q^2 \\ 1 & & 15q & \\ & 1 & & 6q \\ & & 1 & \end{pmatrix},$$

that is,

$$\begin{aligned}\mathrm{h}\partial\psi_0 &= 6q\psi_1 + 36q^2\psi_3 \\ \mathrm{h}\partial\psi_1 &= \psi_0 + 15q\psi_2 \\ \mathrm{h}\partial\psi_2 &= \psi_1 + 6q\psi_3 \\ \mathrm{h}\partial\psi_3 &= \psi_2.\end{aligned}$$

The last three equations permit us to express ψ_0, ψ_1, ψ_2 in terms of $\psi = \psi_3$:

$$\begin{aligned}\psi_0 &= (\mathrm{h}^3\partial^3 - 21q\mathrm{h}\partial - 6\mathrm{h}q)\psi\\ \psi_1 &= (\mathrm{h}^2\partial^2 - 6q)\psi\\ \psi_2 &= \mathrm{h}\partial\psi,\end{aligned}$$

and substitution in the first equation gives

$$\left(\mathrm{h}^4\partial^4 - 27q\mathrm{h}^2\partial^2 - 27\mathrm{h}^2q\partial - 6\mathrm{h}^2q\right)\psi = 0.$$

The equivalence between this scalar equation and the original system is given by

$$\Psi = \begin{pmatrix}(\mathrm{h}^3\partial^3 - 21q\mathrm{h}\partial - 6\mathrm{h}q)\psi\\ (\mathrm{h}^2\partial^2 - 6q)\psi\\ \mathrm{h}\partial\psi\\ \psi\end{pmatrix}.$$

It turns out that the scalar system is given by the simple general formula

$$\left((\mathrm{h}\partial)^{N-1} - k\mathrm{h}^{k-1}q(k\partial + k - 1)(k\partial + k - 2)\dots(k\partial + 1)\right)\psi = 0$$

for any $1 \le k \le N - 1$, although this is not easy to see directly. □

3.3 Intermission

It is now time to make a major decision. The problem is that our example-oriented exposition of quantum cohomology has 'passed the point of diminishing returns'. We have introduced the main ideas without excessive technicalities, and we have worked out a number of examples, and these examples demonstrate that quantum cohomology is more than a simple generalization of ordinary cohomology. But it is clear that any further progress—in particular, the formulation and proof of general theorems—must rest on a careful treatment of the foundations of the theory.

The reader who wishes to study quantum cohomology for its own sake should by now have sufficient motivation to do so. From now on, however, we shall follow a different path. Our objective is to study quantum cohomology in the context of the theory of integrable systems. This will also require a fresh start, again using our informal exposition so far as a source of motivation. The fresh start begins in the next chapter, and represents the true beginning[2] of the book. In this section we shall just make a few

[2] Logically, the first chapter should have been called Chapter −3.

remarks about the new direction, based on the examples of the previous section.

Clearly, the quantum differential equations have special properties, which reflect properties of the original manifold M. A natural question to ask is how the quantum differential equations might be distinguished amongst general systems of differential equations.

Thus, differential equations will form the background to the new approach. Rather than the analytic aspects, however, it is primarily the algebraic aspects that will be needed. A key role is played by the seemingly innocuous relation between the matrix quantum differential equations and the higher-order scalar equations.

Two important points can be seen already from the examples in the previous section. First, various choices are involved, and we need a systematic way of handling these choices. Second, although the quantum differential equations say that 'quantum multiplication corresponds to differentiation', we are dealing with two fundamentally different kinds of object (quantum multiplication is commutative, composition of differential operators is non-commutative), and so the properties of the quantum differential equations are not simple translations of properties of the quantum product.

Let us begin with the first point, which is largely a matter of notation. In all of our examples, we were able to find differential operators $P_0, \dots, P_s$ and $D_1, \dots, D_u$ such that the quantum differential equations $\mathrm{h}\partial_i \Psi = C_i \Psi$, $(1 \le i \le r)$ were equivalent to the system of scalar equations $D_j \psi = 0$, $(1 \le j \le u)$, the relation between Ψ and ψ being of the form

$$\Psi = \begin{pmatrix} \psi_0 \\ \vdots \\ \psi_s \end{pmatrix} = \begin{pmatrix} P_0 \psi \\ \vdots \\ P_s \psi \end{pmatrix}.$$

A different choice of $P_0, \dots, P_s$ gives a different equivalence. How should the roles of $D_1, \dots, D_u$ and $P_0, \dots, P_s$ be described?

We have observed that the quantum differential equations correspond to an underlying differential geometric object, namely a flat connection. It is natural to ask whether the equivalent scalar equations correspond to some intrinsic object. The answer is that they correspond to a 'D-module', which can be represented in the form

$$\mathcal{M} \cong D/(D_1, \dots, D_u) = \frac{\text{all differential operators}}{\text{the left ideal generated by } D_1, \dots, D_u}.$$

A choice of $P_0, \dots, P_s$ is simply a choice of basis of this D-module, considered as a module over the ring of holomorphic functions.

Thus, each example in the previous section illustrates a correspondence between

(1) the representation $d - \frac{1}{\hbar}\sum_1^r C_i dt_i$ of a flat connection, together with a choice of a 'cyclic generator', and
(2) the representation $D/(D_1, \ldots, D_u)$ of a D-module, together with a choice of basis $[P_0], \ldots, [P_s]$.

Modifying any of the choices gives different representations, but the underlying flat connection and D-module remain the same.

The second point is a deeper matter. It can be expressed heuristically as follows. The definition of the quantum differential equations says that the structure constants of multiplication by two-dimensional cohomology classes are the coefficients of a first-order matrix system of differential equations. On the other hand, when we convert the multiplication table to the quantum cohomology algebra QH^*M, and the first-order system to the D-module $\mathcal{M}$, they cannot be expected to match exactly, because multiplication in QH^*M is commutative and 'multiplication' in $\mathcal{M}$ is non-commutative. We have already seen a hint of this in Theorem 3.2, where the passage from the differential operators $D_1, \ldots, D_u$ to the relations $\mathcal{R}_1, \ldots, \mathcal{R}_u$ of QH^*M appears to lose information. Conversely, if we started with an algebra and a D-module which matched exactly, then chose a basis in order to obtain matrix versions of each, these matrix versions would not in general match up.

This 'incompatibility' of the commutative and non-commutative situations is the key to a more abstract approach to quantum cohomology.

4 Linear differential equations in general

In this chapter we start again from the beginning, independently of quantum cohomology. Our objective is to set up some terminology from the algebraic theory of overdetermined systems of partial differential equations, that is, the theory of holonomic D-modules. While the general theory of D-modules is very technical, we only need some basic concepts, which we shall introduce gradually.

4.1 Ordinary differential equations

We use the notation

$$(\partial^{s+1} + a_s\partial^s + \cdots + a_1\partial + a_0)y = 0$$

for an ordinary differential equation (o.d.e.) of order $s + 1$, where

$$\partial = \partial_z = \frac{d}{dz}$$

and the coefficients $a_0, \ldots, a_s$ are functions of the complex variable z.

We shall assume unless stated otherwise that $a_0, \ldots, a_s$ belong to the ring

$$\mathcal{H}_z = \{a : \mathrm{N}_z \to \mathbb{C} \mid a \text{ is holomorphic}\} = \mathrm{Map}(\mathrm{N}_z, \mathbb{C})$$

where N_z is a fixed open disk in $\mathbb{C}$. The suffix z indicates that z is a local coordinate, but we shall write $N_z = N, \mathcal{H}_z = \mathcal{H}$ until later on, when we consider functions of more than one variable.

In the language of classical o.d.e. theory, we are working in a neighbourhood of a regular point. It is a standard result that, for any point $z_0 \in N$ and any values $c_0, \ldots, c_s \in \mathbb{C}$, there is a unique solution $y \in \mathcal{H}$ of the differential equation which satisfies the initial conditions $y(z_0) = c_0, y'(z_0) = c_1, \ldots, y^{(s)}(z_0) = c_s$. It follows that the set of all holomorphic solutions (on N) is a vector space of dimension $s + 1$.

Introducing new variables

$$y_0 = y, \quad y_1 = \partial y = y', \quad \ldots, \quad y_s = \partial^s y = y^{(s)}$$

we may convert the above scalar equation to a system of $s+1$ first-order equations:

$$\partial \begin{pmatrix} y_0 \\ \vdots \\ y_{s-1} \\ y_s \end{pmatrix} = \begin{pmatrix} 0 & 1 & & \\ & \ddots & \ddots & \\ & & 0 & 1 \\ -a_0 & \cdots & -a_{s-1} & -a_s \end{pmatrix} \begin{pmatrix} y_0 \\ \vdots \\ y_{s-1} \\ y_s \end{pmatrix} \quad \text{or, briefly,} \quad \partial Y = AY.$$

This system is equivalent, in an obvious sense, to the original scalar equation.

Let us choose a basis $y_{(0)}, \ldots, y_{(s)}$ of solutions of the scalar equation. The corresponding vector functions

$$Y_{(i)} = \begin{pmatrix} y_{(i)} \\ \partial y_{(i)} \\ \vdots \\ \partial^s y_{(i)} \end{pmatrix}, \qquad 0 \le i \le s$$

constitute a basis for the set of solutions of $\partial Y = AY$ (on the same neighbourhood N).

It will be essential to introduce the 'fundamental solution matrix'

$$H = \begin{pmatrix} | & & | \\ Y_{(0)} & \cdots & Y_{(s)} \\ | & & | \end{pmatrix}.$$

Evidently H satisfies

$$\partial H = AH,$$

and it takes values in the group $\mathrm{GL}_{s+1}\mathbb{C}$ of invertible $(s+1) \times (s+1)$ complex matrices. Conversely, any $\mathrm{GL}_{s+1}\mathbb{C}$-valued solution $\tilde{H}$ of the 'matrix differential equation' $\partial \tilde{H} = A\tilde{H}$ must be related to H by $\tilde{H} = HX$ where X is a constant $(s+1) \times (s+1)$ matrix, because the columns of $\tilde{H}$ are another basis of solutions of the original system.

Thus, the system is equivalent to the matrix equation $\partial H = AH$. Relations between scalar equations such as $(\partial^{s+1} + a_s\partial^s + \cdots + a_1\partial + a_0)y = 0$ and matrix equations such as $\partial H = AH$ will be very important in this book.

Explicit solutions can be expressed locally as power series, whose coefficients can be found term by term, but in general they are not given by elementary functions.

In classical o.d.e. theory, two equations (or systems) are said to be equivalent if their fundamental solutions H_1, H_2 are related by a 'gauge transformation' $X : N \to \mathrm{GL}_{s+1}\mathbb{C}$, that is, $H_2(z) = X(z)H_1(z)$ for all z. The gauge transformation X converts the system

$$\partial H_1 = A_1 H_1$$

to the system

$$\partial H_2 = A_2 H_2, \quad A_2 = XA_1X^{-1} + \partial X X^{-1}.$$

A fundamental solution H itself can be regarded as a gauge transformation which renders the system $\partial Y = AY$ equivalent to the trivial system $\partial Y = 0$. In this sense, differential equations with holomorphic coefficients are trivial—but finding the gauge transformation H is equivalent to solving the original system, which is not trivial if an explicit formula is required. If the coefficients are allowed to have singularities, then the situation is more complicated, as there is no guarantee that a fundamental solution matrix H exists with the same domain of definition as the coefficient functions.

Example 4.1 The second-order o.d.e. $\partial^2 y + a_1 \partial y + a_0 y = 0$ is equivalent to

$$\partial Y = \begin{pmatrix} 0 & 1 \\ -a_0 & -a_1 \end{pmatrix} Y \quad \text{if we define} \quad Y = \begin{pmatrix} y_0 \\ y_1 \end{pmatrix} = \begin{pmatrix} y \\ \partial y \end{pmatrix},$$

but it is equivalent to

$$\partial Y = \begin{pmatrix} 0 & 1/z \\ -za_0 & 1/z - a_1 \end{pmatrix} Y \quad \text{if we define} \quad Y = \begin{pmatrix} y_0 \\ y_1 \end{pmatrix} = \begin{pmatrix} y \\ z\partial y \end{pmatrix}.$$

These two possibilities are related by the gauge transformation

$$\begin{pmatrix} y \\ z\partial y \end{pmatrix} = \begin{pmatrix} 1 & 0 \\ 0 & z \end{pmatrix} \begin{pmatrix} y \\ \partial y \end{pmatrix}$$

(which inadvertently introduces a singularity at $z = 0$). □

It is obvious that the passage from a scalar to a matrix equation depends critically on the definition of $y_0, \dots, y_s$, and one may ask how much freedom there is in this. Instead of using the successive derivatives of y, let us set

$$y_0 = P_0 y, \quad y_1 = P_1 y, \quad \dots, \quad y_s = P_s y$$

where $P_0, \dots, P_s$ are differential operators (polynomials in ∂ with coefficients in the ring of holomorphic functions). This leads to an equivalent system of first-order equations if the P_i are independent in an appropriate sense and if knowledge of $y_0, \dots, y_s$ allows us to recover y.

The latter condition will hold, for example, if one of the P_i is the identity operator 1. An appropriate independence condition is that there do not exist differential operators $S_0, \ldots, S_{s+1}$ such that $S_0P_0 + \cdots + S_sP_s = S_{s+1}T$, where

$$T = \partial^{s+1} + a_s\partial^s + \cdots + a_1\partial + a_0.$$

These two conditions may be expressed by saying that the equivalence classes $[P_0], \ldots, [P_s]$ form a basis of

$$\mathcal{M} = D/(T) = \frac{\text{all differential operators}}{\text{the left ideal generated by } T}$$

when it is regarded as a free module over the ring $\mathcal{H}$ of holomorphic functions. Each choice of basis corresponds to a way of converting the scalar equation to a matrix equation.

Let us review briefly the algebraic terminology used here, which is completely standard. The set D, of all differential operators with coefficients in $\mathcal{H}$, is a ring or algebra, and $\mathcal{M}$ is a module over the ring D, that is, a differential module or D-module. Since D is not commutative, we specify that

$$(T) = \{ST \mid S \in D\} = DT$$

means the *left* ideal, and that the action of an element P of D on an element $[Q] = Q + (T)$ of $\mathcal{M}$ is on the left:

$$P \cdot [Q] = P \cdot (Q + (T)) = PQ + (T) = [PQ].$$

It should be noted that $\mathcal{M}$ is an infinite-dimensional complex vector space, that is, a free module of infinite rank over the field $\mathbb{C}$; indeed, it can be identified with the space $\mathrm{Map}(N, \mathbb{C}^{s+1})$ of (holomorphic) maps from N to $\mathbb{C}^{s+1}$. It is a free module of rank $s+1$ over the ring $\mathcal{H}$ of holomorphic functions; obviously $[1], [\partial], \ldots, [\partial^s]$ constitute a basis. Over D it is a cyclic module, generated by the element $[1]$, but it is not free.

Let $[P_0], \ldots, [P_s]$ be an $\mathcal{H}$-module basis of the D-module $\mathcal{M}$. Then the scalar equation $Ty = 0$ is equivalent to the system

$$\partial Y = AY,$$

where A^t (the transpose of A) is the matrix of ∂, in the sense that

$$\partial \cdot [P_j] = [\partial P_j] = \sum_{k=0}^{s} A_{jk}[P_k].$$

Note that ∂ is a differential operator, not a linear transformation, so it would be more accurate to say 'connection matrix' rather than 'matrix' here.

Example 4.2 Let $T = \partial^2 + a_1\partial + a_0$. The D-module $D/(T)$ has rank 2 over $\mathcal{H}$. With respect to the basis $[1], [\partial]$, we have

$$\begin{aligned}\partial \cdot [1] &= [\partial] = 0[1] + 1[\partial],\\ \partial \cdot [\partial] &= [\partial^2] = [-a_1\partial - a_0] = -a_0[1] - a_1[\partial]\end{aligned}$$

so the matrix of ∂ is

$$\begin{pmatrix} 0 & -a_0 \\ 1 & -a_1 \end{pmatrix},$$

which is the transpose of the matrix A above. With respect to the basis $[1], [z\partial]$, on the other hand, a similar calculation shows that the matrix of ∂ is

$$\begin{pmatrix} 0 & -za_0 \\ 1/z & 1/z - a_1 \end{pmatrix}.$$

These computations represent the D-module version of Example 4.1. □

In D-module terminology, a gauge transformation simply means a change of basis. The gauge transformation

$$X(z) = \begin{pmatrix} a(z) & b(z) \\ c(z) & d(z) \end{pmatrix}$$

converts the basis $[1], [\partial]$ to the basis $[P_0], [P_1]$ where

$$\begin{pmatrix} y_0 \\ y_1 \end{pmatrix} = \begin{pmatrix} a & b \\ c & d \end{pmatrix}\begin{pmatrix} y \\ \partial y \end{pmatrix} = \begin{pmatrix} ay + b\partial y \\ cy + d\partial y \end{pmatrix} = \begin{pmatrix} P_0 y \\ P_1 y \end{pmatrix}.$$

In the example above we have $X = \begin{pmatrix} 1 & \\ & z \end{pmatrix}$, and

$$X\begin{pmatrix} 0 & 1 \\ -a_0 & -a_1 \end{pmatrix}X^{-1} + \partial X X^{-1} = \begin{pmatrix} 0 & 1/z \\ -za_0 & 1/z - a_1 \end{pmatrix}.$$

Remarks on notation In calculations with differential operators, ∂f may be interpreted either as the application of ∂ to the function f, that is, $\partial f/\partial z$, or as the composition of the first-order operator ∂ with the zero-order operator f (i.e. their product in the ring D), which is the same as $f\partial + \partial f/\partial z$. This may lead to confusion. To resolve this without introducing heavy-handed notation, we shall generally rely on the context: the first interpretation always applies in differential equations, while the second is reserved for computations within D-modules. These two situations are essentially distinct and in practice there will be no danger of conflict.

Remark There is a general concept of 'D-module' which can be wider or narrower than in the situation discussed here. First, other coefficient rings may be used, such as polynomials (the 'Weyl algebra'), or rational functions (the 'ring of differential operators'), or smooth functions. Second, the concept may be globalized, using sheaf theory. Third, the concept generalizes to functions of several variables. These theories lead in quite different directions, and the D-module literature is consequently very diverse. For our purposes, suitable references are [25], [100] and [105]. A discussion of the o.d.e. case can be found in [65].

Evidently $\mathcal{M} = D/(T)$ is related in some way to the $(s+1)$-dimensional vector space of solutions of the differential equation $Ty = 0$. But there is a fundamental difference, because $\mathcal{M}$ depends only on T and its coefficients, whereas the solution space depends on the *kind of functions* allowed as solutions. To state the relation precisely, we denote by $\mathcal{F}$ the space of functions allowed as solutions (and we assume that $\mathcal{F}$ is a module over $\mathcal{H}$). For example, $\mathcal{F}$ could be $\mathcal{H}$ itself. Then we have a natural isomorphism of vector spaces

$$\{y \in \mathcal{F} \mid Ty = 0\} \cong \mathrm{Hom}_D(\mathcal{M}, \mathcal{F}),$$

given by the map which assigns to a solution y the D-module homomorphism $[X] \mapsto Xy$. Thus, the solution space has a 'dual nature' to the D-module.

We have seen that any scalar o.d.e. of order $s+1$ can be converted to a system of $s+1$ first-order equations, by choosing a basis $[P_0], \dots, [P_s]$. Conversely, but less obviously, any system can be converted to a scalar equation. To do this, one must produce a cyclic D-module. This procedure will be discussed in detail in the next section. As motivation, we shall give a simple example here without going into the D-module aspects.

Example 4.3 Consider the system

$$\partial \begin{pmatrix} y_0 \\ y_1 \end{pmatrix} = \begin{pmatrix} 0 & u \\ v & 0 \end{pmatrix} \begin{pmatrix} y_0 \\ y_1 \end{pmatrix},$$

where u and v are functions of z, that is,

$$\begin{aligned} y_0' &= u y_1 \\ y_1' &= v y_0. \end{aligned}$$

If we declare that $y = y_0$, and differentiate the first equation, we obtain the scalar equation $y'' - (u'/u)y' - uvy = 0$. If we declare that $y = y_1$, we obtain a different scalar equation, $y'' - (v'/v)y' - uvy = 0$.

When $v = 0$, evidently it is not appropriate to declare that $y = y_1$. In this case the system is

$$\partial \begin{pmatrix} y_0 \\ y_1 \end{pmatrix} = \begin{pmatrix} 0 & u \\ 0 & 0 \end{pmatrix} \begin{pmatrix} y_0 \\ y_1 \end{pmatrix} = \begin{pmatrix} uy_1 \\ 0 \end{pmatrix},$$

which can be solved directly by integration: $y_1 = c =$ constant and $y_0 = c\int u$. This explicit solution is not our immediate concern, but it confirms that $y = y_1 = c$ is inappropriate. The choice $y = y_0$ leads to the scalar equation $y'' - (u'/u)y' = 0$, or more informatively, $\partial(1/u)\partial y = 0$.

A similar phenomenon occurs for the system

$$\partial \begin{pmatrix} y_0 \\ y_1 \end{pmatrix} = \begin{pmatrix} 1 & 0 \\ 0 & 1 \end{pmatrix} \begin{pmatrix} y_0 \\ y_1 \end{pmatrix}.$$

Neither $y = y_0$ nor $y = y_1$ is appropriate here. However, the system is equivalent to the scalar equation $y'' - 2y' + y = 0$ if one declares that $y = y_0 + zy_1$. □

We conclude this section with a very brief comment on the effect of singular points. Let us assume that N is a disk containing the origin, and that we have an o.d.e. whose coefficients a_i are holomorphic on the punctured disk $N^* = N - \{0\}$. The solutions are not holomorphic on N^* in general (they are holomorphic on the universal covering space of N^*). However, analytic continuation of a solution is again a solution, so the analytic continuation around the origin of $H(z)$ must be of the form $H(z)M$ for some matrix M, called the *monodromy matrix*. Writing $M = e^{2\pi i\Gamma}$ for some matrix Γ, we see that $z^\Gamma = e^{\Gamma \log z}$ has exactly the same behaviour. Therefore, $S(z) = H(z)z^{-\Gamma}$ is holomorphic on N^*, and one can say that there is a 'multi-valued' fundamental solution matrix of the form

$$H(z) = S(z)z^\Gamma$$

where the multi-valuedness is entirely described by the function z^Γ.

Let us say that $z = 0$ is a regular singular point of the system $\partial Y = AY$ if $A(z) = \frac{1}{z}B(z)$ for a holomorphic map B on N. (The usual definition is weaker: the system is required to have this form after performing a gauge transformation which is holomorphic on N.) In this case, it can be shown that the map S extends holomorphically to N, and that the monodromy matrix can be determined explicitly from B. For example, in the 'non-resonant' case where no two eigenvalues of $B(0)$ differ by an integer, we have $\Gamma = B(0)$.

This underpins the 'Frobenius method', which consists of using the Ansatz

$$y = z^{\alpha}(c_0 + c_1 z + c_2 z^2 + \cdots)$$

to find a local basis of solutions in a neighbourhood of a regular singular point.

4.2 Partial differential equations

The theory of partial differential equations (p.d.e.) has a very different flavour from the theory of ordinary differential equations, but we shall be concerned only with 'overdetermined' systems of p.d.e., which share many common features with o.d.e., in particular finite dimensionality of the solution space. Let us begin by considering the relation between scalar and matrix equations in the multivariable case.

We introduce the notation

$$\partial_1 = \frac{\partial}{\partial z_1}, \quad \ldots, \quad \partial_r = \frac{\partial}{\partial z_r}$$

and consider a system of p.d.e.

$$T_1 y = 0, \quad \ldots, \quad T_u y = 0$$

for a scalar function $y(z_1, \ldots, z_r)$. The T_i are differential operators, that is, polynomials in $\partial_1, \ldots, \partial_r$ with coefficients in the ring $\mathcal{H} = \mathcal{H}_{z_1,\ldots,z_r}$ (functions of $z_1, \ldots, z_r$ which are holomorphic in some open polydisk $\mathrm{N} = \mathrm{N}_{z_1,\ldots,z_r}$). As in the o.d.e. case where $r = 1$ and $u = 1$, it is possible to convert the scalar system to a *first-order matrix system*

$$\partial_i Y = A_i Y, \quad 1 \le i \le r,$$

by introducing suitable new variables. But, unlike in the o.d.e. case, there is no canonical way to define the new variables. Nor is it obvious how to read off the dimension of the solution space (the size of the matrices A_i) from the degrees of the operators $T_1, \ldots, T_u$. Indeed, *whether* the solution space is finite-dimensional is already difficult to recognize!

It is easy to find examples where the solution space is infinite-dimensional. If $u < r$ this will always be the case, and the system is said to be underdetermined. For example, the wave equation $\partial^2 y/\partial z_1 \partial z_1 = \partial^2 y/\partial z_2 \partial z_2$ is of this type, with $u = 1$ and $r = 2$. But if $u \geq r$, even if the solution space is finite-dimensional (i.e. even if this is not an underdetermined system in disguise), then the 'most likely' outcome is that the solution space consists only of the trivial solution $y = 0$.

The concept of D-module, and the closely related concept of flat connection from differential geometry, are essential for the task of recognizing (and constructing) non-trivial overdetermined systems. We shall therefore start immediately with the D-module point of view.

Let D be the ring of differential operators generated by $\partial_1, \ldots, \partial_r$ with coefficients in the ring $\mathcal{H}$ of holomorphic functions on some open polydisk N in $\mathbb{C}^r$. Let

$$\mathcal{M} = D/(T_1, \ldots, T_u),$$

where $(T_1, \ldots, T_u)$ means the left ideal generated by the differential operators $T_1, \ldots, T_u$. This is a cyclic D-module, generated by the constant differential operator 1.

Assumption $\mathcal{M}$ is a free module of rank $s+1$ over $\mathcal{H}$, and $[P_0], \ldots, [P_s]$ is a basis of $\mathcal{M}$ (over $\mathcal{H}$), with $P_0 = 1$.

The D-module $\mathcal{M}$ is an algebraic version of the system of partial differential equations $T_1 y = \cdots = T_u y = 0$. As we have already explained in the o.d.e. case, it is independent of the kind of functions y under consideration. As in the o.d.e. case, for any D-module of functions $\mathcal{F}$, the vector space $\mathrm{Hom}_D(\mathcal{M}, \mathcal{F})$ is canonically isomorphic to the solution space $\{y \in \mathcal{F} \mid T_1 y = \cdots = T_u y = 0\}$ of the system: to a solution y there corresponds the D-module homomorphism $\mathcal{M} \to \mathcal{F}$ given by $[X] \mapsto Xy$. In particular, the solution space of the original system has dimension $s+1$.

Now we discuss systematically the relation between scalar and matrix equations in the p.d.e. case, under the above assumption.

From scalar systems to matrix systems

Let us construct explicitly the first-order matrix system which corresponds to the basis $[P_0], \ldots, [P_s]$. We define $(s+1) \times (s+1)$ matrices $\Omega_1, \ldots, \Omega_r$ by

$$\partial_i \cdot [P_j] = [\partial_i P_j] = \sum_{k=0}^{s} (\Omega_i)_{kj} [P_k],$$

and we set $\Omega = \sum_{i=1}^{r} \Omega_i dz_i$, a 1-form with values in the space $\mathrm{End}(\mathbb{C}^{s+1})$ of complex $(s+1) \times (s+1)$ matrices. Then $\nabla = d + \Omega$ defines a connection in the trivial vector bundle $N \times \mathbb{C}^{s+1}$, whose space of sections is

$$\mathcal{H}[P_0] \oplus \cdots \oplus \mathcal{H}[P_s] \cong \mathrm{Map}(N, \mathbb{C}^{s+1}).$$

Namely, $\nabla_{\partial_i}[P_j] = \sum_{k=0}^{s}(\Omega_i)_{kj}[P_k]$, and more generally for a section $\sum_{j=0}^{s} f_j[P_j]$ of this bundle, we define

$$\begin{aligned}\nabla_{\partial_i}\left(\sum_{j=0}^{s} f_j[P_j]\right) &= \sum_{j=0}^{s} \partial f_j/\partial z_i[P_j] + \sum_{j=0}^{s} f_j \nabla_{\partial_i}[P_j] \\ &= \sum_{j=0}^{s} \partial f_j/\partial z_i[P_j] + \sum_{j,k=0}^{s} f_j(\Omega_i)_{kj}[P_k].\end{aligned}$$

Appendix 4.5 at the end of this chapter gives a brief review of the theory of connections.

Proposition 4.4 *The connection ∇ is flat.*

Proof We have $\nabla_{\partial_i}\nabla_{\partial_j} = \nabla_{\partial_j}\nabla_{\partial_i}$, since $\partial_i\partial_j = \partial_j\partial_i$. It follows that the curvature of ∇ is zero. □

This implies that the space of covariant constant sections, that is, solutions of

$$\partial_i \begin{pmatrix} f_0 \\ \vdots \\ f_s \end{pmatrix} = -\Omega_i \begin{pmatrix} f_0 \\ \vdots \\ f_s \end{pmatrix}, \quad 1 \le i \le r,$$

has dimension $s + 1$.

However, it turns out that the original system is more closely related to the dual connection ∇^* defined by $(\nabla^*_{\partial_i}[P_j]^*)[P_\alpha] = -[P_j]^*(\nabla_{\partial_i}[P_\alpha])$ where $[P_0]^*, \ldots, [P_s]^*$ is the basis dual to $[P_0], \ldots, [P_s]$. We have

$$\begin{aligned}\left(\nabla^*_{\partial_i} \sum_{j=0}^{s} y_j[P_j]^*\right)[P_\alpha] &= \left(\sum_{j=0}^{s} \partial y_j/\partial z_i[P_j]^* + \sum_{j=0}^{s} y_j \nabla^*_{\partial_i}[P_j]^*\right)[P_\alpha] \\ &= \partial y_\alpha/\partial z_i - \sum_{j=0}^{s} y_j\left([P_j]^* \sum_{\beta=0}^{s} (\Omega_i)_{\beta\alpha}[P_\beta]\right) \\ &= \partial y_\alpha/\partial z_i - \sum_{j=0}^{s} y_j(\Omega_i)_{j\alpha}.\end{aligned}$$

Thus, covariant constant sections $\sum_{j=0}^{s} y_j[P_j]^*$ for the dual connection are solutions of the following system:

$$\partial_i \begin{pmatrix} y_0 \\ \vdots \\ y_s \end{pmatrix} = \Omega_i^t \begin{pmatrix} y_0 \\ \vdots \\ y_s \end{pmatrix}, \quad 1 \le i \le r.$$

Since the dual connection is also flat, the solution space of this system also has dimension $s + 1$.

Proposition 4.5 *The map* $y \longmapsto Y = \begin{pmatrix} P_0 y \\ \vdots \\ P_s y \end{pmatrix}$

from the solution space

$$\{y \mid T_j y = 0, 1 \leq j \leq u\} \; \cong \; \mathrm{Hom}_D(\mathcal{M}, \mathcal{H})$$

to the solution space

$$\{Y \mid \partial_i Y = \Omega_i^t Y, 1 \leq i \leq r\} \; = \; \mathrm{Ker}\nabla^*$$

is an isomorphism of $(s+1)$*-dimensional vector spaces.*

Proof For any solution y of the scalar equation, we have to verify that $y_\alpha = P_\alpha y$ produces a covariant constant section. But this follows immediately from the formula $[\partial_i P_\alpha] = \sum_{j=0}^{s} (\Omega_i)_{j\alpha}[P_j]$ defining Ω. The map in question is therefore a well defined, linear, map between vector spaces of the same dimension. The kernel is zero because $P_0 y = y$. □

This proposition explains the unavoidable appearance of the dual D-module

$$\mathcal{M}^* = \mathrm{Hom}_{\mathcal{H}}(\mathcal{M}, \mathcal{H})$$

(or, more generally, $\mathrm{Hom}_{\mathcal{H}}(\mathcal{M}, \mathcal{F})$, where $\mathcal{F}$ is the ring of functions allowed as solutions). The action of ∂_i on an element π of $\mathcal{M}^*$ is

$$\partial_i \cdot \pi([P]) = -\pi(\partial_i \cdot [P]) + \frac{\partial}{\partial z_i}\pi([P]).$$

This gives rise to the flat connection $\nabla^* = d - \Omega^t$ whose connection form is $-\Omega^t$ with respect to the basis $[P_0]^*, \ldots, [P_s]^*$.

Thus, the solution space of the scalar system is the subspace $\mathrm{Hom}_D(\mathcal{M}, \mathcal{H})$ of $\mathcal{M}^*$, and the solution space of the matrix system is the subspace $\mathrm{Ker}\nabla^*$ of $\mathcal{H}[P_0]^* \oplus \cdots \oplus \mathcal{H}[P_s]^*$. Both solution spaces inhabit a world which is dual to the original D-module $\mathcal{M}$.

Definition 4.6 *Let* $J = (y_{(0)}, \ldots, y_{(s)})$, *where* $y_{(0)}, \ldots, y_{(s)}$ *is any basis of solutions of the scalar system.*

This choice gives also a basis $Y_{(0)},\dots,Y_{(s)}$ of solutions of the matrix system $\partial_i Y = \Omega_i^t Y$, and the fundamental solution matrix can be written

$$H = \begin{pmatrix} | & & | \\ Y_{(0)} & \cdots & Y_{(s)} \\ | & & | \end{pmatrix} = \begin{pmatrix} - & P_0 J & - \\ & \vdots & \\ - & P_s J & - \end{pmatrix}.$$

We have $\Omega^t = dHH^{-1}$, and, up to multiplication on the right by a constant invertible matrix, this equation determines H uniquely.

Finally we point out that our assumptions and choice of basis give a specific identification[1]

$$\mathcal{M} = D/(T_1,\dots,T_u) \cong \mathrm{Map}(N,\mathbb{C}^{s+1}) = \mathcal{H}^{s+1}, \quad [\sum_{i=0}^{s} f_i P_i] \leftrightarrow \begin{pmatrix} f_0 \\ \vdots \\ f_s \end{pmatrix}$$

in which the natural action of ∂_i on the left corresponds to the natural action of $\partial_i + \Omega_i$ on the right. As $P_0 = 1$, the cyclic element $[1]$ on the left corresponds to $(1,0,\dots,0)^t$ on the right. We can regard $\mathcal{H}^{s+1}$ (with D-module structure given by $d+\Omega$) as a concrete realization of the abstract D-module $\mathcal{M}$.

It is worth emphasizing that, under our assumptions, $\mathcal{M}$ is a *trivial* D-module. The equation $\Omega^t = dHH^{-1}$ can be written $H^t(d+\Omega)(H^t)^{-1} = d$, so H provides an isomorphism with the trivial D-module $\mathcal{H}^{s+1}$ (with D-module structure given by d). The cyclic element $(1,0,\dots,0)^t$ is mapped to $H^t(1,0,\dots,0)^t = J^t$. Thus, the concrete realization can also be expressed as

$$\mathcal{M} = D/(T_1,\dots,T_u) \cong \mathrm{Map}(N,\mathbb{C}^{s+1}) = \mathcal{H}^{s+1}, \quad [P] \leftrightarrow PJ$$

where ∂_i acts in the natural way on both sides.

Here are some concrete examples concerning the rank of the D-module given by a scalar system and conversion from a scalar system to a matrix system.

Example 4.7 Let

$$T_1 = \partial_1 + z_2, \quad T_2 = \partial_2 + 1.$$

What is the rank (over $\mathcal{H}$) of $D/(T_1,T_2)$? Since $[\partial_1] = -z_2[1]$ and $[\partial_2] = -[1]$, the rank is at most 1. It is 1 if $[1] \notin (T_1,T_2)$, and 0 if $[1] \in (T_1,T_2)$.

[1] More generally, $\mathcal{M} \otimes_{\mathcal{H}} \mathcal{F} \cong \mathcal{F}^{s+1}$.

But we have

$$\begin{aligned}(T_1, T_2) \ni \partial_2 T_1 - \partial_1 T_2 &= z_2\partial_2 + 1 - \partial_1\\ &= z_2(\partial_2 + 1) + 1 - (\partial_1 + z_2) \in 1 + (T_1, T_2),\end{aligned}$$

so $[1] \in (T_1, T_2)$ and the rank is 0. This implies that the solution space of the system $T_1 y = 0, T_2 y = 0$ consists only of the trivial solution $y \doteq 0$, a fact which may be verified easily by direct computation. □

Example 4.8 More generally, consider

$$T_1 = \partial_1 + f, \quad T_2 = \partial_2 + g$$

where f and g are functions of z_1, z_2. A similar argument shows that the rank of $D/(T_1, T_2)$ is 1 if and only if $\partial f/\partial z_2 = \partial g/\partial z_1$, otherwise it is 0. The system $T_1 y = 0, T_2 y = 0$ is 'consistent' if and only if $\partial f/\partial z_2 = \partial g/\partial z_1$. In this case the space of solutions is one-dimensional. □

Example 4.9 The case of constant coefficient operators reduces to a problem of commutative algebra. Namely, if $T_i = F_i(\partial_1, \ldots, \partial_r)$ where F_i is a polynomial function, then the rank of $D/(T_1, \ldots, T_u)$ is equal to the dimension of the complex vector space $\mathbb{C}[x_1, \ldots, x_r]/(F_1(x), \ldots, F_u(x))$. Whether the rank is finite depends on the nature of the polynomials F_i. For example, if $T_1 = \partial_2^2 - \partial_1\partial_2$, $T_2 = \partial_1^2$ then the rank is 4, and an example of a basis is $[1], [\partial_1], [\partial_2], [\partial_2^2]$. This choice of basis leads to an equivalent matrix system

$$\partial_1 Y = \begin{pmatrix} 0 & 1 & 0 & 0\\ 0 & 0 & 0 & 0\\ 0 & 0 & 0 & 1\\ 0 & 0 & 0 & 0 \end{pmatrix} Y, \quad \partial_2 Y = \begin{pmatrix} 0 & 0 & 1 & 0\\ 0 & 0 & 0 & 1\\ 0 & 0 & 0 & 1\\ 0 & 0 & 0 & 0 \end{pmatrix} Y, \quad \text{where } Y = \begin{pmatrix} y\\ \partial_1 y\\ \partial_2 y\\ \partial_2^2 y \end{pmatrix}.$$

A fundamental solution matrix is $H = e^{z_1 A_1 + z_2 A_2}$, where A_1, A_2 are the above coefficient matrices.

To show directly that the dimension of $\mathbb{C}[x_1, x_2]/(x_2^2 - x_1 x_2, x_1^2)$ is 4, the concept of Gröbner basis may be used. Very briefly, the point is that 'long division' of a polynomial p by polynomials $p_1, \ldots, p_k$ is a well behaved procedure if these polynomials form a Gröbner basis of the ideal $(p_1, \ldots, p_k)$. In this situation, the 'standard monomials' (the monomials which are not divisible by any of the leading terms of $p_1, \ldots, p_k$) provide a basis of the vector space $\mathbb{C}[x_1, \ldots, x_r]/(p_1, \ldots, p_k)$, which may be identified with the 'space of remainders'. This is a natural generalization of the case $r = 1$, where the standard monomials $1, x, \ldots, x^N$ provide a basis of $\mathbb{C}[x]/(p)$ for any polynomial p of degree $N + 1$.

With respect to the 'total degree/lexicographic' monomial order

$$1 < x_2 < x_1 < x_2^2 < x_1x_2 < x_1^2 < \cdots$$

it is possible to show that $x_2^2 - x_1x_2, x_1^2, x_2^3$ is a Gröbner basis of $(x_2^2 - x_1x_2, x_1^2)$. The leading terms of the Gröbner basis elements are x_1x_2, x_1^2. The standard monomials are therefore $1, x_1, x_2, x_2^2$, and the dimension of the vector space they span is 4.

To calculate the above matrices—of the linear transformations given by multiplication by x_1, x_2—it is necessary to multiply the basis vectors by x_1, x_2 then calculate the remainders after division by the elements of the Gröbner basis, which can be done by inspection in this example. In general such computations, and the computation of the Gröbner basis itself, can be carried out by standard algorithms and implemented on a computer. □

Example 4.10 With higher-order differential operators (and two or more variables, non-constant coefficients) the calculations become extremely complicated. However, Gröbner basis theory can be extended to non-commutative rings, and it is an effective tool in the case of operators with polynomial or rational coefficients (see [105]).

For example, if u, v are given functions of z_1, z_2, and

$$T_1 = \partial_2^2 - \partial_1\partial_2 - u, \quad T_2 = \partial_1^2 - v(\partial_1 - \partial_2),$$

then we have a D-module $D/(T_1, T_2)$. The rank of this D-module depends on u, v. It is easy to see that the rank is at most 4, as any operator can be expressed (modulo the relations) in terms of $[1], [\partial_1], [\partial_2], [\partial_2^2]$. It can be verified that the rank is exactly 4 if and only if $u_{z_1} = v_{z_2} = 0$. □

Example 4.11 Let

$$T_1 = \partial_1^2 + u, \quad T_2 = \partial_2 - (\partial_1^3 + \tfrac{3}{2}u\partial_1 + \tfrac{3}{4}u_{z_1}).$$

It is clear that the rank of $D/(T_1, T_2)$ is at most 2, for any particular function u, because ∂_2 is expressed in terms of ∂_1 and T_1 is a quadratic relation in ∂_1. Whether the rank is exactly 2 depends on whether there is a relation between $[1]$ and $[\partial_1]$, and this depends on the nature of u. It turns out (we shall give the proof in section 4.4) that the rank is 2 if and only if u satisfies the condition

$$4u_{z_2} = 6uu_{z_1} + u_{z_1z_1z_1},$$

which arises in the same way as the condition $f_{z_2} = g_{z_1}$ in Example 4.8 or the condition $u_{z_1} = v_{z_2} = 0$ in Example 4.10. This p.d.e. is the famous KdV equation (regarded here, however, in an unorthodox way). We shall see much more of the KdV equation in Chapter 7. The matrix system will be given later in Example 4.15. □

From matrix systems to scalar systems

Given a system

$$\partial_i Y = A_i Y, \quad 1 \le i \le r$$

of *first-order matrix p.d.e.*, it is possible to construct a system of *higher-order scalar p.d.e.*, providing the connection $d - A$ corresponding to the matrix system is flat. In view of the preceding discussion, which we are going to carry out in reverse, let us write $\nabla^* = d - A$; we want to construct a cyclic element of the D-module corresponding to the dual connection $\nabla = d + A^t$.

The connection $d - A$ corresponds to the D-module structure on

$$\mathcal{N} = \mathrm{Map}(N, \mathbb{C}^{s+1}) = \mathcal{H}^{s+1}$$

given by the formula $\partial_i \cdot Y = (\partial_i - A_i)Y$. For an element p of the dual D-module

$$\mathcal{N}^* = \mathrm{Hom}_{\mathcal{H}}(\mathcal{H}^{s+1}, \mathcal{H})$$

we have $\partial_i \cdot p(Y) = -p(\partial_i \cdot Y) + \partial p(Y)/\partial z_i$. With respect to the basis

$$p_0, \dots, p_s$$

of $\mathcal{N}^*$ defined by $p_i(Y) = y_i$, the dual connection is $\nabla = d + A^t$. Writing $p = \sum_{i=0}^{s} f_i p_i$, we have

$$\partial_i \cdot \begin{pmatrix} f_0 \\ \vdots \\ f_s \end{pmatrix} = (\partial_i + A_i^t) \begin{pmatrix} f_0 \\ \vdots \\ f_s \end{pmatrix}.$$

By definition, a cyclic element is an element p_{cyclic} such that $D \cdot p_{\text{cyclic}} = \mathcal{N}^*$. Given such an element we have

$$\mathcal{N}^* \cong D/I$$

for some left ideal I, and if $I = (T_1, \dots, T_u)$ then the required scalar system is $T_1 y = 0, \dots, T_u y = 0$. Moreover, since $p_i = P_i \cdot p_0$ for some $P_i \in D$, we obtain a specific basis $[P_0], \dots, [P_s]$ of D/I. It is known (under our assumptions) that cyclic elements exist.

In practical terms, if $p_{\text{cyclic}} = p_i$, the computation of the ideal amounts to 'declaring that $y = y_i$' and computing the scalar system for y from the matrix system for Y. Of course, the specific operators T_j are not recoverable from this calculation, just the ideal $(T_1, \ldots, T_u)$.

Example 4.12 Assume that $A^t = \Omega$ is the connection form with respect to some basis $[P_0], \ldots, [P_s]$ for a D-module $\mathcal{M} = D/(T_1, \ldots, T_u)$, with $P_0 = 1$. Then, the matrix system $\partial_i Y = A_i Y$ arises from the scalar system $T_1 y = 0, \ldots, T_u y = 0$ (as above) and we have[2] $\mathcal{N} = \mathcal{M}^*$ and $Y = \sum_{i=0}^{s} y_i [P_i]^*$. We recover the scalar system in this case by choosing the cyclic element $p_{\text{cyclic}} = p_0$. □

Example 4.13 Let us examine the role of cyclic elements in Example 4.3, which will enable us to explain the ad hoc computations there. Starting from the matrix system

$$\partial \begin{pmatrix} y_0 \\ y_1 \end{pmatrix} = \begin{pmatrix} 0 & u \\ v & 0 \end{pmatrix} \begin{pmatrix} y_0 \\ y_1 \end{pmatrix},$$

we wish to construct an equivalent scalar system (necessarily an o.d.e., of order 2 in this case).

As a candidate for a cyclic element of the dual D-module we try p_0, defined by

$$p_0(Y) = y_0,$$

then compute $\partial \cdot p_0, \partial^2 \cdot p_0, \ldots$ in the hope of generating the whole D-module. (In this endeavour there is little chance of failure, since all we require are two linearly independent elements.) First, we obtain

$$\begin{aligned} \partial \cdot p_0(Y) &= -p_0(\partial \cdot Y) + p_0(Y)' \\ &= -p_0\left(\begin{pmatrix} y_0 \\ y_1 \end{pmatrix}' - \begin{pmatrix} 0 & u \\ v & 0 \end{pmatrix} \begin{pmatrix} y_0 \\ y_1 \end{pmatrix}\right) + y_0' \\ &= -(y_0' - u y_1) + y_0' \\ &= u y_1 \\ &= u p_1(Y), \end{aligned}$$

that is, $\partial \cdot p_0 = u p_1$. This calculation is sufficient: p_0 and $\partial \cdot p_0$ generate the whole D-module (if u is not identically zero), and this proves that p_0 is a cyclic element.

[2] More generally, $\mathcal{N} = \mathrm{Hom}_{\mathcal{H}}(\mathcal{M}, \mathcal{F})$, and $\mathrm{Hom}_{\mathcal{H}}(\mathcal{N}, \mathcal{F}) = \mathcal{M} \otimes_{\mathcal{H}} \mathcal{F}$.

To find the scalar o.d.e. we compute $\partial^2 \cdot p_0$ in a similar way:

$$\begin{aligned}\partial^2 \cdot p_0(Y) = \partial \cdot (up_1)(Y) &= -up_1(\partial \cdot Y) + (up_1(Y))' \\ &= -u(y_1' - vy_0) + (u'y_1 + uy_1') \\ &= uvy_0 + u'y_1,\end{aligned}$$

that is, $\partial^2 \cdot p_0 = uvp_0 + u'p_1 = uvp_0 + (u'/u)\partial \cdot p_0$. We obtain $(\partial^2 - (u'/u)\partial - uv) \cdot p_0 = 0$, so

$$D \cdot p_0 \cong D/(\partial^2 - (u'/u)\partial - uv),$$

and the scalar o.d.e. is $(\partial^2 - (u'/u)\partial - uv)y = 0$. If $u = 0$, then $\partial \cdot p_0 = 0$, and p_0 is obviously not a cyclic element. Similarly, the element p_1 is cyclic if v is not identically zero, and the D-module can be represented as $D \cdot p_1 \cong D/(\partial^2 - (v'/v)\partial - uv)$.

The dual connection is

$$\nabla = \partial + A^t = \partial + \begin{pmatrix} 0 & v \\ u & 0 \end{pmatrix}$$

and the same result could have been obtained by applying this operator repeatedly. In the case of p_0 we have

$$\nabla \begin{pmatrix} 1 \\ 0 \end{pmatrix} = \begin{pmatrix} 0 \\ u \end{pmatrix}, \quad \nabla^2 \begin{pmatrix} 1 \\ 0 \end{pmatrix} = \nabla \begin{pmatrix} 0 \\ u \end{pmatrix} = \begin{pmatrix} uv \\ u' \end{pmatrix} = uv \begin{pmatrix} 1 \\ 0 \end{pmatrix} + \frac{u'}{u} \nabla \begin{pmatrix} 1 \\ 0 \end{pmatrix},$$

that is, $\nabla^2 - (u'/u)\nabla - uv$ annihilates the vector $\begin{pmatrix} 1 \\ 0 \end{pmatrix}$.

For the system

$$\partial \begin{pmatrix} y_0 \\ y_1 \end{pmatrix} = \begin{pmatrix} 1 & 0 \\ 0 & 1 \end{pmatrix} \begin{pmatrix} y_0 \\ y_1, \end{pmatrix},$$

neither p_0 nor p_1 are cyclic, but $p_0 + zp_0$ (for example) is cyclic. An explicit general construction of cyclic elements in the o.d.e. case appears in [65]. □

4.3 Differential equations with spectral parameter

In the theory of soliton equations (and in differential geometry) one finds partial differential equations with an auxiliary parameter λ. There is no explanation for this, it is just an experimental observation. The parameter sometimes appears most naturally in the scalar form

$$T_j y = 0, \quad 1 \leq j \leq u$$

of the equations, and sometimes in the matrix form

$$\partial_i Y = A_i Y, \quad 1 \le i \le r,$$

depending on the particular example. Even if λ appears in a simple way in the scalar equations, it may appear in a complicated way in the matrix version (or vice versa).

Example 4.14 The quantum differential equations

$$\mathrm{h}\partial_i \Psi = b_i \circ_t \Psi, \quad 1 \le i \le r$$

of Chapter 3 are a matrix system of p.d.e. with very simple dependence on the spectral parameter $\lambda = \mathrm{h}$. □

Example 4.15 The KdV equation

$$4u_t = 6uu_x + u_{xxx}$$

will be discussed in detail in Chapters 7 and 8. We have already seen this equation in Example 4.11 (instead of z_1, z_2 we write x, t and assume these, as well as u, are real, which is the standard convention). It arises from the scalar system

$$Ly = 0, \quad (\partial_t - P)y = 0$$

where $L = \partial_x^2 + u$, $P = \partial_x^3 + \frac{3}{2}u\partial_x + \frac{3}{4}u_x$ (the notation L, P is also standard). Namely, the KdV equation is the condition that the D-module $D/(L, \partial_t - P)$ has rank 2 (we shall prove this in section 4.4).

The KdV equation also arises from the scalar system

$$(L - \lambda)y = 0, \quad (\partial_t - P)y = 0.$$

This is the origin of the term 'spectral parameter', because the first equation says that y is an eigenfunction of the Schrödinger operator L.

The matrix form of this system is less trivial. Let us choose the basis $[1], [\partial_x]$ of $D/(L - \lambda, \partial_t - P)$. Then the matrices Ω_1, Ω_2 of ∂_x, ∂_t (see section 4.4) are

$$\Omega_1 = \begin{pmatrix} 0 & \lambda - u \\ 1 & 0 \end{pmatrix}, \quad \Omega_2 = \begin{pmatrix} -u_x/4 & \lambda^2 - u\lambda/2 - u^2/2 - u_{xx}/4 \\ \lambda + u/2 & u_x/4 \end{pmatrix}.$$

If we introduce the transposed matrices $A_i = \Omega_i^t$, then the original scalar system is equivalent to the matrix system

$$\partial_x Y = A_1 Y, \quad \partial_t Y = A_2 Y \qquad \text{for } Y = \begin{pmatrix} y_0 \\ y_1 \end{pmatrix} = \begin{pmatrix} y \\ \partial_x y \end{pmatrix}.$$

This system is consistent, that is, has a two-dimensional solution space, if and only if u is a solution of the KdV equation.

Note that, if we attempt to return to the scalar system by declaring that $y = y_0$, we do not necessarily obtain the individual operators $L - \lambda$, $\partial_t - P$; we obtain only the left ideal which they generate. □

Example 4.16 The harmonic map equation is, like the KdV equation, a non-linear p.d.e. which can be regarded as the compatibility condition for a system of linear equations with spectral parameter. Unlike the case of the KdV equation, however, the spectral parameter is essential right from the start. Moreover, it arises in a very simple way in the matrix version, whereas the spectral parameter in the KdV case appears most simply in the scalar version.

A harmonic map may be defined as a critical point of the energy functional. For maps $\phi : \mathbb{R} \to M$ where M is a Riemannian manifold, this means that ϕ is a geodesic. For maps $\phi : \mathbb{R}^2 \to G$, where G is a compact Lie group[3] with a bi-invariant Riemannian metric, the critical point condition (Euler–Lagrange equation) is

$$\partial_x(\phi^{-1}\partial_x\phi) + \partial_y(\phi^{-1}\partial_y\phi) = 0.$$

For maps $\phi : \mathbb{R}^{1,1} \to G$, the critical point condition is

$$\partial_x(\phi^{-1}\partial_x\phi) - \partial_y(\phi^{-1}\partial_y\phi) = 0.$$

In fact these hold whether or not G is compact.

The complex equation

$$\partial_1(\phi^{-1}\partial_1\phi) + \partial_2(\phi^{-1}\partial_2\phi) = 0$$

also makes sense, though not necessarily as an Euler–Lagrange equation. Here we assume $\phi : \mathbb{C}^2 \to G^{\mathbb{C}}$, where $G^{\mathbb{C}}$ is the complexification of G. Let $C : G^{\mathbb{C}} \to G^{\mathbb{C}}$ be the conjugation map[4] with respect to the real form G, and let $c : \mathfrak{g}^{\mathbb{C}} \to \mathfrak{g}^{\mathbb{C}}$ be the induced conjugation map of Lie algebras. Evidently, the original harmonic map equation is obtained by putting $z_1 = x, z_2 = y$ and imposing the reality condition $C(\phi) = \phi$.

[3] We are assuming that G is a matrix group here, although the abstract case can be handled easily.

[4] If $G = U_n$, then $G^{\mathbb{C}} = GL_n\mathbb{C}$. The maps $C : G^{\mathbb{C}} \to G^{\mathbb{C}}$ and $c : \mathfrak{g}^{\mathbb{C}} \to \mathfrak{g}^{\mathbb{C}}$ are given respectively by $C(A) = A^{*-1}$ and $c(A) = -A^*$.

However, a different form of the complexified equation is appropriate for the present context. With $z = x + iy$ and

$$\partial = \frac{\partial}{\partial z} = \frac{1}{2}\left(\frac{\partial}{\partial x} - i\frac{\partial}{\partial y}\right), \quad \bar{\partial} = \frac{\partial}{\partial \bar{z}} = \frac{1}{2}\left(\frac{\partial}{\partial x} + i\frac{\partial}{\partial y}\right)$$

the harmonic map equation becomes

$$\partial(\phi^{-1}\bar{\partial}\phi) + \bar{\partial}(\phi^{-1}\partial\phi) = 0.$$

Note that $\phi^{-1}\bar{\partial}\phi, \phi^{-1}\partial\phi$ take values in $\mathfrak{g}^{\mathbb{C}}$ and satisfy $c(\phi^{-1}\partial\phi) = \phi^{-1}\bar{\partial}\phi$. This can be obtained by putting $z_1 = z, z_2 = \bar{z}$ (and $C(\phi) = \phi$) in the complex equation

$$\partial_1(\phi^{-1}\partial_2\phi) + \partial_2(\phi^{-1}\partial_1\phi) = 0.$$

Similarly, putting $z_1 = x + y, z_2 = x - y$ and $C(\phi) = \phi$ gives the equation for harmonic maps $\phi : \mathbb{R}^{1,1} \to G$.

The main observation is that the complex equation (and hence each real version) can be regarded as the compatibility condition for a system of linear equations if we introduce the $\mathfrak{g}^{\mathbb{C}}$-valued 1-form

$$\alpha^\lambda = \tfrac{1}{2}(1 - \tfrac{1}{\lambda})(\phi^{-1}\partial_1\phi)dz_1 + \tfrac{1}{2}(1 - \lambda)(\phi^{-1}\partial_2\phi)dz_2$$

with spectral parameter λ. Namely, a simple calculation, to be given in a more general context in a moment, shows that the connection $d + \alpha^\lambda$ is flat for every (non-zero) value of λ if and only if ϕ satisfies the complexified harmonic map equation $\partial_1(\phi^{-1}\partial_2\phi) + \partial_2(\phi^{-1}\partial_1\phi) = 0$.

The corresponding linear system is

$$\partial_1 Y = \tfrac{1}{2}(1 - \tfrac{1}{\lambda})(\phi^{-1}\partial_1\phi)^t Y$$
$$\partial_2 Y = \tfrac{1}{2}(1 - \lambda)(\phi^{-1}\partial_2\phi)^t Y.$$

Scalar differential equations do not arise naturally at this point, but conversion to scalar form can offer advantages in some situations.

A feature of this example is that the specific ϕ-dependence of the 1-form α^λ is illusory. We state this fact for the complex version of the equations:

Proposition 4.17 *Let*

$$\alpha^\lambda = \tfrac{1}{2}(1 - \tfrac{1}{\lambda})\alpha_1 dz_1 + \tfrac{1}{2}(1 - \lambda)\alpha_2 dz_2$$

be a $\mathfrak{g}^{\mathbb{C}}$-valued 1-form on $\mathbb{C}^2$. If $d + \alpha^\lambda$ is flat for every (non-zero) value of λ, then there exists a map $\phi : \mathbb{C}^2 \to G^{\mathbb{C}}$ such that

$$\alpha_1 = \phi^{-1}\partial_1\phi, \quad \alpha_2 = \phi^{-1}\partial_2\phi$$

and this map satisfies the equation

$$\partial_1(\phi^{-1}\partial_2\phi) + \partial_2(\phi^{-1}\partial_1\phi) = 0.$$

Conversely, let $\phi : \mathbb{C}^2 \to G^{\mathbb{C}}$ *be a map which satisfies the equation* $\partial_1(\phi^{-1}\partial_2\phi) + \partial_2(\phi^{-1}\partial_1\phi) = 0$. *Then the 1-form* $\alpha^\lambda = \frac{1}{2}(1 - \frac{1}{\lambda})(\phi^{-1}\partial_1\phi)dz_1 + \frac{1}{2}(1 - \lambda)(\phi^{-1}\partial_2\phi)dz_2$ *defines a flat connection* $d + \alpha^\lambda$.

Proof The flatness condition for $\alpha^\lambda = \frac{1}{2}(1 - \frac{1}{\lambda})\alpha_1 dz_1 + \frac{1}{2}(1 - \lambda)\alpha_2 dz_2$, that is, $d\alpha^\lambda + \alpha^\lambda \wedge \alpha^\lambda = 0$, breaks down (on taking coefficients of powers of λ) into three equations:

$$\partial_2\alpha_1 - \tfrac{1}{2}[\alpha_1, \alpha_2] = 0$$

$$-\partial_2\alpha_1 + \partial_1\alpha_2 + [\alpha_1, \alpha_2] = 0$$

$$-\partial_1\alpha_2 - \tfrac{1}{2}[\alpha_1, \alpha_2] = 0.$$

The middle equation says that $d + \alpha$ is flat, where $\alpha = \alpha_1 dz_1 + \alpha_2 dz_2$. By the Basic Lemma of section 3.1 (more precisely, Corollary 4.24 of appendix 4.5), there exists a map $\phi : \mathbb{C}^2 \to G^{\mathbb{C}}$ such that $\alpha = \phi^{-1}d\phi$. The other two equations, in the presence of the middle equation, are equivalent to $\partial_1\alpha_2 + \partial_2\alpha_1 = 0$, the complex version of the harmonic map equation. Both implications of the proposition follow from this. □

The harmonic map equations are obtained by imposing the reality conditions

$$z_1 = z, \quad z_2 = \bar{z}, \quad c(\alpha_1) = \alpha_2$$

or

$$z_1 = x + y, \quad z_2 = x - y, \quad c(\alpha_1) = \alpha_1, \quad c(\alpha_2) = \alpha_2.$$

We shall discuss the harmonic map equation in more detail in Chapters 7 and 8. For the moment we just mention one fundamental link with differential geometry (see also appendix 4.5). The Gauss map ϕ of a surface in $\mathbb{R}^3$ may be regarded as a map from (a local chart of) the surface into the Grassmannian $Gr_2(\mathbb{R}^3)$, and this Grassmannian may be regarded as a submanifold of the Lie group $G = O_3$ (the conjugacy class of the matrix $\mathrm{diag}(1, -1, -1)$ with diagonal entries $1, -1, -1$). It is well known that ϕ is harmonic if and only if the surface has constant mean curvature. Thus, harmonic maps are generalizations of such surfaces. Further background information on harmonic maps in differential geometry can be found in [39] and [40]. □

Examples 4.15 and 4.16 illustrate the fundamental concept of an *integrable p.d.e.*—one says (rather optimistically) that a p.d.e. is an integrable p.d.e. if it can be written as the compatibility condition for a system of linear equations, that is, if it can be written as the zero curvature condition for a connection which is given in terms of some auxiliary function(s) $u = u(z_1, \ldots, z_r)$ and possibly a spectral parameter λ. This is too vague to be regarded as a satisfactory definition of integrable system (see section 7.6), but it is a useful starting point.

We shall say more about these examples later. For the moment we just point out that a solution of the KdV equation or the harmonic map equation gives a particular flat connection. Thus, independent methods of constructing solutions of the KdV equation or the harmonic map equation (and there are many such methods) would enable us to construct examples of flat connections in a systematic way. Example 4.14 is different. The quantum cohomology of a manifold does indeed give a particular flat connection, but we have not (yet) seen this as corresponding to a solution of an integrable p.d.e.; we shall return to this matter in Chapter 9.

We should emphasize that, in this chapter, we have been concerned only with the relations between various equations (the integrable p.d.e., the scalar system, the matrix system). We shall discuss solutions of these equations later.

4.4 Flat connections from extensions of D-modules

We have seen, roughly speaking, that solutions of integrable p.d.e. correspond to flat connections. Given one such object, we now ask whether it can be 'extended' to a family of objects of the same type. In the language of connections, the question is whether a given flat connection can be extended to a flat connection over a larger base space. In this section we present a construction method from the D-module point of view.

For simplicity we assume that the original system involves one variable x, and that the extension involves one additional variable t. Thus, we start with a D-module $D_x/(T_0)$ of rank $s+1$, where

$$T_0 = \partial_x^{s+1} + a_s(x)\partial_x^s + \cdots + a_1(x)\partial_x + a_0(x).$$

We wish to extend this to a D-module $D_{x,t}/(T, T_1, T_2, \ldots)$ *of the same rank* $s+1$ by extending T_0 to a t-family T (with $T|_{t=0} = T_0$) and adjoining further partial differential operators $T_1, T_2, \ldots$ as necessary.

As we have seen, in general it is hard to predict the rank of such a D-module, but if we adjoin an operator of the form $\partial_t - P$, where P does not involve ∂_t, then it is obvious that

$$\text{rank } D_{x,t}/(T, \partial_t - P) \leq s+1$$

since the new relation may be used to eliminate ∂_t. We shall derive a condition for the rank to be exactly $s+1$.

We describe the method in the following three steps.

Step 1. Let $T_0 = \partial_x^{s+1} + a_s(x)\partial_x^s + \cdots + a_1(x)\partial_x + a_0(x)$. The D-module $D_x/(T_0)$ may be identified, as a module over the space of functions $\mathcal{H}_x = \text{Map}(N_x, \mathbb{C})$, with the space of sections of the trivial bundle $N_x \times \mathbb{C}^{s+1}$. It is a free module of rank $s+1$, with basis $[1], [\partial_x], \ldots, [\partial_x^s]$, and we have a flat connection ∇ defined by $\nabla_{\partial_x}[\partial_x^i] = [\partial_x^{i+1}]$ for $0 \leq i \leq s$ (and in fact $\nabla_{\partial_x}[X] = [\partial_x X]$ for any $X \in D_x$).

Step 2. Next consider a t-family of operators

$$T = \partial_x^{s+1} + a_s(x,t)\partial_x^s + \cdots + a_1(x,t)\partial_x + a_0(x,t)$$

such that $T(x,0) = T_0(x)$. Then $D_x/(T)$ is a t-family of D-modules of rank $s+1$. We may identify $(D_x \otimes \mathcal{H}_t)/(T)$ with the space of sections of the extended trivial bundle $N_x \times N_t \times \mathbb{C}^{s+1}$ over $N_x \times N_t$. Again $[1], [\partial_x], \ldots, [\partial_x^s]$ provide a local basis of sections at each point.

Step 3. Let P be a fixed element of $D_{x,t}$. We define a connection ∇ on the extended bundle as follows: ∇_{∂_x} is defined in the same way as before, and ∇_{∂_t} is defined by $\nabla_{\partial_t}[\partial_x^i] = [\partial_x^i P]$ for $0 \leq i \leq s$ (this depends on the particular choice of representative ∂_x^i).

Proposition 4.18 *The extended connection ∇ is flat if and only if P satisfies $[\partial_t - P, T] = 0$ mod T.*

The commutator is computed in the ring $D_{x,t}$, but in fact it can be expressed entirely in terms of D_x. To see this, note that $[\partial_t, a_i\partial_x^i] = \partial_t a_i \partial_x^i - a_i\partial_x^i\partial_t = a_i\partial_t\partial_x^i + (\partial a_i/\partial_t)\partial_x^i - a_i\partial_x^i\partial_t = (\partial a_i/\partial_t)\partial_x^i$. Thus, if we define T_t to be the result of differentiating the coefficients of T with respect to t, the condition can be expressed as

$$T_t = [P, T] \text{ mod } T.$$

Before giving the proof we remark that the condition can also be written

$$[\partial_t - P, (T)] \subseteq (T).$$

Proof The connection is flat if and only if $\nabla_{\partial_x}\nabla_{\partial_t} = \nabla_{\partial_t}\nabla_{\partial_x}$ (as operators on sections of the bundle). It suffices to check that $\nabla_{\partial_x}\nabla_{\partial_t}[\partial_x^i] = \nabla_{\partial_t}\nabla_{\partial_x}[\partial_x^i]$

for $0 \le i \le s$. If $0 \le i \le s-1$ the statement follows trivially from the definition of ∇. For $i = s$, we have $\nabla_{\partial_x}\nabla_{\partial_t}[\partial_x^s] = [\partial_x^{s+1}P]$, and $\nabla_{\partial_t}\nabla_{\partial_x}[\partial_x^s] = \nabla_{\partial_t}[\partial_x^{s+1}] = \nabla_{\partial_t}[-a_s\partial_x^s - \cdots - a_0] = [(-a_s\partial_x^s - \cdots - a_0)P - (a_s)_t\partial_x^s - \cdots - (a_0)_t]$. Therefore, the flatness condition is $TP + T_t = 0 \bmod T$, which is equivalent to $T_t = [P, T] \bmod T$, as required. □

Remark The condition $[\partial_t - P, T] = 0 \bmod T$ can be regarded as the intrinsic scalar version of the matrix zero curvature condition $[\partial_x - \Omega_1^t, \partial_t - \Omega_2^t] = 0$. Evidently it is weaker than the condition $[\partial_t - P, T] = 0$, in general. Example 4.15 is very special, as in this case the scalar version *can* be written in the form $[\partial_t - P, T] = 0$.

In $D_{x,t}$ we have $\partial_x\partial_t = \partial_t\partial_x$, so the condition $\nabla_{\partial_x}\nabla_{\partial_t} = \nabla_{\partial_t}\nabla_{\partial_x}$ is simply the condition that the ring $D_{x,t}$ acts on the space of sections of the bundle. Thus, we have investigated whether the space of sections of the extended bundle is a $D_{x,t}$-module. It is possible to take the opposite point of view: we could investigate when the $D_{x,t}$-module $D_{x,t}/(T, \partial_t - P)$ has rank $s+1$.

Corollary 4.19 *The D-module $D_{x,t}/(T, \partial_t - P)$ has rank $s+1$ if and only if $[\partial_t - P, T] = 0$ mod T.*

Proof Assume that the rank of $D_{x,t}/(T, \partial_t - P)$ is $s+1$. Then, the map $D_{x,t}/(T, \partial_t - P) \to (D_x \otimes \mathcal{H}_t)/(T)$, $[\partial_x^i] \mapsto [\partial_x^i]$ is an isomorphism, and the extended connection is flat because $\partial_x\partial_t = \partial_t\partial_x$. Hence $[\partial_t - P, T] = 0 \bmod T$ by the proposition.

Conversely, assume that $[\partial_t - P, T] = 0 \bmod T$. Then the extended connection on $(D_x \otimes \mathcal{H}_t)/(T)$ is flat, hence we obtain an action of $D_{x,t}$ as in the proposition. For this action, $[1]$ is still a cyclic element, and T is still a relation. By construction, $\partial_t - P$ annihilates $[1]$ and is therefore also a relation. It follows that the map $D_{x,t}/(T, \partial_t - P) \to (D_x \otimes \mathcal{H}_t)/(T)$, $[\partial_x^i] \mapsto [\partial_x^i]$ is an isomorphism, hence $\operatorname{rank} D_{x,t}/(T, \partial_t - P) = s+1$. □

The arguments here generalize—at least in one direction—to the following situation:

Theorem 4.20 *Let each T_i be a t-family of differential operators in $z_1, \dots, z_r$ such that the D-module $\mathcal{M} = D_{z_1,\dots,z_r}/(T_1, \dots, T_u)$ has rank $s+1$ for each value of t. Let P be a t-family of differential operators in $z_1, \dots, z_r$ such that*

$$[\partial_t - P, (T_1, \dots, T_u)] \subseteq (T_1, \dots, T_u).$$

Then the extended D-module $D_{z_1,\dots,z_r,t}/(T_1, \dots, T_u, \partial_t - P)$ also has rank $s+1$.

Proof Let $[P_0],\ldots,[P_s]$ be a basis of $D_{z_1,\ldots,z_r}/(T_1,\ldots,T_u)$. The main task is to prove that the extended connection satisfies $\nabla_{\partial_i}\nabla_{\partial_t}[P_j] = \nabla_{\partial_t}\nabla_{\partial_i}[P_j]$ for all i,j. Working in the ring $D_{z_1,\ldots,z_r}$, let us write

$$\partial_i P_j = \sum_{k=0}^{s} (\Omega_i)_{kj} P_k + U$$

where $U \in (T_1,\ldots,T_u)$. Then we have $\nabla_{\partial_i}\nabla_{\partial_t}[P_j] = [\partial_i P_j P]$ with $\partial_i P_j P = \sum_{i=0}^{s} (\Omega_i)_{kj} P_k P + UP$, and

$$\begin{aligned}\nabla_{\partial_t}\nabla_{\partial_i}[P_j] &= \nabla_{\partial_t} \sum_{i=0}^{s} (\Omega_i)_{kj}[P_k] \\ &= \sum_{i=0}^{s} (\Omega_i)_{kj} \nabla_{\partial_t}[P_k] + \sum_{i=0}^{s} \tfrac{\partial}{\partial t}(\Omega_i)_{kj}[P_k] \\ &= \Big[\sum_{i=0}^{s} (\Omega_i)_{kj} P_k P + \tfrac{\partial}{\partial t}(\Omega_i)_{kj} P_k\Big].\end{aligned}$$

Subtracting $\nabla_{\partial_t}\nabla_{\partial_i}[P_j]$ from $\nabla_{\partial_i}\nabla_{\partial_t}[P_j]$, we obtain

$$\begin{aligned}UP - \sum_{i=0}^{s} \tfrac{\partial}{\partial t}(\Omega_i)_{kj} P_k &= [U,P] + PU - (\partial_i P_j - U)_t \\ &= [U,P] + PU + U_t \\ &= [\partial_t - P, U] + PU,\end{aligned}$$

and this is in $(T_1,\ldots,T_u)$, hence $\nabla_{\partial_i}\nabla_{\partial_t}[P_j] = \nabla_{\partial_t}\nabla_{\partial_i}[P_j]$, as required. □

In particular, we may attempt to make a series of extensions of the above type:

Corollary 4.21 *Let T be a t-family of differential operators in x such that the D-module $\mathcal{M} = D_x/(T)$ has rank $s+1$ for each value of t. Let P be a t-family of differential operators in x such that $[\partial_t - P, T] = 0$ mod T. Let Q be an s-family of differential operators in x such that $[\partial_s - Q, T] = 0$ mod T. Assume further that $[\partial_t - P, \partial_s - Q] = 0$ mod T. Then the extended D-module $D_{x,t,s}/(T, \partial_t - P, \partial_s - Q)$ also has rank $s+1$.*

Proof To be able to apply the theorem in two steps, we have to show that $[\partial_t - P, \partial_s - Q] \in (T, \partial_t - P)$. But this follows immediately from the hypothesis $[\partial_t - P, \partial_s - Q] = 0$ mod T. □

In particular, the corollary applies when $[\partial_t - P, T] = 0$, $[\partial_s - Q, T] = 0$, and $[\partial_t - P, \partial_s - Q] = 0$. These are much stronger hypotheses, but we shall meet an important example (the KdV hierarchy) in Chapter 7.

Although our extension procedure appears to produce rather special D-modules of the form $D_{x,t}/(T, \partial_t - P)$, it is in fact rather general. Namely, in a 'generic' D-module of rank $s+1$ of the form $D_{x,t}/I$, the elements

$$[1],\ \partial_x \cdot [1],\ \partial_x^2 \cdot [1],\ \ldots,\ \partial_x^s \cdot [1]$$

will be independent. As the rank is $s+1$, one has a relation of the form $T = \partial_x^{s+1} + a_s \partial_x^s + \cdots + a_0$, that is, $T \cdot [1] = 0$. The element $[\partial_t]$ can be expressed as a linear combination of the above basis vectors, that is, $(\partial_t - P) \cdot [1] = 0$ for some polynomial P in ∂_x. Hence the D-module is of the type constructed in this section.

4.5 Appendix: connections in differential geometry

Connections are fundamental in this book. In the same spirit as Chapter 1 we shall give a brief review here, with motivation and historical comments.

Although connections are usually treated as global objects on manifolds, we shall only need the local theory. Our connections will almost always be flat connections (that is, with zero curvature). A flat connection on a simply connected manifold is trivial, in the sense that it is equivalent to a constant connection. Nevertheless, such connections are of great interest for integrable systems theory because the relation between a connection and its trivialization is essentially the relation between a differential equation and its solution. If further conditions are imposed on the connection, or if the manifold is not simply connected, then the situation becomes even more interesting.

We divide this essay into three parts. First, as motivation, we describe two standard connections which arise in the elementary theory of surfaces in Euclidean space. Then we review briefly the basic notation for abstract connections in vector bundles. Finally we return to surface theory in order to explain how the 'integrable systems approach' arises naturally from the search for differential geometric invariants associated to symmetry groups. The 'obvious' symmetry groups studied by Klein and Cartan are finite-dimensional Lie groups, whereas the 'hidden' symmetry groups of more recent origin are infinite-dimensional Kac–Moody Lie groups. Both arise from the procedure described here.

Part 1: how connections arise in surface theory

It is worth starting with surface theory, because surfaces provide the simplest non-trivial examples of connections, and when they appeared over a century ago they were immediately linked with integrable systems. This link faded away throughout most of the 20th century under the twin pressures of computational messiness and abstract machinery, both of which were necessary steps towards dealing with higher-dimensional submanifolds. Eventually, however, integrable systems theory has been reunited with classical differential geometry, with connections as the basic underlying concept.

Let us consider a surface S in Euclidean space $\mathbb{R}^3$. Vector fields on the surface (sections of the tangent bundle TS) may be regarded as maps

$$v = (v_1, v_2, v_3) : S \to \mathbb{R}^3$$

such that $v(s)$ is tangential to S at s. Let us choose a local coordinate chart $\psi : U \to S$, where U is an open subset of the plane $\mathbb{R}^2$ with standard coordinates x, y. Let $p : U \to \mathbb{R}^3$ be the composition of ψ with the inclusion map $S \to \mathbb{R}^3$:

$$p : U \xrightarrow{\psi} S \xrightarrow{i} \mathbb{R}^3.$$

Then, the maps $p_x, p_y : U \to \mathbb{R}^3$ are (locally defined) vector fields on S. They correspond to the vector fields $\partial_x = \partial/\partial x, \partial_y = \partial/\partial y$ on U. It is common practice to use ∂_x, ∂_y and p_x, p_y interchangeably, and to omit explicit mention of ψ, and we shall follow this convention.

There are two natural connections in this situation. The first is simply the *directional derivative:* the derivative of a map $w : S \to \mathbb{R}^3$ in the direction of the (local) vector field $v = f\partial_x + g\partial_y$ is defined to be

$$\nabla_v^{\mathbb{R}^3} w = f\frac{\partial w}{\partial x} + g\frac{\partial w}{\partial y}.$$

The result is a (locally defined) map $S \to \mathbb{R}^3$.

The second connection is a differential operator on vector fields. If w is a vector field on S, the *covariant derivative* of w in the direction of v (at a point s) is defined to be the orthogonal projection of $(\nabla_v^{\mathbb{R}^3} w)(s)$ on the tangent space T_sS:

$$\nabla_v^{TS} w = \pi_{TS}(\nabla_v^{\mathbb{R}^3} w).$$

Here we regard the tangent space as a linear subspace of $\mathbb{R}^3$ and define orthogonal projection by using the Euclidean inner product $\langle\ ,\ \rangle$.

In fact the operators $\nabla^{\mathbb{R}^3}, \nabla^{TS}$ are both referred to as covariant derivative operators, or connections, and their basic property is that they are differential operators with respect to w (in the sense that $\nabla_v fw = f\nabla_v w + df(v)w$, for any function $f : S \to \mathbb{R}$), yet 'linear over the ring of functions' in v (i.e. $\nabla_{fv} w = f\nabla_v w$). Clearly they have quite different properties: $\nabla_v^{\mathbb{R}^3}$ operates on functions $S \to \mathbb{R}^3$ (sections of the trivial vector bundle $S \times \mathbb{R}^3$), and uses only the vector space structure of $\mathbb{R}^3$, whereas ∇_v^{TS} operates on vector fields (sections of the tangent bundle TS), and depends on the inner product $\langle\ ,\ \rangle$ of $\mathbb{R}^3$ as well as the particular embedding (or immersion) of S in $\mathbb{R}^3$.

From a connection we obtain the concept of curvature. Before giving definitions, we comment briefly on the meaning of curvature for a surface. The Gauss curvature of S measures the extent to which the operators $\nabla_{\partial_x}^{TS}, \nabla_{\partial_y}^{TS}$ fail to commute; the connection ∇^{TS} is said to be flat if and only if

$$\nabla_{\partial_x}^{TS}\nabla_{\partial_y}^{TS} = \nabla_{\partial_y}^{TS}\nabla_{\partial_x}^{TS}.$$

The connection $\nabla^{\mathbb{R}^3}$ is flat in this sense because directional derivatives commute. On the other hand, the connection ∇^{TS} is flat precisely for surfaces which are locally isometric to the Euclidean plane (see Example 4.25). There is another kind of curvature, the mean curvature of S, and this is zero precisely for minimal surfaces, that is, surfaces which are critical points for the area functional.

It is traditional to define Gauss and mean curvature functions K, H in the following way, making use of the 'trivial' connection $\nabla^{\mathbb{R}^3}$. First, we note that p_x, p_y span the tangent space of S at each point, and $\nu = p_x \times p_y / |p_x \times p_y|$ is the unit normal vector. We take the directional derivatives of p_x, p_y, ν with respect to ∂_x, ∂_y. Since p_x, p_y, ν span $\mathbb{R}^3$ the result has the following form

$$\begin{aligned}
\nabla_{\partial_x}^{\mathbb{R}^3} p_x &= \boxed{*}\, p_x + \boxed{*}\, p_y + \Box\, \nu \\
\nabla_{\partial_x}^{\mathbb{R}^3} p_y &= \boxed{*}\, p_x + \boxed{*}\, p_y + \Box\, \nu \\
\nabla_{\partial_x}^{\mathbb{R}^3} \nu &= \Box\, p_x + \Box\, p_y + \Box\, \nu
\end{aligned}$$

with a similar list for derivatives with respect to y. Although the connection $\nabla^{\mathbb{R}^3}$ (and its pullback to U) is trivial, we have expressed it in a non-trivial way by choosing a non-trivial basis of $\mathbb{R}^3$, adapted to the surface S; the 3×3 matrices

$$\begin{pmatrix} \boxed{*} & \boxed{*} & \Box \\ \boxed{*} & \boxed{*} & \Box \\ \Box & \Box & \Box \end{pmatrix}$$

which arise are called the (local) connection matrices of $\nabla^{\mathbb{R}^3}$ with respect to the 'moving frame' p_x, p_y, ν. Evidently the coefficients denoted $\boxed{*}$ form the connection matrices of ∇^{TS}, with respect to p_x, p_y. The definitions of the functions K, H are

$$K = \det T,\ H = \tfrac{1}{2}\operatorname{trace} T, \quad \text{where } T = \begin{pmatrix} E & F \\ F & G \end{pmatrix}^{-1} \begin{pmatrix} L & M \\ M & N \end{pmatrix}$$

and where the matrices whose components are

$$E = \langle p_x, p_x \rangle, F = \langle p_x, p_y \rangle, G = \langle p_y, p_y \rangle$$

and

$$L = \langle p_{xx}, \nu \rangle, M = \langle p_{xy}, \nu \rangle, N = \langle p_{yy}, \nu \rangle$$

represent the first and second fundamental forms of S. These six functions and their derivatives fill up the boxes in the 3×3 connection matrices above, in a somewhat complicated way, and in fact the Gauss and mean curvature functions can be extracted directly from these connection matrices.

The Gauss curvature of S comes entirely from the coefficients $\boxed{*}$. It turns out that these coefficients—and hence the Gauss curvature—can be expressed entirely in terms of E, F, G (a priori, they depend on E, F, G, L, M, N). This fact, Gauss' 'Theorema Egregium', says that the connection ∇^{TS} can be expressed in terms of the Riemannian metric of TS. It is usually said that the Gauss curvature is intrinsic, as it comes entirely from ∇^{TS}, while the mean curvature is extrinsic, being related to the embedding of S in $\mathbb{R}^3$. This is somewhat misleading, as the connection ∇^{TS} came from the embedding as well, but the point is that TS exists independently of any embedding, and it turns out that the Gauss curvature can be defined whenever TS has a Riemannian metric, not just for the Riemannian metric which is inherited from $\mathbb{R}^3$. In fact, for an abstract manifold M with an abstract Riemannian metric g (on TM), there is a *unique* connection ∇^{TM} whose 'torsion tensor' is zero and which is compatible with g in the sense that

$$v_1 g(v_2, v_3) = g(\nabla^{TM}_{v_1} v_2, v_3) + g(v_2, \nabla^{TM}_{v_1} v_3)$$

for all vector fields v_1, v_2, v_3. This connection is called the Riemannian connection, or the Levi–Cività connection. If $M = S$ is a surface in Euclidean space, and g is the induced Riemannian metric, then the Levi–Cività connection is exactly the above connection ∇^{TS}.

Let us return to the 3×3 connection matrices. It turns out that they are more than vehicles carrying the Gauss and mean curvature; they actually characterize the surface. To see this, note that they can be written in the form

$$F^{-1}F_x, \quad F^{-1}F_y$$

where F is the $\mathrm{GL}_3\mathbb{R}$-valued map

$$F = \begin{pmatrix} | & | & | \\ p_x & p_y & \nu \\ | & | & | \end{pmatrix}.$$

(It will be clear from the context whether F refers to the above matrix function or the previously defined scalar function $F = \langle p_x, p_y \rangle$.) The (local) connection 1-form is defined to be

$$\alpha = (F^{-1}F_x)dx + (F^{-1}F_y)dy.$$

It is a 3×3 matrix of 1-forms on U, or 'matrix-valued 1-form'. It contains the same information as the connection itself, that is, the operator $\nabla^{\mathbb{R}^3}$ can be recovered from α. Similarly, the connection ∇^{TS} has a 2×2 matrix-valued local connection 1-form which determines it (the top left 2×2 block of α).

The Fundamental Theorem of Surface Theory says that the connection form α characterizes the surface S, in the following sense: if $\alpha = \alpha_1 dx + \alpha_2 dy$ is a 3×3 matrix-valued 1-form constructed from six arbitrary functions E, F, G, L, M, N in the above manner, which satisfies the obvious necessary conditions, then it must arise from a (piece of a) surface S, and all such (pieces of) surfaces are equivalent under Euclidean motions. The 'obvious' necessary conditions are given by

$$d\alpha + \alpha \wedge \alpha = 0.$$

Indeed, this condition follows immediately by taking the exterior derivative of $\alpha = F^{-1}dF$. From the point of view of classical differential geometry the condition consists of a complicated system of non-linear partial differential equations for the functions E, F, G, L, M, N, known as the Gauss–Codazzi equations, and it seems a small miracle that these equations are also sufficient, that is, that they are precisely the conditions which determine a surface. A proof of the fact that α determines a map F is sketched in the next part of this section. The surface p can then be obtained as $p = \int F_1\, dx + \int F_2\, dy$, where F_1, F_2 are the first two columns of F (see section 3.8 of [84]).

The 2-form $d\alpha + \alpha \wedge \alpha$ represents the curvature of the connection whose local connection form is α (this definition of curvature will be explained in part 2 of this section). In particular, if we define a covariant derivative operator ∇ using α (which is given by the six functions), the condition $\nabla_{\partial_x}\nabla_{\partial_y} = \nabla_{\partial_y}\nabla_{\partial_x}$ turns out to be equivalent to the condition $d\alpha + \alpha \wedge \alpha = 0$.

We conclude that the (non-linear) Gauss–Codazzi equations are the integrability conditions for the (linear) moving frame equations. More simply, the condition for local solvability of the linear equations $F^{-1}dF = \alpha$ is the zero curvature condition $d\alpha + \alpha \wedge \alpha = 0$. The fact that the Gauss–Codazzi equations can be expressed as a zero curvature condition indicates that they have very special properties.

If we impose geometrical conditions on our surface S, for example that its Gauss curvature or mean curvature is constant or zero, then the connection form α and the Gauss–Codazzi equations simplify. However, the relation persists: the simplified Gauss–Codazzi equations are solvable if and only if the simplified connection is flat. Thus, the p.d.e. defining such surfaces can always be expressed as a zero curvature condition (with additional conditions), and that is why flat connections are fundamental tools in modern surface theory.

We shall discuss briefly[5] an important example here, the case of surfaces of constant mean curvature (CMC). In order to simplify the equations as much as possible, one should choose appropriate local coordinates. It can be shown that there exist local coordinates such that $\langle p_x, p_y\rangle = 0$ and $\langle p_x, p_x\rangle = \langle p_y, p_y\rangle$; let us denote the (positive) function $\langle p_x, p_x\rangle$ by e^u. Then direct calculation gives $\alpha = F^{-1}dF = \alpha_1 dx + \alpha_2 dy$ with

$$\alpha_1 = F^{-1}F_x = \begin{pmatrix} \frac{1}{2}u_x & \frac{1}{2}u_y & -e^{-u}L \\ -\frac{1}{2}u_y & \frac{1}{2}u_x & -e^{-u}M \\ L & M & 0 \end{pmatrix}$$

and

$$\alpha_2 = F^{-1}F_y = \begin{pmatrix} \frac{1}{2}u_y & -\frac{1}{2}u_x & -e^{-u}M \\ \frac{1}{2}u_x & \frac{1}{2}u_y & -e^{-u}N \\ M & N & 0 \end{pmatrix}.$$

(As stated earlier, α contains the functions E, F, G, L, M, N—but in a particularly simple way here, because of the choice of coordinates.) The Gauss and mean curvature functions are given by

$$K = \frac{LN - M^2}{e^{2u}}, \quad H = \frac{L + N}{2e^u}.$$

[5] Further details may be found in [66] and [81].

Computing $d\alpha + \alpha \wedge \alpha = 0$ gives the Gauss–Codazzi equations in the form

$$u_{xx} + u_{yy} + 2e^u K = 0$$
$$L_y - M_x - u_y e^u H = 0$$
$$N_x - M_y - u_x e^u H = 0.$$

Now we assume that $H = C$, a constant. There are three essentially different cases to discuss.

(i) Minimal surfaces: If $C = 0$, the surface is a minimal surface. The condition $L + N = 0$ is equivalent to $p_{xx} + p_{yy} = 0$, or $p_{\bar{z}z} = 0$, where $z = x + iy$. In other words, the $\mathbb{C}^3$-valued function

$$\phi = p_z = \tfrac{1}{2}(p_x - ip_y)$$

is holomorphic. From the assumptions $\langle p_x, p_y \rangle = 0$ and $\langle p_x, p_x \rangle = \langle p_y, p_y \rangle$ it follows that $\phi_1^2 + \phi_2^2 + \phi_3^2 = 0$. Conversely, any $\mathbb{C}^3$-valued function ϕ such that $\phi_1^2 + \phi_2^2 + \phi_3^2 = 0$ defines a minimal surface with local parametrization p. This is the *Weierstrass representation* of a minimal surface. It can be made even more explicit by writing

$$\phi = \left(\tfrac{1}{2}f(1 - g^2), \tfrac{1}{2}if(1 + g^2), fg\right)$$

where $f = \phi_1 - i\phi_2$, $g = \phi_3/(\phi_1 - i\phi_2)$. We obtain

$$p(x,y) = 2\mathrm{Re} \int_{z_0}^{z} \left(\tfrac{1}{2}f(1 - g^2), \frac{1}{2}if(1 + g^2), fg\right) dz.$$

Thus, minimal surfaces correspond locally to pairs of arbitrary holomorphic functions (f, g), and this solves explicitly (or, rather, bypasses) the Gauss–Codazzi equations for such surfaces.

(ii) Totally umbilic surfaces: An umbilic point is, by definition, a point where the second fundamental form is proportional to the first fundamental form. With our choice of coordinates, this means $L = N$ and $M = 0$. For a totally umbilic surface, it follows from the Gauss–Codazzi equations that $H_x = H_y = 0$, so H is constant, and also that L/e^u is constant, hence K is constant as well. Such a surface must be part of a plane or a sphere.

We shall see later on in this section that umbilic points of a CMC surface are isolated, if the surface is not totally umbilic. Therefore, the remaining case is the case of surfaces without umbilic points.

(iii) CMC surfaces without umbilic points: If a CMC surface has no umbilic points, it can be proved that local 'line of curvature' coordinates exist, that

is, $M = 0$ as well as $E = G$, $F = 0$. (Necessarily, $L \neq N$ here.) In this case we have $L_y = u_y e^u H$, $N_x = u_x e^u H$, as well as $L + N = 2e^u H$. Taking (for aesthetic reasons) the solutions $L = 2He^{u/2} \cosh u/2 = (e^u + 1)H$, $N = 2He^{u/2} \sinh u/2 = (e^u - 1)H$, we see that the Gauss–Codazzi equations reduce entirely to a single non-linear p.d.e.

$$u_{xx} + u_{yy} + 4C^2 \sinh u = 0,$$

which is known as the elliptic sinh-Gordon equation[6] (when $C \neq 0$). Any solution u gives rise to a map F and hence a surface p. It can be said, with mild exaggeration, that this is 'the equation for CMC surfaces'.

In the third part of this section we shall explain how these three cases may be treated more systematically.

It was mentioned in section 4.3 that a CMC surface is characterized by the condition that its unit normal ν (Gauss map) is a harmonic map to the symmetric space S^2. We shall explain this in Chapter 7 (Example 7.13), where we discuss harmonic maps into symmetric spaces in general.

Part 2: connections in vector bundles

Let us now proceed to the general case of a connection ∇ in a vector bundle $E \to M$ of rank n over an r-dimensional manifold M.

A connection is a covariant derivative operator

$$\Gamma(E) \times \Gamma(TM) \to \Gamma(E), \ (s, v) \mapsto \nabla_v s$$

on sections $\Gamma(E)$ of E which is $\mathbb{R}$-linear as a function of s and v, and satisfies

$$\nabla_v fs = f\nabla_v s + df(v)s, \quad \nabla_{fv} s = f\nabla_v s$$

for functions $f : M \to \mathbb{R}$. We have just seen two examples: $E = S \times \mathbb{R}^3$ and $E = TS$.

If we choose local coordinates $x_1, \ldots, x_r$ on M, we obtain a 'local basis' of vector fields $\partial_1 = \partial/\partial x_1, \ldots, \partial_r = \partial/\partial x_r$, that is, a moving frame for the tangent bundle TM. Let us choose locally defined sections $e_1, \ldots, e_n$ of E which are linearly independent at each point. The local connection form for ∇ is the $n \times n$ matrix-valued 1-form $\alpha = \sum_{i=1}^r \alpha_i dx_i$ where the matrices α_i are defined as follows:

$$\nabla_{\partial_i} e_j = \sum_{k=1}^{n} (\alpha_i)_{kj} e_k.$$

[6] A more direct transformation to the sinh-Gordon equation is given in section 1 of [15].

The connection form α determines ∇ locally, because the covariant derivative $\nabla_v s$ can be calculated from α for any $v = \sum g_i \partial_i$ and $s = \sum f_j e_j$ by applying the rules above:

$$\nabla_v \sum_j f_j e_j = \sum_j f_j \nabla_v e_j + \sum_j df_j(v) e_j$$

and then

$$\nabla_v e_j = \sum_{i,k} (\alpha_i)_{kj} g_i e_k = \sum_{i,k} \alpha(g_i \partial_i)_{kj} e_k = \sum_k \alpha(v)_{kj} e_k = \alpha(v) e_j.$$

If we introduce the notation

$$d_v(\sum_j f_j e_j) = \sum_j df_j(v) e_j,$$

the relation between ∇ and α can be written in the compact form

$$\nabla_v = d_v + \alpha(v),$$

or, even more briefly, as $\nabla = d + \alpha$.

The operators

$$\nabla_{\partial_i} = \partial_i + \alpha_i$$

are useful because $\nabla_{\partial_1}, \ldots, \nabla_{\partial_r}$ determine ∇. For example, a section s is said to be covariant constant, or parallel, if $\nabla_v s = 0$ for all vector fields v, but this condition is equivalent to the system of r *first-order differential equations* $(\partial_i + \alpha_i)s = 0$, $1 \le i \le r$. We can write this system in matrix form as follows:

$$\partial_i \begin{pmatrix} f_1 \\ \vdots \\ f_n \end{pmatrix} + \alpha_i \begin{pmatrix} f_1 \\ \vdots \\ f_n \end{pmatrix} = 0, \quad 1 \le i \le r.$$

Here we regard e_i as the column vector with 1 in the i-th position and 0 elsewhere, but it should be emphasized that e_i is merely a section of E; it is not necessarily 'constant' (nor is it covariant constant in general).

The next result is a consequence of Frobenius' Theorem, and underlies much of the theory of integrable systems:

Theorem 4.22 *Let U be a simply connected local coordinate chart for M, with all notation as above. Then, the following conditions are equivalent.*

(1) The system $(\partial_i + \alpha_i)s = 0$, $1 \le i \le r$ admits a non-trivial (i.e. not identically zero) solution.

(2) *The solution space of the system* $(\partial_i + \alpha_i)s = 0$, $1 \le i \le r$ *is a vector space of dimension* n.
(3) $d\alpha + \alpha \wedge \alpha = 0$.
(4) $[\partial_i + \alpha_i, \partial_j + \alpha_j] = 0$ *for all* i, j *(i.e. the differential operators* $\partial_i + \alpha_i$ *commute).*

Condition (1) or (2) says that the system of differential equations is 'consistent', or that the equations are 'compatible'. Condition (3) says that the connection form is flat, that is, has zero curvature. The curvature 2-form is by definition $d\alpha + \alpha \wedge \alpha$, while in terms of differential operators, the curvature tensor is defined as

$$(v_1, v_2, w) \mapsto \pm(\nabla_{v_1} \nabla_{v_2} w - \nabla_{v_2} \nabla_{v_1} w + \nabla_{[v_1, v_2]} w)$$

(the third term being necessary in order to produce a tensor, that is, a map which is linear over the ring of functions). The curvature 2-form is a local expression for the curvature tensor.

Proof The equivalence of (1) and (2) is elementary, as is the equivalence of (3) and (4). We shall sketch briefly a proof of the remaining equivalence.

The implication (2) $\Rightarrow$ (3) can be proved by choosing a basis of solutions $s_1, \dots, s_n$, defining S to be the matrix with these as its columns (so that $dS = -\alpha S$), then observing that $\alpha = -dSS^{-1}$ implies $d\alpha + \alpha \wedge \alpha = 0$.

Frobenius' Theorem on the integrability of distributions is responsible for the implication (4) $\Rightarrow$ (2); see, for example, Theorem 1.6.10 in [75]. Taking $r = n = 2$ for simplicity, a direct proof may be given as follows. First, the o.d.e. $\nabla_{\dot\gamma} s = 0$ may be solved along any curve γ, and the solution is unique if an initial vector (value of s) is specified. Let us specify the initial condition $s(0, 0) = v$. Taking the curve $\gamma(t) = (t, 0)$ we obtain a section s along this curve which satisfies $(\partial_1 + \alpha_1)s = 0$. For each fixed x_1, taking the curve $\gamma(t) = (x_1, t)$, we extend s to the (x_1, x_2)-plane in such a way that $(\partial_2 + \alpha_2)s = 0$. It remains to prove that $(\partial_1 + \alpha_1)s = 0$ holds for any (x_1, x_2) (not just when $x_2 = 0$). For this, condition (4) is used (cf. the proof of Theorem 11 in Chapter 6 of [115]). □

If we write $F = S^{-1}$, so that $\alpha = F^{-1}dF$, we obtain the following important corollary of the above proof:

Corollary 4.23 *Let* α *be an* $n \times n$ *matrix-valued* 1*-form on a simply connected open subset* U *of* $\mathbb{R}^r$. *Then the following conditions are equivalent.*

(1) $d\alpha + \alpha \wedge \alpha = 0$.
(2) There exists a map $F : U \to GL_n\mathbb{R}$ *such that* $\alpha = F^{-1}dF$.

Moreover, if F_1, F_2 *are two such maps then* $F_1 = XF_2$ *for some* $X \in GL_n\mathbb{R}$.

From this one can deduce a similar result for arbitrary Lie groups:

Corollary 4.24 *Let* α *be a* $\mathfrak{g}$*-valued* 1*-form on a simply connected open subset* U *of* $\mathbb{R}^r$*, where* $\mathfrak{g}$ *is the Lie algebra of a Lie group* G*. Then the following conditions are equivalent.*

(1) $d\alpha + \alpha \wedge \alpha = 0$.
(2) There exists a map $F : U \to G$ *such that* $\alpha = F^{-1}dF$.

Moreover, if F_1, F_2 *are two such maps then* $F_1 = XF_2$ *for some* $X \in G$.

An in-depth discussion of Corollaries 4.23 and 4.24 can be found in [113], where they are regarded as the (non-abelian) 'Fundamental Theorem of Calculus'.

The dual connection ∇^* (in the dual bundle E^*) is defined by $(\nabla^*_{\partial_i} e^*_j)e_\alpha = -e^*_j(\nabla_{\partial_i} e_\alpha)$ where $e^*_1, \dots, e^*_n$ is the dual basis to $e_1, \dots, e_n$. Hence,

$$\begin{aligned}(\nabla^*_{\partial_i} \sum_j y_j e^*_j)e_k &= \big(\sum_j y_j \nabla^*_{\partial_i} e^*_j + \sum_j \partial y_j/\partial x_i e^*_j\big)e_k \\ &= \partial y_k/\partial x_i - \sum_j y_j e^*_j(\nabla_{\partial_i} e_k) \\ &= \partial y_k/\partial x_i - \sum_j y_j(\alpha_i)_{jk}.\end{aligned}$$

Thus, $\nabla^* = d - \alpha^t$. Covariant constant sections correspond to solutions of the following system:

$$\partial_i \begin{pmatrix} y_1 \\ \vdots \\ y_n \end{pmatrix} = \alpha_i^t \begin{pmatrix} y_1 \\ \vdots \\ y_n \end{pmatrix}, \quad 1 \le i \le r.$$

The dual connection is flat if and only if the original connection is flat, since the condition $d(-\alpha^t) + (-\alpha^t) \wedge (-\alpha^t) = 0$ is the transpose of the condition $d\alpha + \alpha \wedge \alpha = 0$. If $F = S^{-1}$ where the columns $s_1, \dots, s_n$ of S are a basis of covariant constant sections of ∇, as above, then the columns of F^t are a basis of covariant constant sections of ∇^*.

The connection matrix is very useful for calculations, but it depends on the choice of local trivialization $e_1, \dots, e_n$ (and, of course, also on the choice of local coordinates $x_1, \ldots, x_r$). If $P = (p_{ij}) : U \to \mathrm{GL}_n\mathbb{R}$ is a map, we can define

$$\hat{e}_i = Pe_i = \sum_j p_{ji} e_j, \quad 1 \le i \le n,$$

and any other local trivialization can be obtained this way. By applying the definitions above, we obtain $\hat{F} = FP$ and

$$\hat{\alpha} = P^{-1}\alpha P + P^{-1}dP$$

for the connection form $\hat{\alpha}$ relative to the new basis $\hat{e}_1, \ldots, \hat{e}_n$. This transformation is called a gauge transformation. The curvature form transforms in the simpler way

$$d\hat{\alpha} + \hat{\alpha} \wedge \hat{\alpha} = P^{-1}(d\alpha + \alpha \wedge \alpha)P,$$

reflecting the fact that it expresses a tensor, rather than a differential operator.

If the domain is not simply connected, or if the connection $d + \alpha$ is not flat, then there is in general no solution F to the equation $F^{-1}dF = \alpha$. Nevertheless, this equation is just an o.d.e. if we restrict to a path in the domain, so a solution certainly exists over any such path. The problem is that the solution depends on the choice of path. If we consider closed loops which begin and end at some point P of the domain, then the initial and final values of F differ by some invertible $n \times n$ matrix, and the set of all such matrices constitutes the *holonomy group* of the connection at P. The monodromy group at a singular point of an o.d.e. on the punctured disk (see the end of section 4.1) is the holonomy group of the corresponding connection.

Example 4.25 As an application of Corollary 4.24, we can give the simple and instructive proof that the Levi–Cività connection of a manifold is flat if and only if the manifold is locally isometric to Euclidean space.

As mentioned in part 1, if g is a Riemannian (or pseudo-Riemannian, real or complex) metric on M, then there is a unique connection on $E = \nabla^{TM}$ whose torsion tensor $\nabla_v w - \nabla_w v - [v, w]$ is zero and which is compatible with g in the sense that $v_1 g(v_2, v_3) = g(\nabla_{v_1} v_2, v_3) + g(v_2, \nabla_{v_1} v_3)$. Let α be the local connection matrix with respect to local coordinates $x_1, \ldots, x_n$ and a local basis of vector fields $e_1, \ldots, e_n$. We shall sketch a proof of the fact that $d\alpha + \alpha \wedge \alpha = 0$ if and only if there exist local coordinates $y_1, \ldots, y_n$ with respect to which

$$g\left(\frac{\partial}{\partial y_i}, \frac{\partial}{\partial y_j}\right) = \pm\delta_{ij}.$$

The signs correspond to the signature of the metric, but let us assume for simplicity that the metric is Riemannian, so that all signs are positive.

First, by the Gram–Schmidt orthogonalization procedure, we may assume that $g(e_i, e_j) = \delta_{ij}$. It follows from the compatibility condition above that α is an $\mathfrak{so}_n$-valued 1-form. Hence, there is an SO_n-valued map F such that $\alpha = F^{-1}dF$. Let $\theta_1, \ldots, \theta_n$ be the 1-forms dual to $e_1, \ldots, e_n$. If the vector-valued 1-form $\theta = (\theta_1, \ldots, \theta_n)^t$ satisfies $d\theta = 0$ (equivalently, $\theta_i = dy_i$ for some local coordinates $y_1, \ldots, y_n$), then we have achieved our objective. In general $d\theta \neq 0$, but the modified 1-form $\tilde{\theta} = F\theta$ satisfies $d\tilde{\theta} = dF \wedge \theta + Fd\theta = F\alpha \wedge \theta + Fd\theta = F(\alpha \wedge \theta + d\theta)$. This is zero because of the 'structure equation' $\alpha \wedge \theta + d\theta = 0$, which follows from the definition of α. The desired local coordinates $y_1, \ldots, y_n$ are given by $\tilde{\theta}_i = dy_i$. □

Part 3: surface theory revisited

In surface theory, it is the connection form $\alpha = F^{-1}dF$, rather than the 'obvious' invariants such as curvature, which is of central importance. We shall reconsider this point now in the spirit of Klein's Erlangen Programme (in which geometry is dictated by a group of allowable transformations) and Cartan's approach to geometry via differential forms (see [75] and [113]). This discussion leads directly to the role of symmetry groups[7] in the theory of integrable systems.

In order to classify geometric objects up to the action of a Lie group G, one strategy is to convert the object into a G-valued map F, then consider $F^{-1}dF$ (which absorbs the action of G, as $(gF)^{-1}d(gF) = F^{-1}dF$). If there is a particularly simple choice of F, the connection form $F^{-1}dF$ may be expected to contain the essential geometric data or invariants of the original object. If the object itself happens to be a G/H-valued map, for some homogeneous space G/H, any G-valued 'lift' would be a candidate for F.

For example, if the geometric objects are curves $c : \mathbb{R} \to \mathbb{R}^2$ parametrized by arc-length, then a natural choice for $F : \mathbb{R} \to SO_2$ is the Frenet matrix of c, and $F^{-1}dF$ contains exactly the 'right' invariant for a plane curve, its curvature function. Alternatively, in the homogeneous space formulation, we regard $\mathbb{R}^2$ as $SO_2 \tilde{\times} \mathbb{R}^2/SO_2$ and use (F, c) instead of F. This gives the curvature function again, which determines the original object up to the action of the Euclidean isometry group $SO_2 \tilde{\times} \mathbb{R}^2$.

Now, this example is deceptively simple, as the domain of F is one-dimensional, where the integrability condition $d\alpha + \alpha \wedge \alpha = 0$ is vacuous. In the case of surfaces $p : \mathbb{R}^2 \to \mathbb{R}^3$, the same strategy produces the right invariants, but they are presented in a complicated way and they are subject to the Gauss–Codazzi equations. Finding an invariant analogous to the

[7] It is usual to say that that symmetry groups are related via Noether's Theorem to conserved quantities of integrable systems. Our discussion here is intended to apply in a wider context.

curvature function of a plane curve is essentially the same as solving the Gauss–Codazzi equations, which is a non-trivial task—and this approach begins to look distinctly unattractive.

Remarkably, however, there is a way to carry out the Cartan/Klein strategy, in the case of CMC surfaces, if we implement it faithfully by searching for a *particularly simple* version of F. In general, when the geometrical object is a G/H-valued map, any map of the form FX represents the same object, if X is H-valued. However, $F^{-1}dF$ changes to $(FX)^{-1}d(FX) = X^{-1}F^{-1}dFX + X^{-1}dX$, and there arises the possibility of simplifying this by choosing X appropriately.

Let us return to the case of surfaces in $\mathbb{R}^3$, with local coordinates chosen as in part 1. If we make a preliminary gauge transformation of F, replacing it by

$$F = \begin{pmatrix} | & | & | \\ e^{-\frac{u}{2}}p_x & e^{-\frac{u}{2}}p_y & \nu \\ | & | & | \end{pmatrix},$$

it takes values in SO_3 (rather than $GL_3\mathbb{R}$), which is the first step in producing a simpler F. As in the discussion of harmonic maps[8] in section 4.3, it is convenient to change coordinates from x, y to $z = x + iy, \bar{z} = x - iy$. The 1-form $\alpha = F^{-1}dF$ can then be written $\alpha' dz + \alpha'' d\bar{z}$ where $\alpha' = F^{-1}\partial F/\partial z$, $\alpha'' = F^{-1}\partial F/\partial \bar{z}$. The functions α', α'' take values in the complexified Lie algebra $\mathfrak{so}_3 \otimes \mathbb{C}$, but they satisfy the reality condition $\alpha'' = \overline{\alpha'}$.

Direct calculation gives

$$\alpha' = \left(\begin{array}{cc|c} 0 & \frac{i}{2}u_z & -(Q + \frac{1}{2}e^u H)e^{-\frac{u}{2}} \\ -\frac{i}{2}u_z & 0 & -i(Q - \frac{1}{2}e^u H)e^{-\frac{u}{2}} \\ \hline (Q + \frac{1}{2}e^u H)e^{-\frac{u}{2}} & i(Q - \frac{1}{2}e^u H)e^{-\frac{u}{2}} & 0 \end{array}\right)$$

where

$$Q = \langle p_{zz}, \nu \rangle = \tfrac{1}{4}(p_{xx} - p_{yy} - 2ip_{xy}) = \tfrac{1}{4}(L - N - 2iM).$$

In terms of Q, the Gauss curvature is given by $K = H^2 - 4|Q|^2/e^{2u}$. The Gauss–Codazzi equations (i.e. $d\alpha + \alpha \wedge \alpha = 0$) take the form

$$u_{z\bar{z}} + \tfrac{1}{2}e^u K = 0$$
$$Q_{\bar{z}} - \tfrac{1}{2}e^u H_z = 0.$$

[8] We shall explain the precise relation between CMC surfaces and harmonic maps in section 7.4.

For a CMC surface, H is constant, so the second equation says that Q is holomorphic. This justifies the earlier assertion that, if the surface is not totally umbilic, its umbilic points are isolated.

The special cases (i) $H = 0$ and (ii) $Q = 0$ correspond, respectively, to minimal surfaces and totally umbilic surfaces. The significance of these conditions becomes clearer if we transform the equations by making use of a Lie algebra isomorphism

$$\mathfrak{so}_3 \to \mathfrak{su}_2, \qquad \left(\begin{array}{cc|c} 0 & c & a \\ -c & 0 & b \\ \hline -a & -b & 0 \end{array}\right) \mapsto \frac{1}{2}\begin{pmatrix} ic & -a+ib \\ a+ib & -ic \end{pmatrix}.$$

(We follow the conventions of the appendix to [30].) Here a, b, c are real, but, taking a, b, c to be complex gives an isomorphism of complexified Lie algebras which preserves the real forms.

This gives the matrix representation

$$\alpha' = \begin{pmatrix} -\frac{1}{4}u_z & Qe^{-\frac{u}{2}} \\ -\frac{1}{2}He^{\frac{u}{2}} & \frac{1}{4}u_z \end{pmatrix}$$

but with the new reality condition $\alpha'' = -\overline{\alpha'}^t$. Following [30], we conjugate everything by $\begin{pmatrix} 0 & 1 \\ 1 & 0 \end{pmatrix}$ and thereby redefine F and $\alpha = \alpha' dz + \alpha'' d\bar{z}$ by

$$F^{-1}dF = \alpha = \begin{pmatrix} \frac{1}{4}u_z & -\frac{1}{2}He^{\frac{u}{2}} \\ Qe^{-\frac{u}{2}} & -\frac{1}{4}u_z \end{pmatrix} dz + \begin{pmatrix} -\frac{1}{4}u_{\bar{z}} & -\bar{Q}e^{-\frac{u}{2}} \\ \frac{1}{2}He^{\frac{u}{2}} & \frac{1}{4}u_{\bar{z}} \end{pmatrix} d\bar{z}.$$

All these modifications of F are merely cosmetic; the main transformation comes next.

(i) *Minimal surfaces:* In the case of minimal surfaces our aim is to arrive at the Weierstrass representation; thus, the 'right' geometric invariant of a minimal surface should be a pair of (unrestricted) holomorphic functions.

When $H = 0$ we have

$$F^{-1}dF = \alpha = \begin{pmatrix} \frac{1}{4}u_z & 0 \\ Qe^{-\frac{u}{2}} & -\frac{1}{4}u_z \end{pmatrix} dz + \begin{pmatrix} -\frac{1}{4}u_{\bar{z}} & -\bar{Q}e^{-\frac{u}{2}} \\ 0 & \frac{1}{4}u_{\bar{z}} \end{pmatrix} d\bar{z}.$$

The key simplification of F is to replace it by F_-, where $F = F_-F_+$ is a lower-triangular/upper-triangular factorization ('Gauss factorization') of the form

$$F = \begin{pmatrix} * & * \\ * & * \end{pmatrix} = \begin{pmatrix} 1 & 0 \\ * & 1 \end{pmatrix} \begin{pmatrix} * & * \\ 0 & * \end{pmatrix} = F_- F_+.$$

That is, we decide to replace F by FX where $X = F_+^{-1}$.

Let us write

$$F_- = \begin{pmatrix} 1 & 0 \\ \delta & 1 \end{pmatrix}.$$

Proposition 4.26 *We have*

$$F_-^{-1} dF_- = \begin{pmatrix} 0 & 0 \\ \delta' & 0 \end{pmatrix} dz + \begin{pmatrix} 0 & 0 \\ 0 & 0 \end{pmatrix} d\bar{z},$$

and $\delta = \delta(z)$ *is holomorphic.*

Proof It follows immediately from the form of F_- that

$$F_-^{-1} dF_- = \begin{pmatrix} 0 & 0 \\ \delta_z & 0 \end{pmatrix} dz + \begin{pmatrix} 0 & 0 \\ \delta_{\bar{z}} & 0 \end{pmatrix} d\bar{z}.$$

We just have to show that $\delta_{\bar{z}} = 0$. Observe that

$$\begin{aligned} F_-^{-1}(F_-)_{\bar{z}} &= (FF_+^{-1})^{-1}(FF_+^{-1})_{\bar{z}} \\ &= F_+(F^{-1}F_{\bar{z}})F_+^{-1} + F_+(F_+^{-1})_{\bar{z}}. \end{aligned}$$

The left and right hand sides are, respectively, of the form

$$\begin{pmatrix} 0 & 0 \\ * & 0 \end{pmatrix}, \quad \begin{pmatrix} * & * \\ 0 & * \end{pmatrix}$$

(here we are using the fact that $H = 0$), hence both must be zero. □

Some explanation is required at this point. First of all, the map F_- determines[9] only the Gauss map ν of the minimal surface, not the map p itself; this is why we obtain only one holomorphic function δ, whereas the Weierstrass representation has two. We shall enhance the procedure (by choosing a slight modification of F) in a moment. The second observation is that the proposition really does solve the Gauss–Codazzi equations; in terms of the data δ, the zero curvature condition is trivial, that is, there is no restriction

[9] The Gauss map is the induced map $[F]$ into $SO_3/SO_2 \cong SU_2/U_1$; it can be identified with either the first or the last column of the SU_2-valued version of F, which in turn may be obtained by orthogonalizing the columns of F_-. It can be further identified with δ, via stereographic projection.

on δ (other than that it is holomorphic). All this rests on the assumption that the original matrix factorization $F = F_- F_+$ can be carried out: it is easy to verify that the necessary and sufficient condition for this is that the $(1, 1)$-entry of F is non-zero. At a point where this condition is not satisfied, the holomorphic function δ may develop a singularity.

To complete the picture, and spare the reader from having to extract the details from the literature, we sketch here a version which produces the minimal surface itself. For this it is necessary to 'enhance' F_-. With respect to the refined factorization

$$F = \begin{pmatrix} 1 & \\ \delta & 1 \end{pmatrix} \begin{pmatrix} e^{i\theta} & \\ & e^{-i\theta} \end{pmatrix} \begin{pmatrix} r & \\ & r^{-1} \end{pmatrix} \begin{pmatrix} 1 & \epsilon \\ & 1 \end{pmatrix}$$

(where δ, ϵ are $\mathbb{C}$-valued and θ, r are $\mathbb{R}$-valued with $r > 0$), we define

$$F_- = \begin{pmatrix} 1 & \\ \delta & 1 \end{pmatrix} \begin{pmatrix} e^{i\theta} & \\ & e^{-i\theta} \end{pmatrix}, \quad F_+ = \begin{pmatrix} r & \\ & r^{-1} \end{pmatrix} \begin{pmatrix} 1 & \epsilon \\ & 1 \end{pmatrix}.$$

The condition $\bar{F}^t = F^{-1}$ gives

$$r = 1/\sqrt{1 + |\delta|^2}$$
$$\epsilon = -\bar{\delta} e^{-2i\theta}$$

so this new choice of F_- determines F and therefore does not lose any information. It follows from the equations $F_-^{-1}(F_-)_{\bar{z}} = (FF_+^{-1})^{-1}(FF_+^{-1})_{\bar{z}}$ that

$$e^{i\theta} = q/|q|$$

for some holomorphic function q. Conversely, if we define F_- (hence F) by the formulae above, it can be verified that the Gauss–Codazzi equations are satisfied. In terms of the classical Weierstrass data (see part 1 of this section) we have

$$f = q^2$$
$$g = -\delta.$$

The 1-form[10] $F_-^{-1} dF_-$ is not itself holomorphic; in addition to the holomorphic function δ' it contains the 1-form $d\theta$.

[10] It is not quite true that the information contained in $F_-^{-1} dF_-$, namely δ' and $d\theta$, specifies (locally) a minimal surface up to Euclidean isometries, because the frame F itself had been orthonormalized, thus losing information equivalent to u or $|q|$. Reinstating $|q|$ gives back δ and q, which are equivalent to the classical Weierstrass data f, g.

In conclusion, we have replaced the differential geometric data u, Q (which is required to satisfy the Gauss–Codazzi equations) by the 'free' holomorphic data δ, q or f, g. These are related explicitly by

$$e^{u/2} = |f|(1+|g|^2) = |q|^2(1+|\delta|^2), \quad Q = -fg' = \delta' q^2.$$

Two obvious questions arise at this point. What is the benefit of this method (after all, the classical Weierstrass representation is easy enough to derive directly), and why should the factorization method work at all? The answer to the first question is that the method is Lie-theoretic: factorizations of the above type exist for arbitrary complex Lie groups, and even Kac–Moody Lie groups. As we shall see, the method is effective in other situations, for example the sine-Gordon equation (Example 7.9 and [86]), harmonic maps (Chapter 7 and [32]), the KdV equation (Chapters 7 and 8), and quantum cohomology (Chapter 6 and [62]). We shall speculate on the second question in section 8.7.

(ii) Totally umbilic surfaces: Essentially the same method works here, as the connection matrices are again triangular. However, the situation is much simpler as $Q = 0$ implies that H is constant (while $H = 0$ implies only that Q is holomorphic).

(iii) CMC surfaces without umbilic points: This is the real test case, and a new idea is needed. Motivated by the existence of the 'associated family' of minimal surfaces given by

$$f \longmapsto \mu f \qquad (\mu \in S^1)$$

in the Weierstrass representation, which is equivalent to multiplying Q by μ (rotation of the second fundamental form of the surface), we consider for *any* surface p the μ-family of 1-forms

$$\alpha^{\mu} = \begin{pmatrix} \frac{1}{4}u_z & -\frac{1}{2}He^{\frac{u}{2}} \\ \mu Qe^{-\frac{u}{2}} & -\frac{1}{4}u_z \end{pmatrix} dz + \begin{pmatrix} -\frac{1}{4}u_{\bar z} & -\frac{1}{\mu}\bar{Q}e^{-\frac{u}{2}} \\ \frac{1}{2}He^{\frac{u}{2}} & \frac{1}{4}u_{\bar z} \end{pmatrix} d\bar z.$$

As a cosmetic modification (following [30] again) we conjugate everything by the constant matrix

$$\begin{pmatrix} \mu^{1/4} & 0 \\ 0 & \mu^{-1/4} \end{pmatrix}$$

and put $\lambda = \mu^{-1/2}$. Thus, we redefine the 1-form as

$$\alpha^{\lambda} = \begin{pmatrix} \frac{1}{4}u_z & -\frac{1}{2}\frac{1}{\lambda}He^{\frac{u}{2}} \\ \frac{1}{\lambda}Qe^{-\frac{u}{2}} & -\frac{1}{4}u_z \end{pmatrix} dz + \begin{pmatrix} -\frac{1}{4}u_{\bar z} & -\lambda\bar{Q}e^{-\frac{u}{2}} \\ \frac{1}{2}\lambda He^{\frac{u}{2}} & \frac{1}{4}u_{\bar z} \end{pmatrix} d\bar z.$$

It is a remarkable fact, though easy enough to verify, that the zero curvature condition $d\alpha^\lambda + \alpha^\lambda \wedge \alpha^\lambda = 0$ is equivalent to the original Gauss–Codazzi equations $d\alpha + \alpha \wedge \alpha = 0$ *together with the condition that* H *be constant.*

We shall say much more about this phenomenon in sections 7.3 and 7.4. For the present discussion, the main lesson is that we should adjust our point of view and consider a family of maps F^λ instead of F alone. Instead of the factorization $F = F_- F_+$ that was used in (i) and (ii) above, we should seek a factorization $F^\lambda = F_-^\lambda F_+^\lambda$, where F^λ is now defined somewhat indirectly as the map satisfying $(F^\lambda)^{-1} dF^\lambda = \alpha^\lambda$. This method works exactly as in (i) and (ii), and *solves* the Gauss–Codazzi equations for CMC surfaces, as we shall see in Example 7.13. There is a substantial technical difference, however, as we have replaced the 'symmetry group' SU_2 by its loop group, an infinite-dimensional group.

4.6 Appendix: self-adjointness

We collect here some further properties of differential equations that will be useful later on, focusing on the notion of self-adjointness.

Frobenius symbolic factors

In a neighbourhood of a regular point z_0, the coefficients of an o.d.e. are related to its solutions by any factorization of the form

$$\partial^{s+1} + a_s \partial^s + \cdots + a_1 \partial + a_0 = (\partial + b_s)(\partial + b_{s-1}) \ldots (\partial + b_0),$$

for the solutions may be obtained by quadrature if the factorized form is given.

To express this more explicitly, let us choose the unique ordered basis of solutions $u_0, \ldots, u_s$ which satisfy the initial conditions $u_i^{(j)}(z_0) = \delta_{ij}$. That is, $H^t(z_0) = I$, where

$$H^t = \begin{pmatrix} | & | & & | \\ u & u' & \cdots & u^{(s)} \\ | & | & & | \end{pmatrix}, \quad u = \begin{pmatrix} u_0 \\ \vdots \\ u_s \end{pmatrix}.$$

By definition H^t satisfies

$$(H^t)^{-1}(H^t)' = \begin{pmatrix} 0 & & & -a_0 \\ 1 & \ddots & & \vdots \\ & \ddots & 0 & -a_{s-1} \\ & & 1 & -a_s \end{pmatrix}.$$

Let us now perform a Gauss factorization[11] $H^t = (H^t)_-(H^t)_+$, where $(H^t)_-$ is lower triangular (with 1's on the diagonal) and $(H^t)_+$ is upper triangular. We obtain

$$(H^t)_- = \begin{pmatrix} 1 & & & & & \\ A_1 & 1 & & & & \\ A_2 & B_2 & 1 & & & \\ A_3 & B_3 & C_3 & 1 & & \\ \vdots & \vdots & \vdots & & \ddots & \\ A_s & B_s & C_s & & \cdots & 1 \end{pmatrix}$$

where

$$A_i = u_i/u_0, \quad B_i = A_i'/A_1, \quad C_i = B_i'/B_2', \quad \ldots$$

This satisfies

$$(H^t)_-^{-1}(H^t)_-' = \begin{pmatrix} 0 & & & & \\ v_1 & 0 & & & \\ & v_2 & 0 & & \\ & & \ddots & \ddots & \\ & & & v_s & 0 \end{pmatrix},$$

where

$$v_1 = A_1', \quad v_2 = B_2', \quad v_3 = C_3', \quad \ldots$$

Let us introduce $v_0 = u_0$ as well. Then the ordered basis $u_0, \ldots, u_s$ satisfying the initial conditions $u_i^{(j)}(z_0) = \delta_{ij}$ determines, and is determined by, the ordered set of functions $v_0, \ldots, v_s$ satisfying $v_i(z_0) = 1$. Namely, the v_i's are obtained from the u_i's as above, and the u_i's are obtained from the v_i's by

$$u_0 = v_0, \quad u_1 = v_0 \textstyle\int v_1, \quad u_2 = v_0 \int(v_1 \int v_2), \quad \ldots, \quad u_s = v_0 \int(v_1 \int \ldots \int v_s).$$

The transformation $H^t \mapsto (H^t)_-$ corresponds to a gauge transformation of the original differential operator, and (from the above matrix) the transformed operator is

$$v_0 v_1 \ldots v_s\, \partial \frac{1}{v_s} \partial \ldots \partial \frac{1}{v_1} \partial \frac{1}{v_0}.$$

11 This amounts to solving a system of equations by Gauss elimination; the initial conditions on the solutions ensure that this is possible. We have already seen a 2×2 example in our description of minimal surfaces in appendix 4.5, and hinted at the role of this factorization in other situations.

(These 'symbolic factors of Frobenius' are classically described using Wronskians—see section 5.2 of [71].) Since $\partial \frac{1}{v} = \frac{1}{v}\partial + (\frac{1}{v})'$, this gives a factorization into linear factors of the type mentioned earlier.

Example 4.27 Any second-order operator may be factorized as

$$v_0 v_1 \partial \frac{1}{v_1} \partial \frac{1}{v_0}.$$

Let us introduce a new variable $\tilde{z} = u_1/u_0$, so that $\tilde{\partial} = (1/v_1)\partial$. Then the operator becomes

$$v_0 v_1^2 \, \tilde{\partial}^2 \frac{1}{v_0}$$

and we obtain the well known (but not very useful) fact that any second-order o.d.e. may be converted to the trivial form

$$\frac{\partial^2 \tilde{y}}{\partial \tilde{z}^2} = 0$$

by introducing a new independent variable $\tilde{z} = u_1/u_0$ and a new dependent variable $\tilde{y} = y/u_0$. □

Self-adjointness

As mentioned in section 4.2, a choice of basis $[P_0], \ldots, [P_s]$ of the D-module $\mathcal{M} = D/(T)$ gives an identification $\mathcal{M} \to \mathcal{H}^{s+1} = \mathrm{Map}(N, \mathbb{C}^{s+1})$ which is an isomorphism of $\mathcal{H}$-modules. It is an isomorphism of D-modules if the action of ∂ on $\mathcal{H}^{s+1}$ is defined by $\partial \cdot f = (\partial + \Omega) f$.

Let us now choose an invertible matrix S and define a bilinear form on $\mathbb{C}^{s+1}$ by

$$(x, y) = x^t S y.$$

We also write $(f, g) = f^t S g$ for the induced pairing $\mathcal{M} \times \mathcal{M} \to \mathcal{H}$.

The map

$$E : \mathcal{M} \to \mathcal{M}^*, \quad f \mapsto (f, \ )$$

is an isomorphism of $\mathcal{H}$-modules, but not necessarily an isomorphism of D-modules.

Lemma 4.28 *The following are equivalent:*

(1) E is an isomorphism of D-modules.

(2) $(f,g)' = (\partial \cdot f, g) + (f, \partial \cdot g)$ *for all* $f, g \in \mathcal{M}$.

(3) $S^{-1}\Omega^t S = -\Omega$.

Proof The map E is an isomorphism of D-modules if and only if $(\partial \cdot (f,\ ))(g) = (\partial \cdot f,\ )(g)$ for all $f, g \in \mathcal{M}$, that is, $-(f, \partial \cdot g) + (f,g)' = (\partial \cdot f, g)$, which is condition (2). Substituting $\partial \cdot f = (\partial + \Omega)f$ gives condition (3). □

Example 4.29 If $S = I$, the bilinear form is the standard (complex) inner product, and the pairing can be regarded as the standard (complex) Riemannian metric on the trivial bundle $N \times \mathbb{C}^{s+1}$. The conditions of the lemma are those for the connection $\nabla = d + \Omega$ to be compatible with this metric. □

Example 4.30 Let S be of the form

$$\begin{pmatrix} & & \pm 1 \\ & \iddots & \\ \pm 1 & & \end{pmatrix}.$$

For example, if

$$S = \begin{pmatrix} & 1 \\ -1 & \end{pmatrix}$$

the bilinear form is skew-symmetric, and condition (3) of the lemma is that Ω has the form

$$\Omega = \begin{pmatrix} a & b \\ c & -a \end{pmatrix}.$$

On the other hand, if

$$S = \begin{pmatrix} & 1 \\ 1 & \end{pmatrix}$$

the bilinear form is symmetric, and condition (3) says

$$\Omega = \begin{pmatrix} a & 0 \\ 0 & -a \end{pmatrix}.$$

□

In order to give a 'scalar version' of the self-adjointness condition, we shall use the concept of (formal) adjoint (see section 5.3 of [71]), and its D-module version (Lemma 4.32 below, which is Lemma 1.5.3 of [79], attributed to O. Gabber). For this reason we focus on the case of ordinary differential equations, postponing comments on the case of two or more variables to the end of the section.

Definition 4.31 *The (formal) adjoint of the differential operator* $T = \sum_{i=0}^{s+1} a_i \partial^i$ *is* $T^* = \sum_{i=0}^{s+1} (-\partial)^i a_i$.

The map $T \mapsto T^*$ satisfies $(T^*)^* = T$ and $(T_1 T_2)^* = T_2^* T_1^*$ (see Chapter 2 of [103]). As in section 4.1 we shall assume that $a_{s+1} = 1$ unless otherwise stated.

Lemma 4.32 *Let* $\mathcal{M} = D/(T)$ *as above. Define* $\delta_i \in \mathcal{M}^*$ *by* $\delta_i([\partial^j]) = \delta_{ij}$. *Then* δ_s *is a cyclic element and it is annihilated by* T^*. *In particular, we have* $\mathcal{M}^* \cong D/(T^*)$.

Proof We use the standard basis $[1], [\partial], \ldots, [\partial^s]$, so that $[P_i] = [\partial^i]$ and $[P_i]^* = \delta_i$. Since the action of ∂ on $\mathcal{M}^*$ is given by $\partial - \Omega^t$ (or by direct calculation), we have

$$\begin{aligned}
\partial \cdot \delta_0 &= a_0 \delta_s \\
\partial \cdot \delta_1 &= a_1 \delta_s - \delta_0 \\
&\cdots \\
\partial \cdot \delta_{s-1} &= a_{s-1} \delta_s - \delta_{s-2} \\
\partial \cdot \delta_s &= a_s \delta_s - \delta_{s-1}.
\end{aligned}$$

This can be rewritten as

$$\begin{aligned}
a_0 \delta_s + (-\partial) \cdot \delta_0 &= 0 \\
a_1 \delta_s + (-\partial) \cdot \delta_1 &= \delta_0 \\
&\cdots \\
a_{s-1} \delta_s + (-\partial) \cdot \delta_{s-1} &= \delta_{s-2} \\
((-\partial) + a_s) \cdot \delta_s &= \delta_{s-1}.
\end{aligned}$$

Proceeding inductively from the last equation, we obtain

$$\left((-\partial)^i + (-\partial)^{i-1} a_s + \cdots + a_{s-i+1} \right) \cdot \delta_s = \delta_{s-i}$$

for $i = 1, \ldots, s$. Substituting the result for $i = s$ into the first equation we obtain $T^* \cdot \delta_s = 0$. □

Remark The equations in the proof can be rewritten formally as

$$\begin{pmatrix} \partial & & & -a_0 \\ 1 & \ddots & & \\ & \ddots & \partial & -a_{s-1} \\ & & 1 & \partial - a_s \end{pmatrix} \begin{pmatrix} \delta_0 \\ \vdots \\ \delta_{s-1} \\ \delta_s \end{pmatrix} = \begin{pmatrix} 0 \\ \vdots \\ 0 \\ 0 \end{pmatrix}.$$

Then (cf. Corollary 2.5 of [57]) we deduce that the equation satisfied by δ_s is obtained by formally 'left-expanding' the determinant

$$|\partial + \Omega| = \begin{vmatrix} \partial & & & -a_0 \\ 1 & \ddots & & \\ & \ddots & \partial & -a_{s-1} \\ & & 1 & \partial - a_s \end{vmatrix}.$$

Lemmas 4.28 and 4.32 suggest that there should be a relation between the matrix condition $S^{-1}\Omega^t S = -\Omega$ and the scalar condition $T^* = T$. Since the first condition depends on a basis of the D-module, it is necessary to make some assumption concerning this, as in the following 'typical' result.

Proposition 4.33 *Assume that* $S^{-1}\Omega^t S = -\Omega$ *with respect to a basis* $[P_0], \ldots, [P_s]$ *where* $P_i = \partial^i +$ *lower-order terms. Assume further (for simplicity) that*

$$S = \begin{pmatrix} & & \pm 1 \\ & \cdot^{\cdot^{\cdot}} & \\ \pm 1 & & \end{pmatrix}.$$

Then $T^* = \pm T$.

Proof By the condition $S^{-1}\Omega^t S = -\Omega$, the map $E : \mathcal{M} \to \mathcal{M}^*$ is a D-module isomorphism. Hence the annihilator of $E([P_0])$ is the same as the annihilator of $[P_0] = [1]$, that is, T. On the other hand, because of the form of S and the form of the basis, we have $E([P_0]) = \delta_s$ (where $\delta_i(\partial^j) = \delta_{ij}$ as in the previous lemma), and we know that the annihilator of δ_s is T^*. Hence T and T^* both annihilate the cyclic element δ_s. Since T is monic, it follows that $T^* = \pm T$. □

The proof could have been carried out in the same way as the proof of the lemma, as that proof works just as well when Ω has the form

$$\begin{pmatrix} * & \cdots & \cdots & * \\ 1 & \ddots & & \vdots \\ & \ddots & \ddots & \vdots \\ & & 1 & * \end{pmatrix}.$$

As in the remark above, T^* may be computed by expanding the analogous determinant. Clearly all these arguments generalize to the case where the elements of the (-1)-diagonal of the matrix are constant (not necessarily all equal to 1).

Example 4.34 Let $T = (\partial + a)(\partial - a) - b$. If we choose the basis of $\mathcal{M} = D/(T)$ given by $P_0 = 1$, $P_1 = \partial - a$, we obtain

$$\Omega = \begin{pmatrix} a & b \\ 1 & -a \end{pmatrix}.$$

Let

$$S = \begin{pmatrix} & 1 \\ -1 & \end{pmatrix}.$$

Then we have $S^{-1}\Omega^t S = -\Omega$, and the conditions of the proposition are satisfied. Direct calculation confirms that $T = T^*$ in this case. □

Example 4.35 Let $T = \partial^2 - (u'/u)\partial - uv$, as in Examples 4.3 and 4.13. If we choose the basis of $\mathcal{M} = D/(T)$ given by $P_0 = 1$, $P_1 = (1/u)\partial$, we obtain

$$\Omega = \begin{pmatrix} & v \\ u & \end{pmatrix}.$$

With the same S as in the previous example we have $S^{-1}\Omega^t S = -\Omega$. However, the basis does not satisfy the conditions of the proposition, and $T \neq T^*$ here. Nevertheless, it is true that the D-modules $D/(T)$ and $D/(T^*)$ are isomorphic, because (1) the map $E : \mathcal{M} \to \mathcal{M}^*$ is an isomorphism, and (2) we have $\mathcal{M}^* \cong D \cdot \delta_1 \cong D/(T^*)$. The proof of the proposition does not apply directly because δ_1 and $E([1])$ are not equal. However, they are related by $u\delta_1 = E([1])$. From this we see that the relation between T and T^* is $T = u\,T^*\,(1/u)$, which is equivalent to $(u^{-1}T)^* = u^{-1}T$. This is the appropriate scalar version of the condition $S^{-1}\Omega^t S = -\Omega$ in this case. □

Lemma 4.32 allows us to introduce a more intrinsic pairing on $\mathcal{M} = D/(T)$:

Definition 4.36 *Assume that $T^* = \pm T$. Then we have a natural pairing*

$$\begin{aligned} \langle\ ,\ \rangle : \mathcal{M} \times \mathcal{M} &\longrightarrow \mathcal{M}^* \times \mathcal{M} \longrightarrow \mathcal{H} \\ ([P],[Q]) &\mapsto ([P\cdot\delta_s],[Q]) \mapsto (P\cdot\delta_s)(Q) \end{aligned}$$

that is, we define $\langle [P],[Q]\rangle = (P\cdot\delta_s)(Q)$.

From its tautological definition the pairing $\langle\ ,\ \rangle$ is non-degenerate. It is $\mathcal{H}$-bilinear and the map

$$\mathcal{M} \to \mathcal{M}^*, \quad [P] \mapsto \langle [P],\ \rangle$$

is an isomorphism of D-modules. In particular, we have

$$\langle [P],[Q]\rangle' = \langle \partial \cdot [P],[Q]\rangle + \langle [P],\partial \cdot [Q]\rangle$$

for all $P,Q \in D$. Thus, the analogues of properties (1) and (2) of Lemma 4.28 are 'built-in'. However, we cannot say anything about property (3) unless we have a suitable basis:

Proposition 4.37 *If there exists a basis $[P_0],\ldots,[P_s]$ of $\mathcal{M}$ such that $\langle [P_i],[P_j]\rangle = s_{ij}$, then $S^{-1}\Omega^t S = -\Omega$.*

Proof If such a basis exists, then the pairing $\langle\ ,\ \rangle$ on $\mathcal{M}$ corresponds exactly to the pairing $(f,g) \mapsto f^t S g$ on $\mathcal{H}^{s+1}$ via the identification

$$\mathcal{M} \cong \mathcal{H}^{s+1}, \quad [\textstyle\sum_{i=0}^{s} f_i P_i] \leftrightarrow \begin{pmatrix} f_0 \\ \vdots \\ f_s \end{pmatrix}$$

of section 4.2. As $[P] \mapsto \langle [P],\ \ \rangle$ is an isomorphism of D-modules, so is $f \mapsto (f,\ \ )$. By Lemma 4.28 we obtain $S^{-1}\Omega^t S = -\Omega$. □

This is analogous to the statement that a self-adjoint linear transformation is represented with respect to an orthonormal basis by a symmetric matrix. As in that case, an 'orthonormal' basis may be constructed by the Gram–Schmidt orthogonalization procedure.

Example 4.38 The operator $T = \partial^2 + a_1\partial + a_0$ is self-adjoint if and only if $a_1 = 0$. In this case it is easy to compute $\langle 1,1\rangle = 0$, $\langle 1,\partial\rangle = 1$, $\langle \partial,1\rangle = -1$, $\langle \partial,\partial\rangle = 0$ (we omit the brackets from $[1]$ and $[\partial]$ for clarity).

Without loss of generality, we may write $T = (\partial + a)(\partial - a) - b$ as in Example 4.34. The basis given by $P_0 = 1, P_1 = \partial - a$ is 'orthonormal' in the sense that $\langle [P_i],[P_j]\rangle = s_{ij}$, where

$$S = \begin{pmatrix} & 1 \\ -1 & \end{pmatrix}.$$

As predicted by Proposition 4.37 we have $S^{-1}\Omega^t S = -\Omega$. The basis $[1],[\partial]$ is also orthonormal in this sense, and again $S^{-1}\Omega^t S = -\Omega$. □

Example 4.39 The case of a third-order operator $T = \partial^3 + a_2\partial^2 + a_1\partial + a_0$ is slightly more interesting. This is anti-self-adjoint (i.e. $T^* = -T$) if and only if $a_2 = 0$ and $a_0 = \frac{1}{2}a_1'$. Let us therefore write $T = \partial^3 - b\partial - \frac{1}{2}b'$. The 'metric' $\langle\ ,\ \rangle$ is given by

$$\langle f_0 + f_1\partial + f_2\partial^2, g_0 + g_1\partial + g_2\partial^2\rangle = f_0g_2 - f_1g_1 + f_2g_0 + f_2g_2b.$$

Thus, the basis $[1], [\partial], [\partial^2]$ is *not* orthonormal with respect to the natural choice

$$S = \begin{pmatrix} & & 1 \\ & -1 & \\ 1 & & \end{pmatrix}$$

(unless $b = 0$). An example of an orthonormal basis here is $[1], [\partial], [\partial^2 - \frac{1}{2}b]$. This arises from the following standard modification of the Gram–Schmidt orthogonalization procedure: first, choose a non-null vector, say $P_1 = \partial$, then obtain its orthogonal complement, which is spanned by $1, \partial^2$, then replace these by any P_0, P_2 with $\langle P_0, P_0 \rangle = \langle P_2, P_2 \rangle = 0$ and $\langle P_0, P_2 \rangle = 1$. Alternatively, taking into account the 'triangular' shape of the metric above, let us try the Ansatz

$$P_0 = 1, \ P_1 = \partial + u, \ P_2 = \partial^2 + v\partial + w.$$

This produces an orthonormal basis if and only if $u = v$ and $2w + b = v^2$. Taking $u = v = 0$ gives the basis above. For this basis we have

$$\Omega = \begin{pmatrix} 0 & \frac{1}{2}b & 0 \\ 1 & 0 & \frac{1}{2}b \\ 0 & 1 & 0 \end{pmatrix}$$

which does indeed satisfy $S^{-1}\Omega^t S = -\Omega$. Conversely, this matrix satisfies the assumptions of Proposition 4.33, which is consistent with $T^* = -T$. □

Example 4.40 The adjoint of the factorized operator

$$T = v_0 \dots v_s \, \partial \, \frac{1}{v_s} \, \partial \, \dots \, \partial \, \frac{1}{v_0}$$

is

$$T^* = \frac{1}{v_0} \, (-\partial) \, \dots \, (-\partial) \, \frac{1}{v_s} \, (-\partial) \, v_0 \dots v_s.$$

Since the ordered set of functions $v_0, \dots, v_s$ (satisfying $v_i(z_0) = 1$ for all i) is unique, we have $T^* = (-1)^{s+1}T$ if and only if

$$\frac{1}{v_0} = v_0 \dots v_s, \quad \frac{1}{v_1} = \frac{1}{v_s}, \quad \frac{1}{v_2} = \frac{1}{v_{s-1}}, \quad \dots$$

With respect to the basis given by

$$\frac{1}{v_0}, \ \frac{1}{v_1}\partial\frac{1}{v_0}, \ \frac{1}{v_2}\partial\frac{1}{v_1}\partial\frac{1}{v_0}, \ \dots$$

we have

$$\Omega = \begin{pmatrix} 0 & & & & \\ v_1 & 0 & & & \\ & v_2 & 0 & & \\ & & \ddots & \ddots & \\ & & & v_s & 0 \end{pmatrix}.$$

The (-1)-diagonal is not constant (cf. Example 4.35), so the hypotheses of Proposition 4.33 are not satisfied, but we do have $S^{-1}\Omega^t S = -\Omega$ with

$$S = \begin{pmatrix} & & & & 1 \\ & & & -1 & \\ & & 1 & & \\ & -1 & & & \\ \iddots & & & & \end{pmatrix}.$$

Conversely, one can verify that the above basis is orthonormal with respect to the metric $\langle\ ,\ \rangle$, which explains why $S^{-1}\Omega^t S = -\Omega$ in this case. □

Finally, we remark that our discussion of $\langle\ ,\ \rangle$ may be generalized to the case where $a_{s+1} \neq 1$, that is, T is not a monic differential operator. To simplify the notation, let us continue to assume that T is monic, but instead of the condition $T^* = \pm T$ we assume that $(f\,T)^* = \pm f\,T$ for some function f (which is holomorphic and never zero on its domain). We still have $\mathcal{M} = D/(T)$ and $\mathcal{M}^* \cong D/(T^*)$, but to relate these one must make use of the isomorphism

$$D/(T^*) \to D/(T^* f), \quad [P] \mapsto [Pf].$$

Then one has

$$D/(T) = D/(f\,T) = D/((f\,T)^*) = D/(T^* f) \cong D/(T^*) \cong (D/(T))^*.$$

This isomorphism, and the associated pairing $\langle\ ,\ \rangle^f$, depend on f. Explicitly, the pairing is given by

$$\langle [P], [Q] \rangle^f = (P\tfrac{1}{f} \cdot \delta_s)(Q).$$

Example 4.41 Consider the operator $T = \partial^2 - (u'/u)\partial - uv$ from Examples 4.13 and 4.35. This is not self-adjoint, but $u^{-1}T$ is, so $f = u^{-1}$ here, and we obtain

$$\langle 1, 1 \rangle^f = 0,\ \langle 1, \partial \rangle^f = u,\ \langle \partial, 1 \rangle^f = -u,\ \langle \partial, \partial \rangle^f = 0.$$

The basis that we found in Example 4.35, given by $P_0 = 1$, $P_1 = (1/u)\partial$, is orthornormal. Thus, the obvious extension of Proposition 4.37 explains why $S^{-1}\Omega^t S = -\Omega$ here. □

Self-adjointness of extensions

While there is a reasonably satisfactory correspondence between self-adjoint scalar and matrix equations in the one variable case, the case of two or more variables is more complicated. For a D-module $\mathcal{M} = D/(T_1, \dots, T_u)$, one cannot expect a direct relation between the conditions $\mathcal{M}^* \cong \mathcal{M}$ and $T_i^* = \pm T_i$ for all i. Apart from the fact that the generators $T_1, \dots, T_u$ are obviously not canonical, there is in general no canonical choice of cyclic generator for $\mathcal{M}^*$ corresponding to δ_s in the o.d.e. case.

In the examples which arise from quantum cohomology theory (which we shall begin to discuss in the next chapter), the situation is not entirely chaotic. First, at least in the case of 'simple' manifolds, natural relations T_i often occur. Second, the quantum cohomology D-module generally admits a grading which reflects the grading of the cohomology algebra—and this has a 'top piece' of dimension one. Projection on this top piece therefore gives a suitable analogue of δ_s.

Even for more general D-modules, one may select a cyclic element and compute generators as in the proof of Lemma 4.32. To obtain meaningful results one should start with a natural choice of cyclic element and make use of whatever additional structure is available.

We have in fact already seen an important family of examples where the self-adjointness property is easy to understand, namely the case of a D-module constructed by successive 'extensions' (section 4.4).

Proposition 4.42 *Consider the D-module $\mathcal{M} = D/(T)$, where $T^* = \pm T$, so that $\mathcal{M}^* \cong \mathcal{M}$. Let us extend $\mathcal{M}$ to a D-module of the form $\tilde{\mathcal{M}} = D/(T, \partial_t - P)$ as in Proposition 4.18. Then $\tilde{\mathcal{M}}^* \cong \tilde{\mathcal{M}}$.*

Proof If $[P_0], \dots, [P_s]$ is an 'orthonormal' basis of $\mathcal{M} = D/(T)$ with respect to $\langle \ , \ \rangle$ (in the sense explained earlier), then it remains orthonormal after the extension process, so the extended D-module is also self-adjoint. □

5 The quantum D-module

In this chapter we shall discuss quantum cohomology using the D-module terminology of Chapter 4. The introduction of D-modules here is not artificial; Floer cohomology gives another approach to quantum cohomology, and a natural D-module structure on Floer cohomology was discovered by A. Givental (see [47] and [48]). This Floer cohomology D-module structure could be regarded as the 'geometric' origin of the D-module structure of quantum cohomology. However, at this point, it is technically easier to proceed by defining the quantum D-module directly in terms of the quantum product.

5.1 The quantum D-module

We introduce the quantum D-module $\mathcal{M}$ in the following steps.

First of all, as a vector space, $\mathcal{M}$ is defined to be the space of sections of the trivial vector bundle

$$H^2M \times H^*M \to H^2M$$

or, more generally, the space of sections over an open subset N of H^2M. This is just the vector space consisting of all H^*M-valued functions on N. We obtain a free $\mathcal{H}$-module of rank $s+1$, where $\mathcal{H} = \mathcal{H}_t$ is the vector space of functions on $N = N_t$, as in Chapter 4. The default version of $\mathcal{H}_t$ will be the space of holomorphic functions. However, in the case of a Fano manifold, which we consider in this chapter, it is sometimes convenient to replace $\mathcal{H}_t$ by the smaller algebra $\mathbb{C}[q]$ consisting of polynomial functions in $q = (q_1, \dots, q_r) = (e^{t_1}, \dots, e^{t_r})$. The space of polynomial sections may be identified with $H^2M \otimes \mathbb{C}[q]$. In other cases, it may be necessary to enlarge $\mathcal{H}_t$, for example, to the space of formal power series.

The quantum product $\circ_t$ on H^*M gives a way of multiplying sections. Thus $\mathcal{M}$ becomes an algebra over $\mathcal{H}_t$. In the Fano case, the quantum product $\circ$ endows $H^*M \otimes \mathbb{C}[q]$ with the structure of an algebra over $\mathbb{C}[q]$.

Next, we introduce the action of a ring of differential operators. Let us provisionally define the action of $\partial_1 = \frac{\partial}{\partial t_1}, \dots, \partial_r = \frac{\partial}{\partial t_r}$ as follows: ∂_i acts on sections as the covariant derivative operator $\nabla_i = \partial_i + \frac{1}{\mathrm{h}} C_i$ (where C_i is the matrix of quantum multiplication by b_i as in Chapter 3). This extends to an action of the ring of all differential operators if and only if the identity $\nabla_i \nabla_j = \nabla_j \nabla_i$ holds for all i, j, and this identity does hold because the connection is flat.

This definition is provisional because we have not yet mentioned h. One approach would be to regard h as a fixed complex number. However, we shall incorporate it into the coefficient ring, by considering the following ring of differential operators:

Definition 5.1 *Let $\mathcal{H}_{\mathrm{h}}^+$ be the ring of $\mathbb{C}$-valued functions of* h *which are holomorphic in a neighbourhood of the origin of $\mathbb{C}$. Let D^{h} be the ring of differential operators with coefficients in $\mathcal{H}_t \otimes \mathcal{H}_{\mathrm{h}}^+$ which is generated by* $\mathrm{h}\partial_1, \dots, \mathrm{h}\partial_r$.

We continue to use the generic term 'D-module', although the quantum D-module will always be considered as a module over the specific ring D^{h}. Note that, according to our standard convention, $\mathcal{H}_{\mathrm{h}}$ denotes the ring of $\mathbb{C}$-valued functions of h which are holomorphic on some open set; the notation $\mathcal{H}_{\mathrm{h}}^+$ is intended to emphasize that this open set contains the point $\mathrm{h} = 0$ (and the use of the symbol '+' is for compatibility with Chapter 8).

We can now state the main definition:

Definition 5.2 *The quantum D-module is the $\mathcal{H}_t \otimes \mathcal{H}_{\mathrm{h}}^+$-module*

$$\mathcal{M} = H^*M \otimes \mathcal{H}_t \otimes \mathcal{H}_{\mathrm{h}}^+ \cong \mathcal{H}_t^{s+1} \otimes \mathcal{H}_{\mathrm{h}}^+$$

together with the action of D^{h} in which $\mathrm{h}\partial_i$ *acts as* $\mathrm{h}\partial_i + C_i$.

In particular, $\mathcal{M}$ is a free module over $\mathcal{H}_t \otimes \mathcal{H}_{\mathrm{h}}^+$, with rank equal to the (vector space) dimension of H^*M.

Remarks on notation (1) It is convenient (for computations) to say that '∂_i acts as $\partial_i + \frac{1}{\mathrm{h}} C_i$, even though ∂_i itself does not belong to D^{h}. (2) We remind the reader that the quantum differential equations were written in Chapter 3 in the form $\mathrm{h}\partial_i \Psi = C_i \Psi$, $1 \le i \le r$. The fundamental solution matrix

$$G = \begin{pmatrix} | & & | \\ \Psi_{(0)} & \cdots & \Psi_{(s)} \\ | & & | \end{pmatrix}$$

satisfies $\frac{1}{\mathrm{h}} C = dGG^{-1}$. From now on (in view of the above definition) our preferred notation for the quantum differential equations will be $\mathrm{h}\partial_i \Phi =$

$C_i^t \Phi$, $1 \le i \le r$. The fundamental solution matrix

$$H = L^t = F^t = \begin{pmatrix} | & & | \\ \Phi_{(0)} & \cdots & \Phi_{(s)} \\ | & & | \end{pmatrix}$$

satisfies $\frac{1}{\hbar} C^t = dHH^{-1}$, that is, $\frac{1}{\hbar} C = L^{-1} dL$.

5.2 The cyclic structure and the *J*-function

The most important property of the quantum D-module is its close relation with the (commutative) quantum cohomology algebra QH^*M. We shall discuss the relation in this section under the assumption that H^2M generates H^*M as an algebra and M is a Fano manifold. Because of this assumption we use the polynomial ring $\mathcal{H}_t = \mathbb{C}[q_1, \dots, q_r]$ (often abbreviated to $\mathbb{C}[q]$).

If H^2M does not generate H^*M, one may obtain similar results for the restricted quantum cohomology algebra $QH^\sharp M$, which is based on the cohomology subalgebra $H^\sharp M$ generated by H^2M. (If the quantum product preserves $H^\sharp M$, then $QH^\sharp M$ is the resulting subalgebra of QH^*M; otherwise $QH^\sharp M$ may be defined using modified Gromov–Witten invariants.)

We shall discuss briefly the situation where M is not Fano in section 6.7.

As mentioned in section 2.4, these hypotheses imply that QH^*M has a presentation

$$QH^*M \cong \mathbb{C}[b_1, \dots, b_r, q_1, \dots, q_r]/(\mathcal{R}_1, \dots, \mathcal{R}_u)$$

and H^*M has a presentation

$$H^*(M; \mathbb{C}) \cong \mathbb{C}[b_1, \dots, b_r]/(R_1, \dots, R_u)$$

where $\mathcal{R}_i|_{q=0} = R_i$. However, there is a more specific connection between QH^*M and H^*M, which generalizes to a precise connection between $\mathcal{M}$ and QH^*M, so let us review this.

First, for any polynomial c in 'abstract variables' $b_1, \dots, b_r, q_1, \dots, q_r$, let us denote by $[c]$ the corresponding element of QH^*M, and by $[[c]]$ the corresponding element of $H^*M \otimes \mathbb{C}[q_1, \dots, q_r]$. We claim that there exist suitable polynomials $c_0, \dots, c_s$ such that, if we identify QH^*M with $H^*M \otimes \mathbb{C}[q_1, \dots, q_r]$ via their respective bases given by $c_0, \dots, c_s$ and $c_0|_{q=0}, \dots, c_s|_{q=0}$, then the natural multiplication in QH^*M corresponds to quantum multiplication in $H^*M \otimes \mathbb{C}[q_1, \dots, q_r]$. That is, for $1 \le i \le r$, we have

$$[b_i][c_j] = \sum_{k=0}^{s} (C_i)_{kj}[c_k]$$

$$[[b_i]] \circ [[c_j|_{q=0}]] = \sum_{k=0}^{s} (C_i)_{kj}[[c_k|_{q=0}]].$$

Furthermore, we may assume that $c_0 = 1$, and $c_i = b_i$ for $1 \le i \le r$.

These facts follows from the observation (Theorem 2.2 of [114]) that 'any quantum polynomial may be written as the same classical polynomial plus lower classical terms, and vice versa'. Namely, if we regard b_j as a polynomial (with respect to the cup product) in $b_1, \ldots, b_r$, then the polynomial c_j is obtained by expressing b_j as a polynomial with respect to the quantum product in $b_1, \ldots, b_r$. The polynomials c_j satisfy $c_j|_{q=0} = b_j$. They are not unique, but they may be found systematically by solving a system of linear equations—that is, by Gauss elimination.

This is an important point, so let us repeat it in different words. Any choice of bases gives a $\mathbb{C}[q_1, \ldots, q_r]$-module isomorphism between QH^*M and $H^*M \otimes \mathbb{C}[q_1, \ldots, q_r]$. The latter has two multiplication operations, its natural multiplication (given by the cup product in H^*M) and the quantum product, and we take here the quantum product. The $\mathbb{C}[q_1, \ldots, q_r]$-module isomorphism is not in general an isomorphism of algebras. However, if we start in each case with the basis given by $b_0, \ldots, b_s$, the above argument produces new polynomials $c_0, \ldots, c_s$ in $b_1, \ldots, b_r$ such that we obtain an algebra isomorphism by choosing the bases $[c_0], \ldots, [c_s]$ and $[[b_0]], \ldots, [[b_s]]$ (and we have $c_i|_{q=0} = b_i$, i.e. each c_i is a 'q-deformation' of b_i).

The fact that QH^*M is a cyclic $\mathbb{C}[b_1, \ldots, b_r, q_1, \ldots, q_r]$-module (with relations which are q-deformations of the relations of H^*M) follows from the same argument. In particular, since 1 is a cyclic generator of the $\mathbb{C}[b_1, \ldots, b_r]$-module H^*M, it is also a cyclic generator of the $\mathbb{C}[b_1, \ldots, b_r, q_1, \ldots, q_r]$-module QH^*M.

Exactly the same method gives the analogous result below for $\mathcal{M}$, because any polynomial in the operators $\mathrm{h}\partial_1 + C_1, \ldots, \mathrm{h}\partial_r + C_r$ can be expressed as the same polynomial in $\mathrm{h}\partial_1, \ldots, \mathrm{h}\partial_r$ plus terms of lower order. Moreover, since the lower-order terms contain 'additional' powers of h, if we replace $\mathrm{h}\partial_i$ by b_i (for each i) then set h equal to 0, these lower-order terms all vanish and we are left with the original polynomial expressed in terms of the variables $b_1, \ldots, b_r$.

Theorem 5.3 *The quantum D-module is cyclic, and (the constant map) 1 is a cyclic generator. It is isomorphic to a D-module of the form* $D^h/(D_1, \ldots, D_u)$, *where* $D_1, \ldots, D_u$ *are converted to* $\mathcal{R}_1, \ldots, \mathcal{R}_u$ *when* $h\partial_i$ *is replaced by* b_i *(for each i) then h set equal to* 0.

Furthermore, there exists a basis $[P_0], \ldots, [P_s]$ *of* $D^h/(D_1, \ldots, D_u)$, *with respect to which the (connection) matrix of* $h\partial_i$ *is* C_i. *In other words, the associated flat connection* $\nabla = d + \Omega$ *(as defined in Chapter 4) is given by* $\Omega = \frac{1}{h}C$. *This basis is converted to* $[c_0], \ldots, [c_s]$ *when* $h\partial_i$ *is replaced by* b_i *(for each i) then* h *set equal to* 0.

We are therefore in the situation of section 4.2, and can write the fundamental solution matrix as

$$H = \begin{pmatrix} | & & | \\ \Phi_{(0)} & \cdots & \Phi_{(s)} \\ | & & | \end{pmatrix} = \begin{pmatrix} - & P_0 J & - \\ & \vdots & \\ - & P_s J & - \end{pmatrix},$$

where $J = (\phi_{(0)}, \ldots, \phi_{(s)})$; here $\phi_{(0)}, \ldots, \phi_{(s)}$ is a solution basis of the scalar system $D_1\phi = 0, \ldots, D_u\phi = 0$. We may assume that $P_0 = 1$ and $P_i = h\partial_i$ for $1 \le i \le r$.

This H is defined only up to multiplication on the right by a constant matrix, but we shall explain in section 5.3 how a natural normalization can be chosen. Remarkably, there is an explicit formula for this normalized H (and hence J) in terms of Gromov–Witten invariants (see [26] and [49]). We shall give this briefly in appendix 5.4. In the quantum cohomology literature, following Givental, J is usually called 'the J-function' of M. It has already made an appearance in Chapter 3, in Theorem 3.2; the functions $(\Psi_{(v)}, 1)$—which form the *last* row of the fundamental solution matrix, because of our conventions there—are the components of J.

We have, essentially, seen several examples of this theorem already in Chapter 3. However, as a different convention was used there ($d - \frac{1}{h}C$ instead of $d + \frac{1}{h}C$), and as we did not use D-module language explicitly, let us work out the case of the degree 3 hypersurface in $\mathbb{C}P^4$ (Example 3.6) again, from the current point of view. We choose this example because it is a very simple example which exhibits non-trivial dependence on the parameter h.

Example 5.4 We begin with the D-module structure on $\mathcal{H}_t^4 \otimes \mathcal{H}_h^+$ defined by

$$h\partial \cdot \begin{pmatrix} f_0 \\ f_1 \\ f_2 \\ f_3 \end{pmatrix} = (h\partial + C) \begin{pmatrix} f_0 \\ f_1 \\ f_2 \\ f_3 \end{pmatrix}$$

where (from Example 2.12)

$$C = \left(\begin{array}{c|c|c|c} & 6q & & 36q^2 \\ \hline 1 & & 15q & \\ \hline & 1 & & 6q \\ \hline & & 1 & \end{array} \right).$$

We calculate

$$\mathrm{h}\partial \cdot \begin{pmatrix} 1 \\ 0 \\ 0 \\ 0 \end{pmatrix} = \begin{pmatrix} 0 \\ 1 \\ 0 \\ 0 \end{pmatrix}, \quad (\mathrm{h}\partial)^2 \cdot \begin{pmatrix} 1 \\ 0 \\ 0 \\ 0 \end{pmatrix} = \begin{pmatrix} 6q \\ 0 \\ 1 \\ 0 \end{pmatrix}, \quad (\mathrm{h}\partial)^3 \cdot \begin{pmatrix} 1 \\ 0 \\ 0 \\ 0 \end{pmatrix} = \begin{pmatrix} 6q\mathrm{h} \\ 21q \\ 0 \\ 1 \end{pmatrix}.$$

This will enable us to find suitable P_0, P_1, P_2, P_3; we need differential operators which produce the standard unit basis vectors (so that the matrix of $\mathrm{h}\partial$ is exactly the above matrix C). By inspection we see that $P_0 = 1$, $P_1 = \mathrm{h}\partial$, $P_2 = (\mathrm{h}\partial)^2 - 6q$, $P_3 = (\mathrm{h}\partial)^3 - 21q\mathrm{h}\partial - 6\mathrm{h}q$ have the desired property.

To obtain a relation D_1 for the D-module $D^{\mathrm{h}}/(D_1)$, that is, a differential operator which annihilates the cyclic element, we compute

$$(\mathrm{h}\partial)^4 \cdot \begin{pmatrix} 1 \\ 0 \\ 0 \\ 0 \end{pmatrix} = \begin{pmatrix} 162q^2 \\ 0 \\ 27q \\ 0 \end{pmatrix} + \mathrm{h} \begin{pmatrix} 0 \\ 27q \\ 0 \\ 0 \end{pmatrix} + \mathrm{h}^2 \begin{pmatrix} 6q \\ 0 \\ 0 \\ 0 \end{pmatrix}$$

$$= \left(27q((\mathrm{h}\partial)^2 - 6q) + 162q^2 + 27\mathrm{h}q(\mathrm{h}\partial) + 6\mathrm{h}^2 q \right) \cdot \begin{pmatrix} 1 \\ 0 \\ 0 \\ 0 \end{pmatrix}.$$

This gives the relation $(\mathrm{h}\partial)^4 - 27q(\mathrm{h}\partial)^2 - 27\mathrm{h}q(\mathrm{h}\partial) - 6\mathrm{h}^2 q$, as expected from Example 3.6.

It would have been possible to write all calculations here in terms of the operators $\mathrm{h}\partial + b\circ$ rather than their matrix versions $\mathrm{h}\partial + C$. We have given the matrix form to emphasize that we are simply dealing with a connection in a vector bundle. Note that the 'commutative version' of these calculations, using the matrix C alone, would produce the polynomials $c_0 = 1$, $c_1 = b$, $c_2 = b^2 - 6q, c_3 = b^3 - 21qb$, and the relation $b^4 - 27qb^2$ for the quantum cohomology algebra. It can be verified that the matrix of multiplication by b on $\mathbb{C}[b, q]/(b^4 - 27qb^2)$ with respect to this basis is exactly C. □

As the theorem illustrates, the D-module language provides a convenient way to discuss the idea of 'matching' commutative and non-commutative objects, mentioned briefly in section 3.3. The action of $\mathrm{h}\partial_i$ on the quantum D-module matches exactly the action of b_i on the quantum cohomology algebra—both are given by the same matrix. However, to accomplish this, a careful choice of basis is necessary in each case, and these bases (like the relations) do not match exactly, *only 'mod* h'. In the simplest examples (such as the quantum cohomology of complex projective space, or the manifolds F_3 and Σ_1 discussed in Chapter 3), such additional 'mod h' terms are not necessary, but in general they cannot be avoided.

This matching property relies heavily on the properties of quantum cohomology. If $*_t$ is *any* family of (commutative, associative) products on the vector space H^*M, we have 'quantum differential equations' which say that multiplication by b_i (in the commutative algebra) has the same effect as multiplication by $\mathrm{h}\partial_i$ in the (hypothetical) 'quantum D-module'. But there is no guarantee that such a D-module is well defined, and, even if it is, there is no guarantee that differential relations exist which match relations of the commutative algebra in the above sense. Conversely, if we start with a D-module $\mathcal{M} = D^{\mathrm{h}}/(D_1, \dots, D_u)$, there will in general be no corresponding products $*_t$. This reflects the phenomenon encountered in section 4.2: a system of matrix equations does not necessarily define a D-module of the same rank, while the rank of the D-module defined by a system of scalar equations is not in general visible. We shall return to this matter in Chapter 6.

5.3 Other properties

First, we discuss a property which the quantum D-module $\mathcal{M}$ inherits from the Frobenius property

$$(b \circ_t x, y) = (x, b \circ_t y)$$

of the quantum product (which in turn comes from the symmetry of the Gromov–Witten invariant $\langle A|B|C\rangle_D$ in A, B, C).

The Frobenius property says that the linear transformation

$$b\circ_t : H^*M \to H^*M, \quad x \mapsto b \circ_t x$$

is symmetric with respect to the Poincaré intersection form

$$(x, y) = x^t \, S y$$

where $S_{ij} = (b_i, b_j)$ and $b_0, \ldots, b_s$ is the usual basis. This condition can be expressed as

$$S^{-1} C_i^t S = C_i,$$

or

$$C_i^{(*)} = C_i,$$

where $C_i^{(*)} = S^{-1} C_i^t S$ is the adjoint of C_i with respect to (,).

Example 5.5 In the case of $M = \mathbb{C}P^n$, the matrix of the intersection form with respect to the usual basis $1, b, \ldots, b^n$ is

$$S = \begin{pmatrix} & & 1 \\ & \cdot^{\cdot^{\cdot}} & \\ 1 & & \end{pmatrix}.$$

The adjoint $X^{(*)}$ is obtained by reflecting X in its NE–SW diagonal, that is, $(X^{(*)})_{ij} = X_{n-j,n-i}$. And indeed (see Example 3.3), the matrix C here is

$$\begin{pmatrix} 0 & & & q \\ 1 & \ddots & & \\ & \ddots & \ddots & \\ & & 1 & 0 \end{pmatrix}$$

which is symmetrical under reflection in the NE–SW diagonal.

The matrices C_1, C_2 in Example 3.4 exhibit the same reflection symmetry, since in that case[1] also we have $(b_i, b_j) = 1$ if $i + j = s + 1$ and $(b_i, b_j) = 0$ otherwise. □

The Frobenius condition can be interpreted as saying that the quantum D-module is self-adjoint, in a certain sense. To explain this, let us (following [9] and [72]) introduce a pairing

$$\mathcal{M} \times \mathcal{M} \to \mathcal{H}_t \otimes \mathcal{H}_\mathrm{h}^+$$

by defining

$$(\!(f, g)\!) = (\bar{f}, g)$$

[1] More generally, the Schubert decomposition of any generalized flag manifold gives rise to such a basis.

where $\bar{f}(t,\mathrm{h}) = f(t,-\mathrm{h})$. The effect of the negative sign is to convert the identity

$$\partial_i((f,g)) = ((\partial_i f,g)) + ((f,\partial_i g))$$

into

$$\mathrm{h}\partial_i((f,g)) = -((\mathrm{h}\partial_i f,g)) + ((f,\mathrm{h}\partial_i g)).$$

This identity for the standard D-module structure has the following analogue for the quantum D-module $\mathcal{M}$, which expresses the Frobenius condition naturally in D-module terms. We write $\partial_i \cdot f = (\partial_i + \frac{1}{\mathrm{h}}C_i)f$ for the quantum D-module action (and $\partial_i f$ for the standard action). □

Proposition 5.6 *For any $f,g \in \mathcal{M}$,*

$$\mathrm{h}\partial_i((f,g)) = -((\mathrm{h}\partial_i \cdot f,g)) + ((f,\mathrm{h}\partial_i \cdot g)).$$

In particular, for constant sections x and y, the identity

$$((\mathrm{h}\partial_i \cdot x,y)) = ((x,\mathrm{h}\partial_i \cdot y))$$

holds.

Proof By definition

$$-((\mathrm{h}\partial_i \cdot f,g)) = -(((\mathrm{h}\partial_i + C_i)f,g)) = -((\mathrm{h}\partial_i f,g)) - ((C_i f,g))$$

and

$$((f,\mathrm{h}\partial_i \cdot g)) = ((f,(\mathrm{h}\partial_i + C_i)g)) = ((f,\mathrm{h}\partial_i g)) + ((f,C_i g)).$$

Hence the formula $\mathrm{h}\partial_i((f,g)) = -((\mathrm{h}\partial_i \cdot f,g)) + ((f,\mathrm{h}\partial_i \cdot g))$ is equivalent to the Frobenius condition $((C_i f,g)) = ((f,C_i g))$. □

The dual D-module $\mathcal{M}^*$ is defined to be the space of $\mathcal{H}_t \otimes \mathcal{H}_{\mathrm{h}}^+$-module homomorphisms $\mathcal{M} \to \mathcal{H}_t \otimes \mathcal{H}_{\mathrm{h}}^+$. The D-module structure of $\mathcal{M}^*$ is specified by

$$(\mathrm{h} \cdot \pi)(f) = \mathrm{h}\pi(f)$$

$$(\partial_i \cdot \pi)(f) = -\pi(\partial_i \cdot f) + \frac{\partial}{\partial t_i}\pi(f)$$

for $\pi \in \mathcal{M}^*$ and $f \in \mathcal{M}$.

Let us denote by $\bar{\mathcal{M}}^*$ the D-module obtained from $\mathcal{M}^*$ by 'reversing the sign in the action of h'. That is, $\bar{\mathcal{M}}^*$ and $\mathcal{M}^*$ are identical as $\mathcal{H}_t$-modules, but the action of D^{h} on $\bar{\mathcal{M}}^*$ is specified by

$$\mathrm{h} \odot \pi = -\mathrm{h}\pi,$$

$$\partial_i \odot \pi = \partial_i \cdot \pi$$

(thus $(\mathrm{h}\partial_i) \odot \pi = -\mathrm{h}\partial_i \cdot \pi$).

We can now interpret the Frobenius property in the following way:

Corollary 5.7 *The map* $E : \mathcal{M} \to \bar{\mathcal{M}}^*, f \mapsto (\!(f, \ )\!)$ *is an isomorphism of D-modules.*

Proof It is clear that $E(f)$ belongs to $\bar{\mathcal{M}}^*$ (the bar is irrelevant for this), and that E is an isomorphism of $\mathcal{H}_t \otimes \mathcal{H}_{\mathrm{h}}^+$-modules. To prove that E is an isomorphism of D-modules we must show (a) $\mathrm{h} \odot E(f)\,(g) = E(\mathrm{h}f)\,(g)$, and (b) $(\mathrm{h}\partial_i) \odot E(f)\,(g) = E(\mathrm{h}\partial_i \cdot f)\,(g)$, for all f and g. Both follow immediately from the formulae above. □

Example 5.8 The case of M_N^k is similar to the case of $\mathbb{C}P^n$ and exhibits the same symmetry. For example, when $k = 3, N = 5$ (Example 3.6) we have

$$C = \left(\begin{array}{c|c|c|c} & 6q & & 36q^2 \\ \hline 1 & & 15q & \\ \hline & 1 & & 6q \\ \hline & & 1 & \end{array}\right).$$

The self-adjointness of the D-module can be seen more directly from the equivalent scalar operator

$$\mathrm{h}^4\partial^4 - 27q\mathrm{h}^2\partial^2 - 27\mathrm{h}^2 q\partial - 6\mathrm{h}^2 q,$$

which is self-adjoint in the sense of appendix 4.6. For general k and N we have the operator

$$(\mathrm{h}\partial)^{N-1} - k\mathrm{h}^{k-1}q(k\partial + k - 1)(k\partial + k - 2)\dots(k\partial + 1).$$

This is always self-adjoint *providing we incorporate* h *into the definition of adjoint* by using

$$\partial_i \mapsto -\partial_i, \quad \mathrm{h} \mapsto -\mathrm{h},$$

as $(q\mathrm{h}(k\partial + i))^* = q\mathrm{h}(k\partial + k - i)$. (The same applies to the operator for $\mathbb{C}P^n$, which is self-adjoint in the sense of appendix 4.6 only when n is odd.) We shall discuss this extended definition systematically in section 6.3. □

The grading of the cohomology and quantum cohomology algebras gives rise to a similar grading of the quantum D-module. This implies that the connection matrix C acquires a homogeneity property. To describe this, we write the (α, β) block of the matrix C_i corresponding to $\mathrm{Hom}(H^{2\alpha}M, H^{2\beta}M)$ as $(C_i)_{\alpha,\beta}$. Then,

(1) for $\alpha \geq \beta + 2$ we have $C_{\alpha,\beta} = 0$,
(2) each non-zero entry of the block $(C_i)_{\alpha,\beta}$ has degree $2(\beta - \alpha) + 2$.

For example, the entries of the matrices C_i for the quantum cohomology of the flag manifold F_3 (Example 3.4) are as shown in the boxes below:

$$\left(\begin{array}{c|cc|cc|c}
\boxed{2} & \boxed{4} & \boxed{4} & \boxed{6} & \boxed{6} & \boxed{8} \\
\hline
\boxed{0} & \boxed{2} & \boxed{2} & \boxed{4} & \boxed{4} & \boxed{6} \\
\boxed{0} & \boxed{2} & \boxed{2} & \boxed{4} & \boxed{4} & \boxed{6} \\
\hline
 & \boxed{0} & \boxed{0} & \boxed{2} & \boxed{2} & \boxed{4} \\
 & \boxed{0} & \boxed{0} & \boxed{2} & \boxed{2} & \boxed{4} \\
\hline
 & & & \boxed{0} & \boxed{0} & \boxed{2}
\end{array}\right).$$

We mention here some further properties of the quantum D-module. These play an important role in quantum cohomology theory, but as they do not generalize directly to other integrable systems we shall not go into any detail.

The first property is the nature of the singularity of the quantum differential equations at the point $q = 0$ (or, more precisely, on the union of the divisors $q_i = 0$).

For Fano manifolds, the C_i are polynomial in $q_1, \ldots, q_r$. However, even in this case, the quantum differential equations will have a singular point at $q = (q_1, \ldots, q_r) = 0$, because

$$\partial_i = \frac{\partial}{\partial t_i} = q_i \frac{\partial}{\partial q_i}.$$

This means that the fundamental solution matrix will not be single-valued at $q = 0$, in general. It is clear that the singularity is a regular singularity, so, at least in the case $r = 1$, $L\ (= H^t)$ is expected to be of the form

$$L = q^{\Gamma} S$$

where Γ is a matrix independent of q and[2] S is holomorphic at $q = 0$ (cf. the remarks at the end of section 4.1).

[2] The notation S is standard in the quantum cohomology literature. As this conflicts with our matrix S in section 5.3, we shall be careful to use $X^{(*)}$ for $S^{-1} X^t S$ when necessary.

Theorem 5.9 *It is possible to normalize*[3] L *so that*

$$L(q_1,\ldots,q_r,\mathrm{h}) = \exp\left(\frac{1}{\mathrm{h}}\sum_{i=1}^{r} t_i(C_i|_{q=0})\right) S(q_1,\ldots,q_r,\mathrm{h})$$

with $S|_{q=0} = I$.

For the proof see section 3.2 of [72].

Symbolically, it is usual to write

$$L = \mathrm{e}^{t/\mathrm{h}} S$$

with the understanding that $t = \sum_{i=1}^{r} t_i(C_i|_{q=0}) = \sum_{i=1}^{r} t_i b_i$ here means the multiplication operator on cohomology.

By making use of S, we obtain the following version of the Frobenius condition:

Proposition 5.10 *The normalized map* L *satisfies*

$$L^{(*)^{-1}} = L(-\mathrm{h})$$

that is, $((Lf, Lg)) = ((f, g))$ *for all* $f, g \in \mathcal{M}$.

Proof The Lie algebra condition on $\alpha = F^{-1}dF$ corresponding to the Lie group condition $F^{(*)^{-1}} = F(-\mathrm{h})$ is

$$\begin{aligned}
-\alpha^{(*)} &= -(F^{-1}dF)^{(*)} \\
&= -dF^{(*)}F^{(*)^{-1}} \\
&= -dF^{-1}(-\mathrm{h})F(-\mathrm{h}) \\
&= F(-\mathrm{h})^{-1}dF(-\mathrm{h})F(-\mathrm{h})^{-1}F(-\mathrm{h}) \\
&= \alpha(-\mathrm{h}),
\end{aligned}$$

which is equivalent to the earlier condition $C^{(*)} = C$, when $\alpha = \frac{1}{\mathrm{h}}C$. (It should be noted that the apparently simpler condition $\alpha^{(*)} = \alpha$, which $\alpha = \frac{1}{\mathrm{h}}C$ also satisfies, is not a Lie subalgebra condition.)

We would like to use the principle (cf. Corollary 4.24) that a map F takes values in a Lie subgroup if and only if the 1-form $\alpha = F^{-1}dF$ takes values in the corresponding Lie subalgebra (where the map F is assumed to be defined on a simply connected domain and to take the value I at some point).

[3] Recall that L is defined up to a multiplicative constant by $L^{-1}dL = \frac{1}{\mathrm{h}}C$.

This principle does not apply directly to $F = L$. However, we have $L = e^{t/\mathrm{h}}S$ where $S|_{q=0} = I$, and an analogous principle applies to S. Namely, the equation

$$\Omega = S^{-1}\,\tfrac{1}{\mathrm{h}}C|_{q=0}\,S + S^{-1}dS$$

satisfied by S is also satisfied by $S(-\mathrm{h})^{(*)^{-1}}$. Since both take the value I at $q = 0$, we obtain $S^{(*)^{-1}} = S(-\mathrm{h})$, from which the result for L follows. □

Another aspect of the quantum D-module, though one which is less easy to describe, is its 'arithmetic' nature, arising from the definition of quantum multiplication in terms of intersections of cycles in a certain moduli space. The Gromov–Witten invariants that we have used so far are integers, and so are the coefficients of the quantum differential equations (in all our examples, at least). Moreover, the explicit formula (to be given in the next section) for the fundamental solution matrix shows that the *solutions* of the quantum differential equations have series expansions whose coefficients are generalized Gromov–Witten invariants.

5.4 Appendix: explicit formula for the *J*-function

Givental's explicit formula for the fundamental solution matrix referred to in section 5.2 involves the (gravitational) 'descendant' Gromov–Witten invariants

$$\langle \tau_{d_1}X_1|\tau_{d_2}X_2|\ldots|\tau_{d_i}X_i\rangle_D.$$

These reduce to the primary Gromov–Witten invariants $\langle X_1|X_2|\ldots|X_i\rangle_D$ when $d_1 = d_2 = \cdots = d_i = 0$. Whereas the primary invariants are defined using cycles in the moduli space which come from cycles $X_1, X_2, \ldots, X_i$ in M, the descendant invariants incorporate additional cycles representing the first chern classes of certain line bundles $\mathcal{L}_1^{d_1}, \mathcal{L}_2^{d_2}, \ldots, \mathcal{L}_i^{d_i}$ on the moduli space. (The $d_1, d_2, \ldots, d_i$ are non-negative integers.) The definition of $\langle \tau_{d_1}X_1|\tau_{d_2}X_2|\ldots|\tau_{d_i}X_i\rangle_D$ is

$$\int_{[\overline{M}(D)]^{\mathrm{virt}}} c_1(\mathcal{L}_1^{d_1}) \wedge ev_1^* x_1 \wedge c_1(\mathcal{L}_2^{d_2}) \wedge ev_2^* x_2 \wedge \cdots \wedge c_1(\mathcal{L}_i^{d_i}) \wedge ev_i^* x_i$$

(see section 10.1 of [26]). When $\langle \tau_{d_1}X_1|\tau_{d_2}X_2|\ldots|\ \tau_{d_i}X_i\rangle_D \neq 0$, the numerical condition

$$\sum_{j=1}^{i} |x_j| + 2\sum_{j=1}^{i} d_j = 2(n + i - 3) + 2\langle c_1(TM), D\rangle$$

holds, where n is as usual the complex dimension of M.

Theorem 5.11 *For $t \in H^2M$ and $a \in H^*M$, a solution of the system*

$$\mathrm{h}\partial_j G = b_j \circ_t G, \quad 1 \le j \le r,$$

is given by

$$G(t)(a) = a\mathrm{e}^{t/\mathrm{h}} + \sum_{D\neq 0,\ l\geq 0,\ j=0,\ldots,s} \frac{1}{\mathrm{h}^{l+1}} \langle \tau_l \mathrm{e}^{T/\mathrm{h}} A | B_j \rangle_D \, \mathrm{e}^{\langle t,D\rangle} \, a_j.$$

Here, $\mathrm{e}^{T/\mathrm{h}}A$ means $\sum_{k\geq 0} T^k A/(\mathrm{h}^k k!)$, where $T^k A$ is the Poincaré dual homology class to $t^k a$. As in the previous section, $\mathrm{e}^{t/\mathrm{h}}$ means $\exp\left(\frac{1}{\mathrm{h}} \sum_{i=1}^r t_i (C_i|_{q=0})\right)$. This theorem was proved in [49]. A detailed explanation can be found in [26], with a brief sketch of the proof (using the current notation) in [63].

In particular, the theorem gives an explicit formula for $J = (\psi_{(0)}, \ldots, \psi_{(s)})$:

Corollary 5.12

$$J(t) = \mathrm{e}^{t/\mathrm{h}} \left(1 + \sum_{D\neq 0,\ l\geq 0,\ j=0,\ldots,s} \frac{1}{\mathrm{h}^{l+1}} \langle \tau_l B_j | M \rangle_D \, \mathrm{e}^{\langle t,D\rangle} \, a_j \right).$$

Example 5.13 It is easy to solve the quantum differential equations directly for $M = \mathbb{C}P^n$. We shall do this and compare the result with the formulae above. Using the notation of Chapter 3, the quantum differential equation is $(\mathrm{h}\partial_1)^{n+1}\psi = q_1\psi$, and we need $n+1$ linearly independent solutions in order to construct J. The rows of G are then given by J and its successive derivatives.

Substituting $\psi(q_1) = a_0 + a_1 q_1 + a_2 q_1^2 + \cdots$ into the equation, we discover that the series

$$\psi(t) = \sum_{d\geq 0} \frac{q_1^d}{(d!)^{n+1} \mathrm{h}^{d(n+1)}},$$

is a solution (recall that $t = t_1 b_1 = t_1 x_1$ and $q_1 = \mathrm{e}^{t_1}$). It is less easy to discover, but easy to verify, that the *cohomology-valued* series

$$J(t) = \sum_{d\geq 0} \frac{q_1^{x_1/\mathrm{h}+d}}{[(x_1+\mathrm{h})(x_1+2\mathrm{h})\ldots(x_1+d\mathrm{h})]^{n+1}}$$

is a solution. Indeed, since $\mathrm{h}q_1 \frac{d}{dq_1} q_1^{x_1/\mathrm{h}+d} = (x_1 + d\mathrm{h}) q_1^{x_1/\mathrm{h}+d}$, it follows immediately that

$$\left(\mathrm{h}q_1 \frac{d}{dq_1}\right)^{n+1} J = q_1 J.$$

Thus, writing $J = \psi_{(0)} + \psi_{(1)}x_1 + \cdots + \psi_{(n)}x_1^n$ we immediately obtain $n+1$ solutions, and it turns out[4] that they are linearly independent.

The expression for J from the corollary is

$$J(t) = e^{t/h}\left(1 + \sum_{d\geq 1,\ l\geq 0,\ j=0,\ldots,n} \frac{1}{h^{l+1}}\langle \tau_l X_1^j | \mathbb{C}P^n\rangle_d e^{t_1 d} x_1^{n-j}\right).$$

We shall verify that this coincides with the formula in the case $n = 1$, by making use of the following known descendant Gromov–Witten invariants (see [26]):

$$\langle \tau_{2d-1}\,\text{point}\,|\mathbb{C}P^1\rangle_d = \frac{1}{(d!)^2},\quad \langle \tau_{2d}\mathbb{C}P^1|\mathbb{C}P^1\rangle_d = \frac{-2}{(d!)^2}\left(1+\cdots+\frac{1}{d}\right).$$

All other invariants of the form $\langle \tau_l\,\text{point}\,|\mathbb{C}P^1\rangle_d$, $\langle \tau_l\mathbb{C}P^1|\mathbb{C}P^1\rangle_d$ are zero, because of the numerical condition mentioned earlier.

To indicate clearly the degrees of the cohomology classes, we shall write x_0 for 1 here. Noting that $B_0 = \mathbb{C}P^1$, $B_1 =$ point, and $a_0 = x_1$, $a_1 = x_0$, we have

$$\begin{aligned}
J(t) &= e^{t_1x_1/h}\left(x_0 + \sum_{d\geq 1,\ l\geq 0}\frac{1}{h^{l+1}}\left(\langle \tau_l\mathbb{C}P^1|\mathbb{C}P^1\rangle_d x_1 + \langle \tau_l\,\text{point}\,|\mathbb{C}P^1\rangle_d x_0\right)e^{t_1 d}\right)\\
&= e^{t_1x_1/h}\left(x_0 + \sum_{d\geq 1} e^{t_1 d}\left(\frac{1}{h^{2d}}\frac{1}{(d!)^2}x_0 + \frac{-2}{h^{2d+1}(d!)^2}\left(1+\ldots+\frac{1}{d}\right)x_1\right)\right)\\
&= e^{t_1x_1/h}\left(x_0 + \sum_{d\geq 1}\frac{e^{t_1 d}}{h^{2d}}\frac{1}{(d!)^2}\left(x_0 - \frac{2x_1}{h}\left(1+\cdots+\frac{1}{d}\right)\right)\right).
\end{aligned}$$

Since $x_1^2 = 0$, we have

$$\begin{aligned}
x_0 - \frac{2x_1}{h}\left(1+\frac{1}{2}+\cdots+\frac{1}{d}\right) &= \left(x_0 + \frac{2x_1}{h}\left(1+\cdots+\frac{1}{d}\right)\right)^{-1}\\
&= \left(x_0 + \frac{x_1}{h}\left(1+\cdots+\frac{1}{d}\right)\right)^{-2}.
\end{aligned}$$

[4] They are a 'Frobenius basis'. See appendix 4.6 and [116].

Hence

$$J(t) = e^{t_1 x_1/h}\left(x_0 + \sum_{d\ge 1}\frac{e^{t_1 d}}{\left(h^d d!\left(x_0 + \frac{x_1}{h}\left(1+\cdots+\frac{1}{d}\right)\right)\right)^2}\right)$$

$$= e^{t_1 x_1/h}\sum_{d\ge 0}\frac{e^{t_1 d}}{\left(d!h^d x_0 + d!h^{d-1}\left(1+\cdots+\frac{1}{d}\right)x_1\right)^2}$$

which is the same as

$$e^{t_1 x_1/h}\sum_{d\ge 0}\frac{e^{t_1 d}}{\left((x_1 + hx_0)(x_1+2hx_0)\ldots(x_1+dhx_0)\right)^2}$$

(the linear term in x_0, x_1 is the only non-zero part of the expansion of the denominator).

Conversely, starting with this formula, one could deduce the formulae for the descendant Gromov–Witten invariants given earlier.

Let us write out the formula for G explicitly. In the notation of Chapter 3 we have

$$G = \begin{pmatrix} | & | \\ \Psi_{(0)} & \Psi_{(1)} \\ | & | \end{pmatrix} = \begin{pmatrix} - & h\partial_1 J & - \\ - & J & - \end{pmatrix},$$

and this is a fundamental solution matrix of the system

$$hq_1\frac{d}{dq_1}\begin{pmatrix}\psi_0\\ \psi_1\end{pmatrix} = \begin{pmatrix}0 & q_1\\ 1 & 0\end{pmatrix}\begin{pmatrix}\psi_0\\ \psi_1\end{pmatrix}.$$

The resulting formula for $G = Se^{t/h}$ is given by

$$S = \sum_{d\ge 0}\begin{pmatrix} 1-2d\left(1+\cdots+\frac{1}{d}\right) & dh \\ -2\left(1+\cdots+\frac{1}{d}\right)\frac{1}{h} & 1\end{pmatrix}\frac{q_1^d}{(d!)^2 h^{2d}},$$

with

$$e^{t/h} = \begin{pmatrix}1 & 0\\ t_1/h & 1\end{pmatrix} = I + \frac{1}{h}\begin{pmatrix}0 & 0\\ t_1 & 0\end{pmatrix}.$$

In the notation of Chapter 5 we have $L = e^{t/h}S$ where S is obtained by reflecting the above matrix S in its NE-SW diagonal. □

6 Abstract quantum cohomology

We shall describe a general method of producing D-modules which 'resemble' quantum cohomology D-modules, based on [62] and [72]. This could be considered as a method of construction of Frobenius manifolds (see Chapter 9). But it differs from other approaches in the literature both in its starting point (systems of scalar equations of a rather specific type) and its goal (a wider class of integrable systems).

6.1 The Birkhoff factorization

Motivated by the properties of the quantum cohomology D-module described in the previous chapter, we shall begin by showing how to construct h-families of connections $\nabla = d + \Omega$ such that

(1) ∇ has zero curvature (i.e. $d\Omega + \Omega \wedge \Omega = 0$ for all h),
(2) Ω has the special form $\frac{1}{\mathrm{h}}\omega$ (where ω is independent of h).

Our starting point will be a system of (scalar) partial differential equations of a certain type. Such systems produce flat connections immediately, but do not in general satisfy (2). We shall give a modification procedure, the result of which satisfies both of these conditions.

Special case: ordinary differential equations

Consider the ordinary differential equation

$$\left((\mathrm{h}\partial)^{s+1} + a_s(\mathrm{h}\partial)^s + \cdots + a_1\mathrm{h}\partial + a_0\right)y = 0$$

for a function $y = y(q, \mathrm{h})$.

Assumption All coefficient functions $a_i(q, \mathrm{h})$ are holomorphic functions of t (or $q = e^t$) on an open set $N = N_t$, and holomorphic functions of h in a neighbourhood of $\mathrm{h} = 0$. That is, $a_i \in \mathcal{H}_t \otimes \mathcal{H}_{\mathrm{h}}^+$ for all i.

If we convert to a system of first-order o.d.e.

$$\partial \begin{pmatrix} y_0 \\ \vdots \\ y_{s-1} \\ y_s \end{pmatrix} = \frac{1}{\mathrm{h}} \begin{pmatrix} 0 & 1 & & \\ & \ddots & \ddots & \\ & & 0 & 1 \\ -a_0 & -a_1 & \cdots & -a_s \end{pmatrix} \begin{pmatrix} y_0 \\ \vdots \\ y_{s-1} \\ y_s \end{pmatrix} \quad \text{or} \quad \partial Y = AY$$

by introducing

$$y_i = (\mathrm{h}\partial)^i y, \quad i = 1, \ldots, r$$

the assumption implies that the coefficient matrix A has (at worst) a simple pole at $\mathrm{h} = 0$. With $\Omega = A^t$, we can write

$$\Omega = \tfrac{1}{\mathrm{h}}\omega + \theta^{(0)} + \mathrm{h}\theta^{(1)} + \mathrm{h}^2\theta^{(2)} + \cdots$$

where $\omega, \theta^{(0)}, \theta^{(1)}, \ldots$ are independent of h.

General case: partial differential equations

Next consider a system of p.d.e.

$$T_1 y = 0, \ldots, T_u y = 0$$

for a function $y = y(q_1, \ldots, q_r, \mathrm{h})$.

Assumption Each differential operator T_i is a polynomial in $\mathrm{h}\partial_1, \ldots, \mathrm{h}\partial_r$, with coefficient functions in $\mathcal{H}_t \otimes \mathcal{H}_{\mathrm{h}}^+$. The D-module

$$\mathcal{M} = D^{\mathrm{h}}/(T_1, \ldots, T_u),$$

is free of rank $s+1$ over $\mathcal{H}_t \otimes \mathcal{H}_{\mathrm{h}}^+$.

With respect to any basis $[P_0], \ldots, [P_s]$ of $\mathcal{M}$ we obtain a first-order system

$$\partial_i Y = A_i Y, \quad 1 \le i \le r,$$

where $A = \Omega^t$, and where Ω is defined as follows:

Definition 6.1 *For $i = 1, \ldots, r$ let Ω_i denote 'the matrix of the action of ∂_i' on $\mathcal{M}$, that is, $[\partial_i P_j] = \sum_{k=0}^{s} (\Omega_i)_{kj} [P_k]$.*

Then Ω is of the form

$$\Omega = \frac{1}{\mathrm{h}}\omega + \theta^{(0)} + \mathrm{h}\theta^{(1)} + \mathrm{h}^2\theta^{(2)} + \cdots .$$

As in Chapter 5, we extend the coefficients to $\mathcal{H}_t \otimes \mathcal{H}_\mathrm{h}$ when it is convenient to do so; in particular Ω has coefficients in this ring.

As explained in Chapter 4, the connection $\nabla = d + \Omega$ is flat, for any value of h. If $\theta^{(i)} = 0$ for all i, then we have found a family of connections satisfying (1) and (2) above, and thereby achieved our objective. In general, we shall modify Ω by choosing a new basis $[\hat{P}_0], \ldots, [\hat{P}_s]$ in such a way that the new connection form $\hat{\Omega}$ is of the required type

$$\hat{\Omega} = \tfrac{1}{\mathrm{h}}\hat{\omega},$$

that is, $\hat{\theta}^{(i)} = 0$ for all i. The modification depends on a result in the theory of loop groups called the *Birkhoff factorization.*

We have $\Omega = L^{-1}dL$ where L is the transpose of the fundamental solution matrix H. This is a map $N \times S^1 \to \mathrm{GL}_{s+1}\mathbb{C}$, if we restrict to values of h in the unit circle. When we write $L(q,\mathrm{h})$ we have in mind this interpretation, that is, $L(q,\mathrm{h})$ is regarded as a matrix. Alternatively, L (or H) can be regarded as a map $N \to \Lambda\mathrm{GL}_{s+1}\mathbb{C}$, where $\Lambda\mathrm{GL}_{s+1}\mathbb{C}$ is the loop group of $\mathrm{GL}_{s+1}\mathbb{C}$, that is, the space of all (smooth) maps $S^1 \to \mathrm{GL}_{s+1}\mathbb{C}$. When we write $L(q)$ we mean the map whose value at $\mathrm{h} \in S^1$ is $L(q,\mathrm{h})$.

The Birkhoff factorization says that 'almost every' loop $\gamma \in \Lambda\mathrm{GL}_{s+1}\mathbb{C}$ may be factorized in the form

$$\gamma(\mathrm{h}) = \underbrace{(a_0 + \tfrac{1}{\mathrm{h}}a_1 + \tfrac{1}{\mathrm{h}^2}a_2 + \cdots)}_{\gamma_-(\mathrm{h})}\ \underbrace{(b_0 + \mathrm{h}b_1 + \mathrm{h}^2 b_2 + \cdots)}_{\gamma_+(\mathrm{h})}.$$

Moreover, the factorization is unique apart from the ambiguity of being able to replace a_i, b_i by $a_i x, x^{-1}b_i$ for an invertible matrix x.

This result is best stated, proved, and understood by considering an infinite-dimensional Grassmannian

$$\mathrm{Gr}^{(s+1)} \cong \Lambda\mathrm{GL}_{s+1}\mathbb{C}/\Lambda_+\mathrm{GL}_{s+1}\mathbb{C}$$

which will be discussed in more detail in section 8.1. Here $\Lambda_+\mathrm{GL}_{s+1}\mathbb{C}$ is the subgroup of $\Lambda\mathrm{GL}_n\mathbb{C}$ consisting of maps which, together with their inverses, extend holomorphically to the unit disk $\{\mathrm{h} \mid 0 \le |\mathrm{h}| \le 1\}$. The (isomorphic) subgroup $\Lambda_-\mathrm{GL}_{s+1}\mathbb{C}$ may be defined as a similar way with respect to the disk $\{\mathrm{h} \mid 1 \le |\mathrm{h}| \le \infty\}$. Now, $\Lambda\mathrm{GL}_{s+1}\mathbb{C}$ acts naturally on the homogeneous space $\mathrm{Gr}^{(s+1)}$, and one has:

Theorem 6.2 *The orbit of* $[I]$ *(where I is the constant loop) under the action of* $\Lambda_-GL_{s+1}\mathbb{C}$ *is a dense open subset of* $Gr^{(s+1)}$.

This orbit (or its identity component, which is a dense open subset of the identity component of $\mathrm{Gr}^{(s+1)}$) is called the 'big cell', and it is analogous to the big cell in a finite-dimensional Grassmannian.

Theorem 6.2 implies the Birkhoff factorization: if $\gamma \in \Lambda\mathrm{GL}_{s+1}\mathbb{C}$ is such that $[\gamma]$ belongs to the big cell of $\mathrm{Gr}^{(s+1)}$, then there exist loops $\gamma_- \in \Lambda_-\mathrm{GL}_{s+1}\mathbb{C}$, $\gamma_+ \in \Lambda_+\mathrm{GL}_{s+1}\mathbb{C}$, such that $\gamma = \gamma_-\gamma_+$, as stated earlier. Further information on the Birkhoff factorization can be found in appendix 8.8, and full details in [102].

Proposition 6.3 *Assume that Ω is holomorphic in $q = (q_1, \ldots, q_r)$, for q in some open set. Then, for any point q_0 in this set, there is a neighbourhood U of q_0 on which the connection $\nabla = d + \Omega$ is gauge equivalent to a connection $\hat{\nabla} = d + \hat{\Omega}$ with $\hat{\Omega} = \frac{1}{\mathrm{h}}\hat{\omega}$. Moreover, we have $\hat{\omega} = Q_0\omega Q_0^{-1}$, for some holomorphic map $Q_0 : U \to GL_{s+1}\mathbb{C}$.*

Proof Let $L(q) = L_-(q)L_+(q)$ be the Birkhoff factorization of $L(q)$, where we normalize L_- by imposing the condition $L_-|_{\mathrm{h}=\infty} = I$. For any given point q_0, we may choose $\gamma \in \Lambda\mathrm{GL}_{s+1}\mathbb{C}$ so that $[\gamma L(q_0)]$ belongs to the big cell. Replacing L by γL, we obtain a factorization at q_0, and hence on a neighbourhood U of this point. Let us write

$$L_-(q,\mathrm{h}) = I + \tfrac{1}{\mathrm{h}}R_1(q) + \tfrac{1}{\mathrm{h}^2}R_2(q) + \cdots$$

$$L_+(q,\mathrm{h}) = Q_0(q)(I + \mathrm{h}Q_1(q) + \mathrm{h}^2Q_2(q) + \cdots)$$

for some R_i, Q_j.

The gauge transformation $L \mapsto \hat{L} = L(L_+)^{-1} = L_-$ transforms $\Omega = L^{-1}dL$ into $\hat{\Omega} = \hat{L}^{-1}d\hat{L} = L_-^{-1}dL_-$, and the Fourier expansion of the latter manifestly contains only negative powers of h. But we have the alternative expression

$$\begin{aligned} L_-^{-1}dL_- &= (LL_+^{-1})^{-1}d(LL_+^{-1}) = L_+L^{-1}dLL_+^{-1} + L_+d(L_+^{-1}) \\ &= L_+\big(\tfrac{1}{\mathrm{h}}\omega + \theta^{(0)} + \mathrm{h}\theta^{(1)} + \cdots\big)L_+^{-1} + L_+(dL_+^{-1}), \end{aligned}$$

whose only negative power of h occurs in the term $\frac{1}{\mathrm{h}}Q_0\omega Q_0^{-1}$. It follows that $\hat{\Omega} = \frac{1}{\mathrm{h}}Q_0\omega Q_0^{-1}$, as required. □

Another way to express this modification of Ω is to say that we replace the original basis $[P_0], \ldots, [P_s]$ of $\mathcal{M}$ by the new basis $[\hat{P}_0], \ldots, [\hat{P}_s]$, where $\hat{P}_i = \sum_{j=0}^{s}(L_+)_{ji}^{-1}P_j$. Then $\hat{\Omega}_i$ is the matrix of the action of ∂_i with respect to the basis $[\hat{P}_0], \ldots, [\hat{P}_s]$. Roughly speaking, the Birkhoff factorization generalizes (to the case of h-dependent matrices, i.e. loops) the Gauss factorization of a matrix into a lower triangular matrix times an upper triangular matrix.

The above result is, therefore, a generalization of the 'Gauss elimination' procedure which we encountered in section 5.2 in the case of quantum cohomology. We shall take this 'change of basis' point of view in the first example below.

Example 6.4 We begin with a simple example based on operators of the form

$$T = (\mathrm{h}\partial)^2 - a_1 q \mathrm{h}\partial - a_2 q^2 - a_3 \mathrm{h} q - a_4 \mathrm{h}^2$$

where a_1, a_2, a_3, a_4 are constants and $\partial = q\partial/\partial q$. If $|q| = |\mathrm{h}| = 2$ this operator is homogeneous of degree 4, although the homogeneity will play only a background role in this example.

The D-module $\mathcal{M} = D^{\mathrm{h}}/(T)$ has rank 2. With respect to the monomial basis $[1], [\mathrm{h}\partial]$ we find

$$\Omega = \frac{1}{\mathrm{h}}\begin{pmatrix} 0 & a_2 q^2 \\ 1 & a_1 q \end{pmatrix} + \begin{pmatrix} 0 & a_3 q \\ 0 & 0 \end{pmatrix} + \mathrm{h}\begin{pmatrix} 0 & a_4 \\ 0 & 0 \end{pmatrix}.$$

Let us consider a new basis of the form $[1], [\mathrm{h}\partial - uq - v\mathrm{h}]$ where u, v are constant. (This particular form is chosen so that each basis element is homogeneous.) Using the new basis we obtain

$$\Omega = \frac{1}{\mathrm{h}}\begin{pmatrix} uq & (a_1 u - u^2 + a_2)q^2 \\ 1 & (a_1 - u)q \end{pmatrix} + \begin{pmatrix} v & (a_1 v - 2uv + a_3 - u)q \\ 0 & -v \end{pmatrix}$$
$$+ \mathrm{h}\begin{pmatrix} 0 & a_4 - v^2 \\ 0 & 0 \end{pmatrix}.$$

There are two essentially different cases, depending on a_4. If $a_4 = 0$, we can achieve the form $\frac{1}{\mathrm{h}}\omega$ by taking $u = a_3$ and $v = 0$. Then

$$\hat{\Omega} = \frac{1}{\mathrm{h}}\begin{pmatrix} a_3 q & (a_1 a_3 - a_3^2 + a_2)q^2 \\ 1 & (a_1 - a_3)q \end{pmatrix}.$$

Since the Birkhoff factorization is essentially unique, it must give the same result, so we have succeeded in carrying out this factorization indirectly. In particular, from the form of the new basis $[1], [\mathrm{h}\partial - a_3 q]$, we see that

$$L_+ = \begin{pmatrix} 1 & a_3 q \\ 0 & 1 \end{pmatrix}.$$

When $a_4 \neq 0$, the special form $\frac{1}{\mathrm{h}}\hat{\omega}$ cannot be achieved by a basis of the above type, since one of $a_4 - v^2$ and v must be non-zero. The Birkhoff factorization (and basis change) exists, but is more complicated in this case. We shall give a method to compute it in the next example. □

Example 6.5 The most general second-order o.d.e. of the permitted type has

$$T = (\mathrm{h}\partial)^2 - (u_0 + \mathrm{h}u_1 + \mathrm{h}^2 u_2 + \cdots)\mathrm{h}\partial - (v_0 + \mathrm{h}v_1 + \mathrm{h}^2 v_2 + \cdots)$$

with all u_i, v_i holomorphic in q on some open set.

With respect to $[1], [\mathrm{h}\partial]$ we have

$$\Omega = \frac{1}{\mathrm{h}}\begin{pmatrix} 0 & u_0 \\ 1 & v_0 \end{pmatrix} + \begin{pmatrix} 0 & u_1 \\ 0 & v_1 \end{pmatrix} + \mathrm{h}\begin{pmatrix} 0 & u_2 \\ 0 & v_2 \end{pmatrix} + \cdots$$

Even here, to perform the Birkhoff factorization and find the gauge transformation $L_+ = Q_0(I + \mathrm{h}Q_1 + \mathrm{h}^2 Q_2 + \cdots)$ is not easy. Let us consider the special case where

$$u_i = 0 \text{ for } i \geq 3, \quad v_i = 0 \text{ for } i \geq 2.$$

The significance of these conditions will become clear in sections 6.6 and 6.7, from which it follows that L_+ must have the special form

$$L_+ = \begin{pmatrix} \alpha_0 & \beta \\ 0 & \alpha_1 \end{pmatrix}\left(\begin{pmatrix} 1 & 0 \\ 0 & 1 \end{pmatrix} + \mathrm{h}\begin{pmatrix} 0 & \gamma \\ 0 & 0 \end{pmatrix}\right),$$

where the functions $\alpha_0, \alpha_1, \beta, \gamma$ satisfy

$$\begin{aligned} \alpha_0' &= \alpha_0 \gamma \\ \alpha_1' &= \alpha_1 v_1 - \alpha_1 \gamma \\ \beta' &= \alpha_0(u_1 + \gamma v_0) + \beta(v_1 - \gamma) \\ \gamma' &= u_2 + \gamma v_1 - \gamma^2. \end{aligned}$$

If the Ricatti equation for γ can be solved, then the remaining equations for $\alpha_0, \alpha_1, \beta$ may be solved by quadrature. The constants of integration here (for L_+) reflect the constants of integration in solving the differential equation $L^{-1}dL = \Omega$ for L.

The previous example has $u_0 = a_2 q^2$, $u_1 = a_3 q$, $u_2 = a_4$, and $v_0 = a_1 q$, $v_1 = 0$, so it leads to the Ricatti equation

$$\gamma' = a_4 - \gamma^2.$$

This can be solved explicitly; the solution is

$$\gamma = \sqrt{a_4}\,\frac{(Cq^{\sqrt{a_4}})^2 - 1}{(Cq^{\sqrt{a_4}})^2 + 1}.$$

If $a_4 = 0$ we have $\gamma' = -\gamma^2$. The solution $\gamma = 0, \alpha_0 = \alpha_1 = 1, \beta = a_4 q$ gives

$$L_+ = \begin{pmatrix} 1 & a_3 q \\ 0 & 1 \end{pmatrix},$$

as in the previous example. □

Example 6.6 This example is simply the quantum D-module for the flag manifold F_3 (Example 3.4), but worked out in reverse, starting from the D-module $\mathcal{M} = D^{\mathrm{h}}/(T_1, T_2)$ where

$$T_1 = \mathrm{h}^2\partial_1^2 + \mathrm{h}^2\partial_2^2 - \mathrm{h}^2\partial_1\partial_2 - q_1 - q_2$$
$$T_2 = \mathrm{h}^3\partial_1\partial_2^2 - \mathrm{h}^3\partial_1^2\partial_2 - q_2\mathrm{h}\partial_1 + q_1\mathrm{h}\partial_2.$$

These operators arose from the quantum differential equations in Example 3.4, but it is of interest that they also arise independently, as the 'quantum integrals of motion' for the Toda lattice (see [55] and the remarks at the end of Chapter 2).

It is not immediately obvious that $\mathcal{M}$ has rank 6, but this can be verified by a Gröbner basis calculation (cf. the examples in section 4.2). A monomial basis is given by $1, \mathrm{h}\partial_1, \mathrm{h}\partial_2, \mathrm{h}^2\partial_1^2, \mathrm{h}^2\partial_2^2, \mathrm{h}^3\partial_1\partial_2^2$. The matrices Ω_1, Ω_2 are

$$\Omega_1 = \frac{1}{\mathrm{h}}\left(\begin{array}{c|c|c|cc|c} 0 & 0 & -q_1-q_2 & \mathrm{h}q_1 & 0 & 0 \\ \hline 1 & 0 & 0 & q_1 & 0 & 0 \\ 0 & 0 & 0 & q_1 & 0 & 0 \\ \hline 0 & 1 & 1 & 0 & 0 & q_2 \\ 0 & 0 & 1 & 0 & 0 & q_1 \\ \hline 0 & 0 & 0 & 0 & 1 & 0 \end{array}\right),$$

$$\Omega_2 = \frac{1}{\mathrm{h}}\left(\begin{array}{c|cc|cc|c} 0 & -q_1-q_2 & 0 & 0 & \mathrm{h}q_2 & -(q_1+q_2)q_2 \\ \hline 0 & 0 & 0 & -q_2 & q_2 & \mathrm{h}q_2 \\ 1 & 0 & 0 & q_1 & q_2 & 0 \\ \hline 0 & 1 & 0 & 0 & 0 & 2q_2 \\ 0 & 1 & 1 & 0 & 0 & q_2 \\ \hline 0 & 0 & 0 & 1 & 0 & 0 \end{array}\right).$$

Thus $\Omega = \frac{1}{\mathrm{h}}\omega + \theta^{(0)}$ where $\theta^{(0)} = \theta_1^{(0)}dt_1 + \theta_2^{(0)}dt_2$, and

$$\theta_1^{(0)} = \left(\begin{array}{c|cc|cc|c} 0 & 0 & 0 & q_1 & 0 & 0 \\ \hline 0 & 0 & 0 & 0 & 0 & 0 \\ 0 & 0 & 0 & 0 & 0 & 0 \\ \hline 0 & 0 & 0 & 0 & 0 & 0 \\ 0 & 0 & 0 & 0 & 0 & 0 \\ \hline 0 & 0 & 0 & 0 & 0 & 0 \end{array}\right), \quad \theta_2^{(0)} = \left(\begin{array}{c|cc|cc|c} 0 & 0 & 0 & 0 & q_2 & 0 \\ \hline 0 & 0 & 0 & 0 & 0 & q_2 \\ 0 & 0 & 0 & 0 & 0 & 0 \\ \hline 0 & 0 & 0 & 0 & 0 & 0 \\ 0 & 0 & 0 & 0 & 0 & 0 \\ \hline 0 & 0 & 0 & 0 & 0 & 0 \end{array}\right).$$

It is easily verified that

$$L_+ = Q_0 = \left(\begin{array}{c|cc|cc|c} 1 & 0 & 0 & q_1 & q_2 & 0 \\ \hline 0 & 1 & 0 & 0 & 0 & q_2 \\ 0 & 0 & 1 & 0 & 0 & 0 \\ \hline 0 & 0 & 0 & 1 & 0 & 0 \\ 0 & 0 & 0 & 0 & 1 & 0 \\ \hline 0 & 0 & 0 & 0 & 0 & 1 \end{array}\right)$$

produces new 1-forms $\hat{\omega}$, $\hat{\Omega}$ satisfying $\hat{\Omega} = \frac{1}{\mathrm{h}}\hat{\omega}$. This amounts to using the new basis given by $1, \mathrm{h}\partial_1, \mathrm{h}\partial_2, \mathrm{h}^2\partial_1^2 - q_1, \mathrm{h}^2\partial_2^2 - q_2, \mathrm{h}^3\partial_1\partial_2^2 - q_2\mathrm{h}\partial_1$. Computing $\hat{\omega}_i = Q_0\omega_i Q_0^{-1}$ explicitly, we find

$$\hat{\omega}_1 = \left(\begin{array}{c|cc|cc|c} 0 & q_1 & 0 & 0 & 0 & q_1q_2 \\ \hline 1 & 0 & 0 & 0 & 0 & 0 \\ 0 & 0 & 0 & q_1 & 0 & 0 \\ \hline 0 & 1 & 1 & 0 & 0 & 0 \\ 0 & 0 & 1 & 0 & 0 & q_1 \\ \hline 0 & 0 & 0 & 0 & 1 & 0 \end{array}\right), \quad \hat{\omega}_2 = \left(\begin{array}{c|cc|cc|c} 0 & 0 & q_2 & 0 & 0 & q_1q_2 \\ \hline 0 & 0 & 0 & 0 & q_2 & 0 \\ 1 & 0 & 0 & 0 & 0 & 0 \\ \hline 0 & 1 & 0 & 0 & 0 & q_2 \\ 0 & 1 & 1 & 0 & 0 & 0 \\ \hline 0 & 0 & 0 & 1 & 0 & 0 \end{array}\right).$$

Of course, we expected this answer (the quantum multiplication matrices for F_3), but it should be noted that the calculation we have just performed is essentially different from that of Example 3.4. In that example, we produced a cyclic element of the D-module, namely the identity element of cohomology, and we also produced a basis for the D-module by applying ∂_1 and ∂_2 repeatedly to that cyclic element, namely, $[1], [\mathrm{h}\partial_1], [\mathrm{h}\partial_2]$, $[\mathrm{h}^2\partial_1^2 - q_1], [\mathrm{h}^2\partial_1\partial_2 - \mathrm{h}^2\partial_1^2 + q_1], [\mathrm{h}^3\partial_1^2\partial_2 - q_1\mathrm{h}\partial_2]$. In the current example we arrived at the same basis by a different route, by 'normalizing' an arbitrarily chosen monomial basis. □

6.2 Quantization of an algebra

In this section, we rephrase the construction of the previous section.

Let us begin with an abstract (commutative) graded algebra of the form

$$QA = \mathcal{H}_t[b_1,\dots,b_r]/(\mathcal{R}_1,\dots,\mathcal{R}_u),$$

where the relations $\mathcal{R}_1,\dots,\mathcal{R}_u$ are weighted homogeneous with respect to a fixed assignment of degrees $|b_i|$, $|q_j|$. We assume that QA is a free $\mathcal{H}_t$-module of rank $s+1$. In addition, we assume that QA is a deformation of an algebra

$$A = \mathbb{C}[b_1,\dots,b_r]/(\mathcal{R}_1|_{q=0},\dots,\mathcal{R}_u|_{q=0})$$

with $\dim_{\mathbb{C}} A = s+1$. We have in mind the situation where QA, A are the quantum cohomology and cohomology algebras of a Fano manifold.

Recall that D^{h} denotes the ring of differential operators with coefficients in $\mathcal{H}_t \otimes \mathcal{H}_{\mathrm{h}}^{+}$ which is generated by $\mathrm{h}\partial_1,\dots,\mathrm{h}\partial_r$.

Definition 6.7 *A quantization of QA is a D-module $\mathcal{M} = \mathrm{D}^{\mathrm{h}}/(T_1,\dots,T_u)$ such that*

(1) $\mathcal{M}$ is free over $\mathcal{H}_t \otimes \mathcal{H}_{\mathrm{h}}^{+}$ of rank $s+1$
(2) $S(T_i)|_{\mathrm{h}=0} = \mathcal{R}_i$, where $S(T_i)$ is the result of replacing $\mathrm{h}\partial_1,\dots,\mathrm{h}\partial_r$ by $b_1,\dots,b_r$ in T_i (for $i=1,\dots,u$).

There is no guarantee that such a quantization exists, but it is sometimes possible to produce a quantization simply by replacing $b_1,\dots,b_r$ by $\mathrm{h}\partial_1,\dots,\mathrm{h}\partial_r$ in each $\mathcal{R}_i$. When this works, that is, when the resulting D-module is free of rank $s+1$, we refer to it as the *naive quantization.*

Example 6.8 The one variable case is straightforward, as the rank of the cyclic D-module $D^{\mathrm{h}}/(T)$ is simply the order of the ordinary differential operator T. In Example 6.4 the D-module

$$\mathcal{M} = \mathrm{D}^{\mathrm{h}}/(\,(\mathrm{h}\partial)^2 - a_1 q\mathrm{h}\partial - a_2 q^2 - a_3 q\mathrm{h} - a_4\mathrm{h}^2\,)$$

is a quantization of

$$QA = \mathcal{H}_t[b]/(b^2 - a_1 q b - a_2 q^2)$$

and this is a deformation of

$$A = \mathcal{H}_t[b]/(b^2).$$

This simple example illustrates that the quantization of a given QA is certainly not unique. □

Example 6.9 The quantum D-module

$$D^{\mathrm{h}}/\big((\mathrm{h}\partial)^4 - 27q(\mathrm{h}\partial)^2 - 27\mathrm{h}q(\mathrm{h}\partial) - 6\mathrm{h}^2 q\big)$$

is a particular quantization of $QH^*M_5^3$ for the Fano hypersurface M_5^3 (see Example 5.4). The naive quantization

$$D^{\mathrm{h}}/\big((\mathrm{h}\partial)^4 - 27q(\mathrm{h}\partial)^2\big)$$

is also a valid quantization of $QH^*M_5^3$. Later (Example 6.24) we shall consider the effect of these different quantizations. □

Example 6.10 Here is an example where the naive quantization procedure fails (Example 5.4 of [62]). The rank 4 D-module

$$D^{\mathrm{h}}/\big((\mathrm{h}\partial_1)^2 - q_1(\mathrm{h}\partial_2 - 2\mathrm{h}\partial_1)(\mathrm{h}\partial_2 - 2\mathrm{h}\partial_1 - \mathrm{h}), \mathrm{h}\partial_2(\mathrm{h}\partial_2 - 2\mathrm{h}\partial_1) - q_2\big)$$

is a quantization of

$$QA = \mathcal{H}_{t_1,t_2}[b_1, b_2]/\big(b_1^2 - q_1(b_2 - 2b_1)(b_2 - 2b_1), b_2(b_2 - 2b_1) - q_2\big),$$

but the 'naive quantization'

$$D^{\mathrm{h}}/\big((\mathrm{h}\partial_1)^2 - q_1(\mathrm{h}\partial_2 - 2\mathrm{h}\partial_1)(\mathrm{h}\partial_2 - 2\mathrm{h}\partial_1), \mathrm{h}\partial_2(\mathrm{h}\partial_2 - 2\mathrm{h}\partial_1) - q_2\big)$$

has rank 0, and is therefore *not* a quantization of QA. We shall say more about this example in section 6.7. □

When a quantization of QA exists, the Birkhoff factorization produces a D-module $\mathcal{M}$ which 'matches' QA in the sense of section 5.2: the matrix of $\mathrm{h}\partial_i$ on $\mathcal{M}$ is exactly the matrix of b_i on QA. The 1-form ω, associated to a commutative algebra QA, always satisfies $\omega \wedge \omega = 0$, but it will not in general satisfy $d\omega + \omega\wedge\omega = 0$. The 1-form Ω, associated to a D-module $\mathcal{M}$, always satisfies $d\Omega + \Omega\wedge\Omega = 0$, but it will not in general satisfy $\Omega\wedge\Omega = 0$. The main point of our discussion in this section is that, for $\hat{\Omega} = \frac{1}{\mathrm{h}}\hat{\omega}$, both sets of conditions hold.

6.3 Digression on D^h-modules

The definition of the quantum D-module (Chapter 5) involves a parameter h, and we have seen in the previous two sections that this parameter plays an important role. We shall explain briefly how to modify the theory of D-modules in Chapter 4 in order to take account of h, that is, when the ring of differential operators from Chapter 4 is replaced by the ring D^{h}.

The relation between scalar systems and matrix systems (section 4.2) generalizes in a straightforward way. The quantum D-module in section 5.1 is an example of passing from a matrix system to a scalar system. In the present chapter we are primarily interested in the reverse procedure, which we can carry out under the assumptions of section 6.1.

As in section 4.2, we regard the matrix system as a concrete realization of the D-module. With the parameter h, the identification takes the following form:

$$\mathcal{M} = D^{\mathrm{h}}/(T_1, \dots, T_u) \to \mathcal{H}_t^{s+1} \otimes \mathcal{H}_{\mathrm{h}}^+, \quad [P] = [\textstyle\sum_{i=0}^s f_i P_i] \mapsto \begin{pmatrix} f_0 \\ \vdots \\ f_s \end{pmatrix}.$$

This is an isomorphism of free $\mathcal{H}_t \otimes \mathcal{H}_{\mathrm{h}}^+$-modules of rank $s+1$, and also an isomorphism of cyclic D-modules. The ring D^{h} acts on $\mathcal{M}$ in the standard way, and its action on $\mathcal{H}_t^{s+1} \otimes \mathcal{H}_{\mathrm{h}}^+$ is given by the connection $\nabla = d + \Omega$. We assume that $P_0 = 1$; this corresponds to the cyclic element $(1, 0, \dots, 0)^t$ on the right hand side. In Chapter 8 we shall see how this concrete realization of $\mathcal{M}$ is related to the infinite-dimensional Grassmannian of section 6.1.

Self-adjointness (appendix 4.6) has to be handled somewhat differently when the parameter h is involved. As in section 5.3, let us introduce a pairing

$$\mathcal{M} \times \mathcal{M} \to \mathcal{H}_t \otimes \mathcal{H}_{\mathrm{h}}^+$$

by defining

$$((f, g)) = (\bar{f}, g)$$

where $\bar{f}(t, \mathrm{h}) = f(t, -\mathrm{h})$ and $(f, g) = f^t S g$ for some invertible matrix S. As in section 5.3, we write $\partial_i \cdot f = (\partial_i + \Omega_i) f$ for the quantum D-module action and $\partial_i f$ for the standard action. We then have the following version of Lemma 4.28:

Lemma 6.11 *The following are equivalent:*

(1) The map $E : \mathcal{M} \to \bar{\mathcal{M}}^*, f \mapsto ((f, \ ))$ *is an isomorphism of D-modules.*
(2) $\mathrm{h}\partial_i((f, g)) = -((\mathrm{h}\partial \cdot f, g)) + ((f, \mathrm{h}\partial \cdot g))$ *for all* $f, g \in \mathcal{M}$.
(3) $S^{-1}\bar{\Omega}^t S = -\Omega$.

Proof Condition (1) is equivalent to (a) $\mathrm{h} \odot E(f)(g) = E(\mathrm{h}f)(g)$ and (b) $(\mathrm{h}\partial_i) \odot E(f)(g) = E(\mathrm{h}\partial_i \cdot f)(g)$, for all f, g. But (a) is automatic from the definition of E, and (b) may then be written $\partial_i \odot E(f)\,(g) = E(\partial_i \cdot f)\,(g)$. This means $-E(f)(\partial_i \cdot g) + \partial_i E(f)(g) = E(\partial_i \cdot f)(g)$, that is, $\partial_i (\!(f,g)\!) = (\!(\partial_i \cdot f, g)\!) + (\!(f, \partial_i \cdot g)\!)$, which is equivalent to condition (2). Substituting $\mathrm{h}\partial_i.f = (\mathrm{h}\partial_i + \mathrm{h}\Omega_i)f$ gives condition (3). □

Note that the condition $S^{-1}\bar{\Omega}^t S = -\Omega$ reduces to the Frobenius condition $S^{-1}\omega^t S = \omega$ of section 5.3 when $\Omega = \frac{1}{\mathrm{h}}\omega$ and S is the matrix of the intersection form.

As in appendix 4.6 we shall discuss self-adjointness for scalar systems mainly in the o.d.e. context.

Definition 6.12 *The adjoint of the differential operator* $T = \sum_{i=0}^{s+1} a_i(\mathrm{h}\partial)^i$ *is* $\bar{T}^* = \sum_{i=0}^{s+1}(\mathrm{h}\partial)^i \bar{a}_i$.

This extends the classical definition in appendix 4.6; in that case $T = \bar{T}$. Note the absence of signs here.

For a D-module $\mathcal{M} = \mathrm{D}^{\mathrm{h}}/(T)$, with T of this form (and $a_{s+1} = 1$), we have the following version of Lemma 4.32:

Lemma 6.13 *Define* $\delta_i \in \bar{\mathcal{M}}^*$ *by* $\delta_i([(\mathrm{h}\partial)^j]) = \delta_{ij}$. *Then* δ_s *is a cyclic element and it is annihilated by* $\bar{T}^*$. *In particular, we have* $\bar{\mathcal{M}}^* \cong \mathrm{D}/(\bar{T}^*)$.

Proof The proof is a straightforward modification of Lemma 4.32, taking appropriate care with the notation. Direct computation gives

$$\begin{aligned}
(\mathrm{h}\partial) \odot \delta_0 &= -a_0\delta_s \\
(\mathrm{h}\partial) \odot \delta_1 &= -a_1\delta_s + \delta_0 \\
&\cdots \\
(\mathrm{h}\partial) \odot \delta_{s-1} &= -a_{s-1}\delta_s + \delta_{s-2} \\
(\mathrm{h}\partial) \odot \delta_s &= -a_s\delta_s + \delta_{s-1}.
\end{aligned}$$

This can be rewritten as

$$\begin{aligned}
\bar{a}_0 \odot \delta_s + (\mathrm{h}\partial) \odot \delta_0 &= \delta_0 \\
\bar{a}_1 \odot \delta_s + (\mathrm{h}\partial) \odot \delta_1 &= \delta_0 \\
&\cdots \\
\bar{a}_{s-1} \odot \delta_s + (\mathrm{h}\partial) \odot \delta_{s-1} &= \delta_{s-2} \\
((\mathrm{h}\partial) + \bar{a}_s) \odot \delta_s &= \delta_{s-1}.
\end{aligned}$$

Proceeding inductively from the last equation, we obtain

$$\big((\mathrm{h}\partial)^i + (\mathrm{h}\partial)^{i-1}\bar{a}_s + \cdots + \bar{a}_{s-i+1})\big) \odot \delta_s = \delta_{s-i}$$

for $i = 1, \ldots, s$. Substituting the result for $i = s$ into the first equation we obtain $\bar{T}^* \odot \delta_s = 0$. □

As in appendix 4.6, we can use this to introduce an intrinsic pairing on $\mathcal{M} = D^{\mathrm{h}}/(T)$ when T is self-adjoint in the above sense, by making use of the isomorphism $\mathcal{M} \to \bar{\mathcal{M}}^*$, $[P] \mapsto P \odot \delta_s$:

Definition 6.14 *Assume that $\bar{T}^* = T$. Then we have a natural pairing*

$$\begin{array}{rcccc} \langle\langle\, ,\, \rangle\rangle : \mathcal{M} \times \mathcal{M} & \longrightarrow & \bar{\mathcal{M}}^* \times \mathcal{M} & \longrightarrow & \mathcal{H}_t \otimes \mathcal{H}_{\mathrm{h}}^+ \\ ([P],[Q]) & \mapsto & ([P \odot \delta_s],[Q]) & \mapsto & (P \odot \delta_s)(Q) \end{array}$$

that is, we define $\langle\langle [P],[Q] \rangle\rangle = (P \odot \delta_s)(Q)$.

This pairing is non-degenerate, and it satisfies

$$\langle\langle f[P],[Q] \rangle\rangle = \bar{f}\langle\langle [P],[Q] \rangle\rangle, \quad \langle\langle [P], f[Q] \rangle\rangle = f\langle\langle [P],[Q] \rangle\rangle.$$

As the map $\mathcal{M} \to \bar{\mathcal{M}}^*$, $[P] \mapsto \langle\langle [P],\ \rangle\rangle$ is an isomorphism of D-modules, we always have the analogue of the Frobenius property:

$$\mathrm{h}\langle\langle [P],[Q] \rangle\rangle' = -\langle\langle \mathrm{h}\partial \cdot [P],[Q] \rangle\rangle + \langle\langle [P], \mathrm{h}\partial \cdot [Q] \rangle\rangle.$$

The results of appendix 4.6 relating self-adjointness for scalar and matrix systems extend naturally to the current situation:

Proposition 6.15 *Assume that $S^{-1}\bar{\Omega}^t S = -\Omega$ with respect to a basis $[P_0], \ldots, [P_s]$ of $D^{\mathrm{h}}/(T)$, where $P_i = (\mathrm{h}\partial)^i +$ lower-order terms. Assume further that*

$$S = \begin{pmatrix} & & 1 \\ & \cdot^{\cdot^{\cdot}} & \\ 1 & & \end{pmatrix}.$$

Then $\bar{T}^* = T$.

Proof Exactly as in the proof of Proposition 4.33, we find that $\bar{T}^*$ and T both annihilate δ_s. Since T and $\bar{T}^*$ are of the form $(\mathrm{h}\partial)^{s+1} +$ lower-order terms, this implies $\bar{T}^* = T$. □

Proposition 6.16 *Let S be the matrix of Proposition 6.15. If there exists a basis $[P_0], \ldots, [P_s]$ of $\mathcal{M}$ such that $\langle\langle [P_i],[P_j] \rangle\rangle = s_{ij} = \delta_{i+j,s+1}$, then $S^{-1}\bar{\Omega}^t S = -\Omega$.*

Proof The proof of Proposition 4.37 extends directly to this situation. □

The condition $\langle\!\langle [P_i],[P_j]\rangle\!\rangle = s_{ij}$ says that $\langle\!\langle\ ,\ \rangle\!\rangle$ corresponds to $((\ ,\ ))$ under the isomorphism $\mathcal{M} \to \mathcal{H}_t^{s+1} \otimes \mathcal{H}_{\mathrm{h}}^{+}$.

Example 6.17 Even for a second-order o.d.e., the presence of h permits non-trivial examples. For

$$T = (\mathrm{h}\partial)^2 + a_1\mathrm{h}\partial + a_0$$

we have $\bar{T}^* = T$ if and only if $\bar{a}_1 = a_1$ and $a_0 - \bar{a}_0 = \mathrm{h}a_1'$. The pairing is then given by

$$\langle\!\langle f_0 + f_1\mathrm{h}\partial, g_0 + g_1\mathrm{h}\partial\rangle\!\rangle = \bar{f}_0 g_1 + \bar{f}_1 g_0 + \bar{f}_1 g_1(-a_1)$$

(for clarity we omit the square brackets). The basis of $D^{\mathrm{h}}/(T)$ given by 1, $\mathrm{h}\partial + \frac{1}{2}a_1$ is orthonormal in the sense that $\langle\!\langle 1,1\rangle\!\rangle = 0$, $\langle\!\langle 1, \mathrm{h}\partial + \frac{1}{2}a_1\rangle\!\rangle = \langle\!\langle \mathrm{h}\partial + \frac{1}{2}a_1, 1\rangle\!\rangle = 1$, $\langle\!\langle \mathrm{h}\partial + \frac{1}{2}a_1, \mathrm{h}\partial + \frac{1}{2}a_1\rangle\!\rangle = 0$. With respect to this basis we have

$$\Omega = \frac{1}{\mathrm{h}}\begin{pmatrix} -\frac{1}{2}a_1 & \frac{1}{4}a_1^2 - a_1 + \frac{1}{2}\mathrm{h}a_1' \\ 1 & -\frac{1}{2}a_1 \end{pmatrix}$$

which satisfies $S^{-1}\bar{\Omega}^t S = -\Omega$. □

Example 6.18 For

$$T = (\mathrm{h}\partial)^3 + a_2(\mathrm{h}\partial)^2 + a_1\mathrm{h}\partial + a_0$$

we have $\bar{T}^* = T$ if and only if

$$\bar{a}_2 = a_2,\ a_1 - \bar{a}_1 = 2\mathrm{h}a_2',\ a_0 - \bar{a}_0 = \mathrm{h}\bar{a}_1' + \mathrm{h}^2 a_2''.$$

In this case, $\langle\!\langle f_0 + f_1\mathrm{h}\partial + f_2(\mathrm{h}\partial)^2, g_0 + g_1\mathrm{h}\partial + g_2(\mathrm{h}\partial)^2\rangle\!\rangle$ is equal to

$$\begin{aligned} &\bar{f}_0 g_2 + \bar{f}_1 g_2(-a_2) + \bar{f}_2 g_2(a_2^2 - \tfrac{1}{2}(a_1 + \bar{a}_1)) \\ &+ \bar{f}_1 g_1 \qquad + \bar{f}_2 g_1(-a_2) \\ &\qquad\qquad\quad + \bar{f}_2 g_0. \end{aligned}$$

An example of an orthonormal basis is $1, \mathrm{h}\partial, (\mathrm{h}\partial)^2 + a_2\mathrm{h}\partial + \frac{1}{2}a_1$. With respect to this we have

$$\Omega = \frac{1}{\mathrm{h}}\begin{pmatrix} 0 & -\frac{1}{2}a_1 & \frac{1}{2}\mathrm{h}a_1' - a_0 \\ 1 & -a_2 & -\frac{1}{2}\bar{a}_1 \\ 0 & 1 & 0 \end{pmatrix},$$

which satisfies $S^{-1}\bar{\Omega}^t S = \Omega$. Note that the conditions of Propositions 6.15 and 6.16 hold here. □

The shape of the matrix $\langle\langle (h\partial)^i, (h\partial)^j \rangle\rangle$ in Example 6.18 shows that an orthonormal basis of the form

$$P_0 = 1, P_1 = h\partial + x, P_2 = (h\partial)^2 + yh\partial + z$$

may always be found. Namely, since $\langle\langle [P_i], [P_j] \rangle\rangle = \delta_{i+j,2}$ when $i + j \le 2$, it suffices to choose x, y, z so that $\langle\langle [P_i], [P_j] \rangle\rangle = 0$ for $i + j \ge 3$. These conditions are $\bar{x} + y - a_2 = 0, z + \bar{z} + y\bar{y} + (\bar{y} + y)(-a_2) - \frac{1}{2}(\bar{a}_1 + a_1) + a_2^2 = 0$, which may be solved for x, y, z. Taking $x = 0$ gives $y = a_2$ and $z + \bar{z} = \frac{1}{2}(\bar{a}_1 + a_1)$; $z = \frac{1}{2}a_1$ gives the basis chosen in Example 6.18.

An induction argument shows that this procedure works in general.

Proposition 6.19 *Let $T = (h\partial)^{s+1} + a_s(h\partial)^s + \cdots + a_0$, and assume that $\bar{T}^* = T$. Then there exists a basis $[P_0], \ldots, [P_s]$ of $D^h/(T)$, where $P_i = (h\partial)^i +$ lower-order terms, such that $\langle\langle [P_i], [P_j] \rangle\rangle = \delta_{i+j,s}$.*

The discussion in this section, as for the pairing $\langle\ ,\ \rangle$ in appendix 4.6, may be generalized to the case where $(\overline{fT})^* = fT$ for some function f (we continue to assume that T is monic, i.e. $a_{s+1} = 1$). Using $\bar{\mathcal{M}}^* \cong \mathcal{M}$ we define a pairing $\langle\langle\ ,\ \rangle\rangle^f$ by

$$\langle\langle [P], [Q] \rangle\rangle^f = (P\tfrac{1}{f} \odot \delta_s)(Q).$$

Example 6.20 Generalizing Example 6.17, let us consider

$$fT = f\big((h\partial)^2 + a_1 h\partial + a_0\big).$$

We have $(\overline{fT})^* = fT$ if and only if $f = \bar{f}$, $f(a_1 - \bar{a}_1) = 2hf'$, and $f(a_0 - \bar{a}_0) = h(\bar{a}_1 f)' + h^2 f''$. The pairing is then given by

$$\langle\langle f_0 + f_1 h\partial, g_0 + g_1 h\partial \rangle\rangle^f = \tfrac{1}{f}(\bar{f}_0 g_1 + \bar{f}_1 g_0) + \bar{f}_1 g_1 \left(\tfrac{hf'}{f^2} - \tfrac{a_1}{f}\right).$$

The basis $f, h\partial + \frac{1}{2}(a_1 - hf'/f)$ is orthonormal. □

6.4 Abstract quantum cohomology

Let us assume now that $\mathcal{M} = D^h/(T_1, \ldots, T_u)$ is a quantization of $QA = \mathcal{H}_t[b_1, \ldots, b_r]/(\mathcal{R}_1, \ldots, \mathcal{R}_u)$ as in section 6.2. Let $[P_0], \ldots, [P_s]$ be a basis of $\mathcal{M}$ such that $[c_0], \ldots, [c_s]$ is a basis of QA and $[c_0|_{q=0}], \ldots, [c_s|_{q=0}]$ is a basis of A, where $c_i = S(P_i)|_{h=0}$. We assume in addition that $P_0 = 1$ and $P_i = h\partial_i$ for $1 \le i \le r$.

With respect to these bases,

$$\Omega_i = \frac{1}{\hbar}\omega_i + \theta_i^{(0)} + \hbar\theta_i^{(1)} + \hbar^2\theta_i^{(2)} + \cdots$$

is the matrix of the action of ∂_i on $\mathcal{M}$, and ω_i is the matrix of the action of b_i on QA.

Let us apply the Birkhoff factorization method. We replace the original basis $[P_0], \ldots, [P_s]$ of $\mathcal{M}$ by a new basis $[\hat{P}_0], \ldots, [\hat{P}_s]$, where $\hat{P}_i = \sum_{j=0}^{s} (L_+)_{ji}^{-1} P_j$. At the same time, we replace the original basis $[c_0], \ldots, [c_s]$ of QA by a new basis $[\hat{c}_0], \ldots, [\hat{c}_s]$, where $\hat{c}_i = \sum_{j=0}^{s} (Q_0^{-1})_{ji} c_j$. The matrices $\hat{\Omega}_i, \hat{\omega}_i$ are the matrices of ∂_i, b_i with respect to these new bases.

Assumption From now on we assume that $L_+|_{q=0} = I$ and $\hat{c}_0 = 1$, $\hat{c}_i = b_i$ for $1 \le i \le r$.

These hold under reasonable conditions, although we shall occasionally meet examples where they do not hold. The property $L_+|_{q=0} = I$ holds, for example, if $\Omega|_{q=0} = \omega|_{q=0}$ as in the case of quantum cohomology. (To prove this, we use the fact that $L = e^{t/\hbar} S$ with $S|_{q=0} = I$—see section 3.2 of [72], and the remarks before Theorem 5.9.) Then $S = S_- S_+$ with $S_-|_{q=0} = S_+|_{q=0} = I$, and $L_+ = S_+$.) As we shall see in section 6.6, we always have $\hat{c}_0 = 1$, $\hat{c}_i = b_i$ for $1 \le i \le r$ when $|q_i| \ge 4$.

Under these assumptions, the choice of quantization leads to an 'abstract quantum product' which has many of the properties of the usual quantum product. First of all, the above choice of bases of QA and A gives an identification

$$\delta : QA \to A \otimes \mathcal{H}_t, \quad [\hat{c}_i] \mapsto [[\hat{c}_i|_{q=0}]] \quad (0 \le i \le r).$$

Definition 6.21 *We define a product operation* $\circ$ *(abstract quantum product) on* $A \otimes \mathcal{H}_t$ *by* $x \circ y = \delta(\delta^{-1}(x)\delta^{-1}(y))$.

It follows that the matrix of the operator $b_i\circ$ on $A \otimes \mathcal{H}_t$, with respect to the basis $[[\hat{c}_0|_{q=0}]], \ldots, [[\hat{c}_s|_{q=0}]]$, is $\hat{\omega}_i$.

The usual quantum product can be described in the same way, and in this case the map δ is the 'quantum evaluation map'.

To summarize, we have the following theorem:

Theorem 6.22 *Under the above assumptions we obtain an abstract quantum product operation* $\circ$ *on* $A \otimes \mathcal{H}_t$ *with the property that the connection* $d + \frac{1}{\hbar}C$ *is flat for all* $\hbar$, *where* $C = \sum_{i=1}^{r} C_i dt_i$ *and* C_i *is the matrix of multiplication by* b_i.

Evidently this product depends on the choice of quantization and the choice of basis.

Any choice of basis—or any $\mathcal{H}_t$-module identification of QA with $A \otimes \mathcal{H}_t$—would give a product in the same way. But the special property of our abstract quantum product is that the connection $d + \frac{1}{\mathrm{h}}C$ is flat. Of course, this is not the only property satisfied by the usual quantum product, and we shall consider other properties after looking at some concrete examples.

In these examples we use the polynomial ring $\mathbb{C}[q]$ instead of $\mathcal{H}_t$, and take D^{h} to mean operators whose coefficients are polynomial in q. The fact that L_+ is polynomial in q follows from our assumption that it is holomorphic at $q = 0$ combined with the fact that it is homogeneous (see section 6.5).

Example 6.23 Consider the algebras

$$QA = \mathbb{C}[b,q]/(b^3 - \beta qb), \quad A = \mathbb{C}[b]/(b^3)$$

where β is a constant. It is a trivial matter to find a quantization. For example, the D-module

$$\mathcal{M} = D^{\mathrm{h}}/((\mathrm{h}\partial)^3 - \beta q \mathrm{h}\partial - \alpha q \mathrm{h})$$

is a quantization of QA for any constant α.

Let us compute the abstract quantum products of $[[1]], [[b]], [[b^2]] \in A \otimes \mathbb{C}[q]$. First, with respect to the monomial basis given by $P_0 = 1$, $P_1 = \mathrm{h}\partial$, $P_2 = (\mathrm{h}\partial)^2$, we have

$$\Omega = \frac{1}{\mathrm{h}} \begin{pmatrix} 0 & 0 & 0 \\ 1 & 0 & \beta q \\ 0 & 1 & 0 \end{pmatrix} + \begin{pmatrix} 0 & 0 & \alpha q \\ 0 & 0 & 0 \\ 0 & 0 & 0 \end{pmatrix}.$$

It is easy to see that

$$L_+ = Q_0 = \begin{pmatrix} 1 & 0 & \alpha q \\ 0 & 1 & 0 \\ 0 & 0 & 1 \end{pmatrix}.$$

We obtain a new basis given by $\hat{P}_0 = 1$, $\hat{P}_1 = \mathrm{h}\partial$, $\hat{P}_2 = (\mathrm{h}\partial)^2 - \alpha q$. The matrix of ∂ with respect to this new basis is

$$\hat{\Omega} = \frac{1}{\mathrm{h}} Q_0 \omega Q_0^{-1} = \frac{1}{\mathrm{h}} \begin{pmatrix} 0 & \alpha q & 0 \\ 1 & 0 & (\beta - \alpha)q \\ 0 & 1 & 0 \end{pmatrix}.$$

Hence $[[b]] \circ [[1]] = [[b]]$, $[[b]] \circ [[b]] = [[b^2]] + \alpha q[[1]]$, and $[[b]] \circ [[b^2]] = (\beta - \alpha)q[[b]]$. Thus, we see how the abstract quantum products depend on

α (i.e. on the choice of quantization). Of course, the identity

$$[[b]] \circ [[b]] \circ [[b]] - \beta q[[b]] = 0$$

holds, for any value of α, because this is a property of QA. □

Example 6.24 Let us consider the quantization

$$\mathcal{M} = D^{\mathrm{h}}/((\mathrm{h}\partial)^4 - a_1 q(\mathrm{h}\partial)^2 - a_2\mathrm{h}q(\mathrm{h}\partial) - a_3\mathrm{h}^2 q)$$

of the algebra $QA = \mathbb{C}[b,q]/(b^4 - a_1 q b^2)$, with $A = \mathbb{C}[b]/(b^4)$. This family includes the quantum D-module for the hypersurface M_5^3 (Example 6.9), which is the particular case $a_1 = 27, a_2 = 27, a_3 = 6$.

With respect to the monomial basis we have

$$\Omega = \frac{1}{\mathrm{h}}\begin{pmatrix} & & & \\ 1 & & & \\ & 1 & & a_1 q \\ & & 1 & \end{pmatrix} + \begin{pmatrix} & & & \\ & & & a_2 q \\ & & & \\ & & & \end{pmatrix} + \mathrm{h}\begin{pmatrix} & & & a_3 q \\ & & & \\ & & & \\ & & & \end{pmatrix}.$$

The method of section 6.6 (see Example 6.36) gives $L_+ = Q_0(I + \mathrm{h}Q_1)$ with

$$Q_0 = \begin{pmatrix} 1 & & a_3 q & \\ & 1 & & (a_2 - a_3)q \\ & & 1 & \\ & & & 1 \end{pmatrix}, \quad Q_1 = \begin{pmatrix} & & & a_3 q \\ & & & \\ & & & \\ & & & \end{pmatrix},$$

hence

$$\hat{\omega} = Q_0 \omega Q_0^{-1} = \begin{pmatrix} & a_3 q & & a_3(a_1 - a_2 + a_3)q^2 \\ 1 & & (a_2 - 2a_3)q & \\ & 1 & & (a_1 - a_2 + a_3)q \\ & & 1 & \end{pmatrix}.$$

We obtain the following abstract quantum products: $[[b]] \circ [[1]] = [[b]]$, $[[b]] \circ [[b]] = [[b^2]] + a_3 q[[1]]$, $[[b]] \circ [[b^2]] = [[b^3]] + (a_2 - 2a_3)q[[b]]$, $[[b]] \circ [[b^3]] = (a_1 - a_2 + a_3)q[[b^2]] + a_3(a_1 - a_2 + a_3)q^2[[1]]$.

When $a_1 = 27, a_2 = 27, a_3 = 6$ these are exactly the quantum products for M_5^3 (Example 2.12). In Example 3.6 we used the known quantum products to obtain the quantum D-module. Here, in contrast, we have used the quantum D-module to recover the quantum products. This result holds in general for Fano hypersurfaces (see [49], [76], and [106]), that is, for any M_N^k the abstract quantum product obtained from the differential operator

$$T_N^k = (\mathrm{h}\partial)^{N-1} - k\mathrm{h}^{k-1}q(k\partial + k - 1)(k\partial + k - 2)\ldots(k\partial + 1)$$

is in fact the usual quantum product.

Finally, we note that the naive quantization

$$\mathcal{M} = D^{\mathrm{h}}/((\mathrm{h}\partial)^4 - a_1 q(\mathrm{h}\partial)^2)$$

of $QA = \mathbb{C}[b,q]/(b^4 - a_1 qb^2)$ gives quite different abstract quantum products (and thus quite different abstract Gromov–Witten invariants). This is the case $a_1 = 27, a_2 = 0, a_3 = 0$ so we obtain $[[b]] \circ [[1]] = [[b]]$, $[[b]] \circ [[b]] = [[b^2]]$, $[[b]] \circ [[b^2]] = [[b^3]]$, $[[b]] \circ [[b^3]] = 27[[b^2]]$. However, since the matrices $\hat{\omega}$ for different quantizations are all conjugate—we are just changing the basis of QA—the algebraic structure remains the same.

This (and the previous) example illustrates clearly how 'Gromov–Witten invariants contain more information than the quantum cohomology algebra'; however, the extra information is contained in the quantum D-module.

As we pointed out in Example 5.8, the differential operator T_N^k satisfies $(\bar{T}_N^k)^* = T_N^k$. However, the naive quantization $(\mathrm{h}\partial)^{N-1} - k^k q(\mathrm{h}\partial)^{k-1}$ is not self-adjoint in this sense. □

Example 6.25 Let

$$QA = QH^*F_3, \quad A = H^*F_3.$$

As we have seen in Example 6.6, the 'usual' quantization (the quantum D-module, or the quantum Toda lattice) produces $\hat{\omega} = C = C_1 dt_1 + C_2 dt_2$ where C_1, C_2 are the usual quantum multiplication matrices (Example 3.4). A similar result holds for the flag manifold F_n, for any n, as was shown in [4]. □

Example 6.26 The following simple example illustrates the problems that occur when the assumptions at the beginning of this section do not hold.

We have investigated (in Example 6.8) the D-module

$$\mathcal{M} = D^{\mathrm{h}}/(\,(\mathrm{h}\partial)^2 - a_1 q\mathrm{h}\partial - a_2 q^2 - a_3 q\mathrm{h} - a_4 \mathrm{h}^2\,),$$

which is a quantization of $QA = \mathcal{H}_t[b]/(b^2 - a_1 qb - a_2 q^2)$. The monomial basis gives the obvious identification $QA \to A \otimes \mathbb{C}[q]$, $[b^i] \mapsto [[b^i]]$, with the obvious products $[[b]] \circ [[1]] = [[b]]$, $[[b]] \circ [[b]] = [[b^2]] + a_1 q[[b]] + a_2 q^2[[1]]$.

However, the Birkhoff factorization does not produce a product of the expected form. First, if $a_4 \neq 0$, then L_+ cannot be the identity matrix at $q = 0$ (see Examples 6.4, and 6.5), and $\hat{c}_i|_{q=0}$ is not well defined. And even if $a_4 = 0$, then we obtain $\hat{c}_0 = 1$, $\hat{c}_1 = b - a_3 q$, which contravenes our assumption that $\hat{c}_i = b_i$ for $1 \le i \le r$. The products are well defined in this case but we have $[[b]] \circ [[1]] = [[b]] + a_3 q[[1]]$, so $[[1]]$ is no longer the identity element. □

A more general theory of 'abstract quantum cohomology D-modules' has been developed by H. Iritani in [72].

6.5 Properties of abstract quantum cohomology

Let us now consider some further properties of the abstract quantum product, following section 5.3 for the quantum product. We assume that $\mathcal{M}$ is a quantization of QA and fix bases as above.

Let S be the matrix of a bilinear form $(\ ,\ )$ on the vector space A, thus $(x,y) = x^t S y$ where x, y are vectors expressed relative to the chosen basis. We extend this to a $\mathbb{C}[q]$-linear pairing on QA and (as in section 5.3) define a pairing

$$\mathcal{M} \times \mathcal{M} \to \mathbb{C}[q,\mathrm{h}], \quad ((f,g)) = (\bar{f},g) = \bar{f}^t S g$$

where $\bar{f}(t,\mathrm{h}) = f(t,-\mathrm{h})$.

Proposition 6.27 *If $S^{-1}\bar{\Omega}^t S = -\Omega$ for some basis $[P_0],\dots,[P_s]$, then the abstract quantum product satisfies the Frobenius property $(z \circ x, y) = (x, z \circ y)$.*

Proof The method of the proof of Proposition 5.10 shows that the Lie algebra condition $S^{-1}\bar{\Omega}^t S = -\Omega$ implies the Lie group condition $S^{-1}\bar{L}_+^t S = L_+^{-1}$. From this we obtain $S^{-1}\bar{\hat{\Omega}}^t S = -\hat{\Omega}$, that is, $S^{-1}\bar{\hat{\omega}}^t S = \hat{\omega}$. As in section 5.3, this is equivalent to the Frobenius property. □

Example 6.28 The abstract quantum product in Example 6.23 satisfies the Frobenius condition with

$$S = \begin{pmatrix} & & 1 \\ & 1 & \\ 1 & & \end{pmatrix}$$

if and only if $\beta = 2\alpha$. In this case we have

$$\hat{\Omega} = \frac{1}{\mathrm{h}} \begin{pmatrix} 0 & \alpha q & 0 \\ 1 & 0 & \alpha q \\ 0 & 1 & 0 \end{pmatrix}.$$

The condition for $T = (\mathrm{h}\partial)^3 - \beta q \mathrm{h}\partial - \alpha q \mathrm{h}$ to satisfy $\bar{T}^* = T$ is also $\beta = 2\alpha$.

This example is a special case of Example 6.18. The pairing is

$$\begin{aligned}&\langle\!\langle f_0+f_1\mathrm{h}\partial+f_2(\mathrm{h}\partial)^2, g_0+g_1\mathrm{h}\partial+g_2(\mathrm{h}\partial)^2\rangle\!\rangle\\&=\bar{f}_0g_2+\bar{f}_1g_1+\bar{f}_2g_0+\bar{f}_2g_2(\beta q).\end{aligned}$$

The orthonormal basis $1, \mathrm{h}\partial, (\mathrm{h}\partial)^2-\alpha q$ found in Example 6.18 is exactly the same as the basis found in Example 6.23. Propositions 6.15 and 6.16 both apply here. □

Example 6.29 The abstract quantum product in Example 6.24 satisfies the Frobenius condition with

$$S=\begin{pmatrix} & & & 1\\ & & 1 & \\ & 1 & & \\ 1 & & & \end{pmatrix}$$

if and only if $a_3=a_1-a_2+a_3$, that is, $a_1=a_2$. (The quantum D-module of M_5^3 satisfies this condition, but the naive quantization does not.) In this case we have

$$\hat{\omega}=\left(\begin{array}{c|c|c|c} & a_3q & & a_3^2q^2\\ \hline 1 & & (a_1-2a_3)q & \\ \hline & 1 & & a_3q\\ \hline & & 1 & \end{array}\right).$$

This condition for $T=(\mathrm{h}\partial)^4-a_1q(\mathrm{h}\partial)^2-a_2q\mathrm{h}(\mathrm{h}\partial)-a_3\mathrm{h}^2q$ to satisfy $\bar{T}^*=T$ is also $a_1=a_2$.

In this case the pairing $\langle\!\langle\ ,\ \rangle\!\rangle$ is given by $\bar{f}_0g_3+\bar{f}_1g_2+\bar{f}_2g_1+\bar{f}_3g_0+\bar{f}_2g_3(a_1q)+\bar{f}_3g_2(a_1q)$. The basis found in Example 6.24 is given by $1, \mathrm{h}\partial, (\mathrm{h}\partial)^2-a_3q, (\mathrm{h}\partial)^3-(a_2-a_3)q\mathrm{h}\partial-\mathrm{h}a_3q$, and this is orthonormal. □

These two examples are rather special as the particular choice of orthonormal basis happens to coincide with the basis from section 6.1 (giving $\hat{\Omega}$). However, the result of applying the procedure of section 6.1 to any orthonormal basis will again be an orthonormal basis; this is the 'scalar version' of Proposition 6.27. Thus, the coincidence is not misleading.

Next we turn to the homogeneity property.

Assumption Each differential operator T_i is homogeneous if we assign degrees as follows: $|b_i|=2, |q_i|>0, |\mathrm{h}|=2, |\partial_i|=0$.

Let $\mathcal{M}_i$ be the subspace of $\mathcal{M}$ which is spanned (over $\mathcal{H}_t$) by the P_j of degree i in $\mathrm{h}\partial_1,\dots,\mathrm{h}\partial_r$. Then we have a decomposition $\mathcal{M}=\mathcal{M}_0\oplus\mathcal{M}_1\oplus$

$\cdots \oplus \mathcal{M}_v$, with respect to which the (α,β)-th block of the matrix Ω_i will be denoted $(\Omega_i)_{\alpha,\beta}$.

Proposition 6.30 *(1) For $\alpha \geq \beta+2$ we have $(\Omega)_{\alpha,\beta} = 0$. (2) Each non-zero entry of the block $(\Omega_i)_{\alpha,\beta}$ has degree $2(\beta-\alpha)$. In particular, each non-zero entry of $\theta^{(i)}_{\alpha,\beta}$ has degree $2(\beta-\alpha-i)$.*

Let us write $L_+ = Q_0(I + \mathrm{h}Q_1 + \mathrm{h}^2 Q_2 + \cdots)$ as usual.

Proposition 6.31 *Each non-zero entry of the block $(L_+)_{\alpha,\beta}$ has degree $2(\beta-\alpha)$. In particular, each non-zero entry of $(Q_i)_{\alpha,\beta}$ has degree $2(\beta-\alpha-i)$.*

Proof The homogeneity condition for a matrix-valued function X can be expressed as

$$
\begin{aligned}
&X(\epsilon^{-|q_1|}q_1,\ldots,\epsilon^{-|q_r|}q_r,\epsilon^{-2}\mathrm{h})\\
&\quad = \mathrm{diag}(1,\epsilon^2,\ldots,\epsilon^{2v})X(q_1,\ldots,q_r,\mathrm{h})\mathrm{diag}(1,\epsilon^2,\ldots,\epsilon^{2v})^{-1}
\end{aligned}
$$

where $\mathrm{diag}(1,\epsilon^2,\ldots,\epsilon^{2v})$ denotes a matrix in block diagonal form (this applies equally well to $X=\Omega$ or $X=L_+$). Let us assume that this condition holds for $X=\Omega$. Then the method of the proof of Proposition 5.10 shows that the condition holds for $X=L_+$. □

Corollary 6.32 *Under the above assumptions, the abstract quantum product preserves the grading of $A \otimes \mathbb{C}[q]$.*

Example 6.33 The degrees of the entries of $\Omega = \frac{1}{\mathrm{h}}\omega + \theta^{(0)} + \mathrm{h}\theta^{(1)} + \mathrm{h}^2\theta^{(2)} + \cdots$ in Example 6.8 (where $|q| = |\mathrm{h}| = 2$) are as shown in the boxes below:

$$
\Omega = \frac{1}{\mathrm{h}}\begin{pmatrix} \boxed{2} & \boxed{4} \\ \boxed{0} & \boxed{2} \end{pmatrix} + \begin{pmatrix} \boxed{0} & \boxed{2} \\ & \boxed{0} \end{pmatrix} + \mathrm{h}\begin{pmatrix} & \boxed{0} \\ & \end{pmatrix}
$$

(the empty spaces indicate entries which are zero by Proposition 6.30). If $a_4 \neq 0$, then L_+ is not homogeneous in the above sense (the above argument fails as L_+ is not the identity matrix at $q=0$). If $a_4 = 0$ then the components of L_+ have the following degrees:

$$
\begin{pmatrix} \boxed{0} & \boxed{2} \\ & \boxed{0} \end{pmatrix}.
$$

We shall make use of the block structure of Ω and L_+ exhibited here in the next section. □

6.6 Computations for Fano type examples

Our definition of abstract quantum product in section 6.4 is somewhat indirect as it involves performing the Birkhoff factorization $L = L_- L_+$. However, the 'Fano type' assumption[1] that $|q_i| > 0$ leads to a simple and explicit algorithm for the computation of L_+, and L_+ is all we need to obtain the abstract quantum product.

We shall describe this algorithm, from [4] and [62]. There are two aspects to the computation. First (step 1 below), Gröbner basis techniques are used to obtain an explicit expression for $\Omega = \frac{1}{\mathrm{h}}\omega + \theta^{(0)} + \mathrm{h}\theta^{(1)} + \cdots$. Then (step 2), L_+ is obtained by solving a system of differential equations. The differential equations are solvable by 'quadrature', that is, successive integration, and their solutions are polynomial functions (under our assumptions), so this procedure is essentially algebraic.

Step 1: P_i and Ω_i. For computational purposes, let us choose $P_0, \ldots, P_s$ to be the 'standard monomials' in $\mathrm{h}\partial_1, \ldots, \mathrm{h}\partial_r$ with respect to a choice of Gröbner basis for the ideal $(T_1, \ldots, T_u)$. The basic principle of (commutative or non-commutative) Gröbner basis theory was mentioned in Examples 4.9 and 4.10, and the theory for our (non-commutative) situation is explained in detail in [105]. Explicit computation of a Gröbner basis and its standard monomials, and then the computation of Ω itself, may be carried out using the Ore algebra package of MAPLE.

Step 2: L_+ and $\hat{\Omega}_i$. Recall from the proof of Proposition 6.3 that $L_+ = Q_0(I + \mathrm{h}Q_1 + \mathrm{h}^2 Q_2 + \cdots)$ satisfies

$$\begin{aligned} L_-^{-1} dL_- &= (LL_+^{-1})^{-1} d(LL_+^{-1}) = L_+ L^{-1} dL L_+^{-1} + L_+ d(L_+^{-1}) \\ &= L_+ \Omega L_+^{-1} + L_+ (dL_+^{-1}), \end{aligned}$$

from which it follows that $L_-^{-1} dL_- = \frac{1}{\mathrm{h}} Q_0 \omega Q_0^{-1}$. Hence, we obtain a system of differential equations $\frac{1}{\mathrm{h}} Q_0 \omega Q_0^{-1} = L_+ \Omega L_+^{-1} - dL_+ L_+^{-1}$, or

$$\frac{1}{\mathrm{h}} Q_0 \omega Q_0^{-1} L_+ = L_+ \Omega - dL_+,$$

[1] As explained in Chapter 2, $|q_i| = 2\langle c_1(TM), A_i \rangle$ for the quantum cohomology of a manifold M, and this is positive if M is a Fano manifold. If M is a Calabi–Yau manifold (see section 10.2) then $c_1(TM) = 0$ and so $|q_i| = 0$. The term nef (numerically effective) covers both cases, that is, $|q_i| \geq 0$.

for L_+. As explained earlier, we may impose the condition $L_+|_{q=0} = I$, that is,

$$Q_0|_{q=0} = I, \quad Q_i|_{q=0} = 0 \text{ for } i \geq 1.$$

It can be shown (as we shall confirm now, by a direct argument) that these conditions determine L_+.

First, by comparing coefficients of powers of h, we can rewrite the system as

$$Q_0^{-1}dQ_0 = \theta^{(0)} + [Q_1, \omega], \; dQ_1 = \theta^{(1)} + [Q_1, \theta^{(0)}] + [Q_2, \omega] - [Q_1, \omega]Q_1,$$

and, for $i \geq 2$,

$$dQ_i = \theta^{(i)} + Q_1\theta^{(i-1)} + \cdots + Q_{i-1}\theta^{(1)} + [Q_i, \theta^{(0)}] + [Q_{i+1}, \omega] - [Q_1, \omega]Q_i.$$

For any $(s+1) \times (s+1)$ matrix $A = (a_{ij})$, let us say that the *k-diagonal* of A is the matrix $A^{[k]}$ whose (i,j)-th entry is a_{ij} when $j = i + k$ and zero when $j \neq i + k$. Thus, $A^{[0]}$ is the usual diagonal part of A, that is, the matrix obtained from A by deleting all non-diagonal entries. Because of the homogeneity property (Propositions 6.30 and 6.31), all our matrices will have a natural block structure $A = (A_{\alpha,\beta})$ where $A_{\alpha,\beta}$ is a square submatrix. For such matrices we define the term *block k-diagonal* in the obvious way. As there will be no danger of confusion, we use the same notation $A^{[k]}$ for the block k-diagonal.

In view of the remarks before Example 6.23 we may work over the polynomial ring $\mathbb{C}[q]$.

Thus, for some m, we have

$$\begin{aligned}
Q_0 &= I + Q_0^{[1]} + Q_0^{[2]} + \cdots + Q_0^{[m]} \\
Q_i &= Q_i^{[i]} + Q_i^{[i+1]} + \cdots + Q_i^{[m]} \quad (1 \leq i \leq m-1) \\
\omega &= \omega^{[-1]} + \omega^{[0]} + \omega^{[1]} + \cdots + \omega^{[m]} \\
\theta^{(i)} &= \theta^{(i),[i]} + \theta^{(i),[i+1]} + \cdots + \theta^{(i),[m]} \quad (0 \leq i \leq m-1)
\end{aligned}$$

and in particular $\Omega = \frac{1}{\mathrm{h}}\omega + \theta^{(0)} + \mathrm{h}\theta^{(1)} + \cdots + \mathrm{h}^{m-1}\theta^{(m-1)}$.

The case where $|q_i| \geq 4$ is easier to deal with, so let us assume this for the moment. Then we have

$$
\begin{aligned}
Q_0 &= I + Q_0^{[2]} + Q_0^{[3]} + \cdots \\
Q_i &= Q_i^{[i+2]} + Q_i^{[i+3]} + \cdots \quad (i \geq 1) \\
\omega &= \omega^{[-1]} + \omega^{[1]} + \omega^{[2]} + \cdots \\
\theta^{(i)} &= \theta^{(i),[i+2]} + \theta^{(i),[i+3]} + \cdots \quad (i \geq 0)
\end{aligned}
$$

and we see immediately from the form of Q_0 that we will have $\hat{c}_0 = c_0 = 1$ and $\hat{c}_i = c_i = b_i$ for $i = 1, \ldots, r$.

Since

$$
(XY)^{[j]} = \sum_k X^{[j-k]} Y^{[k]}, \quad [X, Y]^{[j]} = \sum_k [X^{[j-k]}, Y^{[k]}]
$$

the equation for dQ_i becomes

$$
\begin{aligned}
dQ_i^{[j]} &= \theta^{(i),[j]} + \sum_{k=i+1}^{j-3} Q_1^{[j-k]} \theta^{(i-1),[k]} + \cdots + \sum_{k=3}^{j-i-1} Q_{i-1}^{[j-k]} \theta^{(1),[k]} \\
&+ \sum_{k=2}^{j-i-2} [Q_i^{[j-k]}, \theta^{(0),[k]}] + \sum_{k=-1}^{j-i-3} [Q_{i+1}^{[j-k]}, \omega^{[k]}] - \sum_{k=-1}^{m-1} \sum_{l=3}^{m} [Q_1^{[l]}, \omega^{[k]}] Q_i^{[j-k-l]},
\end{aligned}
$$

and the equations for Q_0 and Q_1 decompose in a similar way.

Let us now define a total ordering on the (symbols) $Q_i^{[j]}, 2 \leq i+2 \leq j \leq m$ by $Q_{i_1}^{[j_1]} < Q_{i_2}^{[j_2]}$ if and only if (a) $j_1 - i_1 < j_2 - i_2$ or (b) $j_1 - i_1 = j_2 - i_2$ and $j_2 < j_1$.

The key observation concerning the above system is that it is of the form

$$
dQ_i^{[j]} = \text{an expression involving } Q_l^{[k]} \text{ with } Q_l^{[k]} < Q_i^{[j]}.
$$

Starting with the smallest term $Q_{m-2}^{[m]}$, we may, therefore, integrate successively to find all $Q_i^{[j]}$. The constants of integration are determined by $L_+|_{q=0} = I$, that is, $Q_0|_{q=0} = I$ and $Q_i|_{q=0} = 0$ for $i \geq 1$.

Example 6.34 The calculations in Example 6.6 were carried out using the above algorithm. Let us write down the form of the equations in step 2, in order to see how L_+ is calculated.

First, since we are dealing with 4×4 block matrices, we have $\Omega = \frac{1}{\mathrm{h}}\omega + \theta^{(0)} + \mathrm{h}\theta^{(1)} + \mathrm{h}^2\theta^{(2)} + \mathrm{h}^3\theta^{(3)}$. Since $|q_1| = |q_2| = 4$, however, only the entries indicated below can be non-zero. As usual, the numbers in the boxes are the degrees:

$$\frac{1}{\mathrm{h}}\left(\begin{array}{c|c|c|c} & \boxed{4} & & \boxed{8} \\ \hline \boxed{0} & & \boxed{4} & \\ \hline & \boxed{0} & & \boxed{4} \\ \hline & & \boxed{0} & \end{array}\right) + \left(\begin{array}{c|c|c|c} & & \boxed{4} & \\ \hline & & & \boxed{4} \\ \hline & & & \\ \hline & & & \end{array}\right) + \mathrm{h}\left(\begin{array}{c|c|c|c} & & & \boxed{4} \\ \hline & & & \\ \hline & & & \\ \hline & & & \end{array}\right).$$

In this situation we have

$$\begin{aligned} Q_0 &= I + Q_0^{[2]} \\ Q_1 &= Q_1^{[3]} \\ \omega &= \omega^{[-1]} + \omega^{[1]} + \omega^{[3]} \\ \theta^{(0)} &= \theta^{(0),[2]} \\ \theta^{(1)} &= \theta^{(1),[3]}. \end{aligned}$$

It turns out (after computation of Ω) that $\theta^{(1)} = 0$, so $Q_1 = 0$, and the system reduces to the equation $Q_0^{-1} dQ_0 = \theta^{(0)}$ for Q_0. This is just

$$d\left(I + Q_0^{[2]}\right) = \left(I + Q_0^{[2]}\right)\theta^{(0),[2]} = \theta^{(0),[2]},$$

in other words

$$q_1 \frac{\partial}{\partial q_1} Q_0^{[2]} = \theta_1^{(0),[2]}, \quad q_2 \frac{\partial}{\partial q_2} Q_0^{[2]} = \theta_2^{(0),[2]}.$$

Since $\theta_1^{(0),[2]}$ is linear in q_1 here, and $\theta_2^{(0),[2]}$ is linear in q_2, we obtain $Q_0^{[2]} = \theta_1^{(0),[2]} + \theta_2^{(0),[2]}$, which gives the formula for L_+ in Example 6.6. This produces the quantum product for the flag manifold F_3.

For F_n the degrees of all matrix entries are again divisible by 4. For example, when $n = 4$ we have 7×7 block matrices, and the equations for $L_+ = Q_0(I + \mathrm{h}Q_1 + \mathrm{h}^2 Q_2 + \mathrm{h}^3 Q_3 + \mathrm{h}^4 Q_4)$ with

$$\begin{aligned} Q_0 &= I + Q_0^{[2]} + Q_0^{[4]} + Q_0^{[6]} \\ Q_1 &= Q_1^{[3]} + Q_1^{[5]} \\ Q_2 &= Q_2^{[4]} + Q_2^{[6]} \\ Q_3 &= Q_3^{[5]} \\ Q_4 &= Q_4^{[6]} \end{aligned}$$

are

$$
\begin{aligned}
dQ_1^{[3]} &= \theta^{(1),[3]} + [Q_2^{[4]}, \omega^{[-1]}] \\
dQ_1^{[5]} &= \theta^{(1),[5]} + [Q_1^{[3]}, \theta^{(0),[2]}] + [Q_2^{[4]}, \omega^{[1]}] + [Q_2^{[6]}, \omega^{[-1]}] - [Q_1^{[3]}, \omega^{[-1]}]Q_1^{[3]} \\
dQ_2^{[4]} &= \theta^{(2),[4]} + [Q_3^{[5]}, \omega^{[-1]}] \\
dQ_2^{[6]} &= \theta^{(2),[6]} + \left[Q_1^{[3]}, \theta^{(1),[3]}\right] + [Q_2^{[4]}, \theta^{(0),[2]}] + [Q_3^{[5]}, \omega^{[1]}] - [Q_1^{[3]}, \omega^{[-1]}]Q_2^{[4]} \\
dQ_3^{[5]} &= \theta^{(3),[5]} + [Q_4^{[6]}, \omega^{[-1]}] \\
dQ_4^{[6]} &= \theta^{(4),[6]}.
\end{aligned}
$$

The ordering of the unknown matrices here is $Q_1^{[5]} > Q_2^{[6]} > Q_1^{[3]} > Q_2^{[4]} > Q_3^{[5]} > Q_4^{[6]}$. Beginning with $Q_4^{[6]}$, one can integrate the equations successively, then substitute Q_1 into $Q_0^{-1}dQ_0 = \theta^{(0)} + Q_1\omega$ and obtain Q_0 in a similar manner. For the answer in this case see [4].

We should make some comments on the significance of this calculation. First of all the calculation does not produce the Gromov–Witten invariants of the flag manifold 'from scratch'; we are starting from non-trivial information, the ideal of relations of a D-module. This D-module could be regarded (in the spirit of this chapter) as having nothing to do with quantum cohomology, since it arises naturally from the Hamiltonian system known as the Toda lattice. However, the most straightforward point of view is that the calculation gives a method to extract the Gromov–Witten invariants from the quantum D-module (which we assume is given). The key point is that, in the case of a Fano manifold like F_n, a simple algorithm exists.

This may well have other benefits. For example, one side effect is that the procedure can be used to define (and compute) 'quantum Schubert polynomials' (see [4]). Such polynomials were introduced in [82] by performing a Gram–Schmidt orthogonalization procedure within the quantum cohomology algebra; in view of section 6.3, our procedure involves a Gram–Schmidt orthogonalization within the quantum D-module. □

Example 6.35 The computation of L_+ in Example 6.23 is very easy because $\Omega = \frac{1}{\hbar}\omega + \theta^{(0)}$, and whenever Ω has this form we must have $L_+ = Q_0$. The equation for Q_0 is then $Q_0^{-1}dQ_0 = \theta^{(0)}$. Here we have $|q| = 4$, and $Q_0 = I + Q_0^{[2]}$, and we obtain $Q_0^{[2]} = \theta^{(0),[2]}$. □

Example 6.36 Computations for the D-module associated to the Fano hypersurfaces M_N^k were carried out in [106]. Let us give the computation of L_+ in Example 6.24 for the D-module

$$\mathcal{M} = D^{\mathrm{h}}/((\mathrm{h}\partial)^4 - a_1 q(\mathrm{h}\partial)^2 - a_2 \mathrm{h} q(\mathrm{h}\partial) - a_3 \mathrm{h}^2 q)$$

(a special case of which is the quantum cohomology of M_5^3). Here we have 4×4 block matrices (with 1×1 blocks), and again $|q| = 4$. We have

$$\Omega = \tfrac{1}{\mathrm{h}}\left(\omega^{[-1]} + \omega^{[1]} + \omega^{[3]}\right) + \theta^{(0),[2]} + \mathrm{h}\,\theta^{(1),[3]}$$

with

$$\begin{aligned} Q_0 &= I + Q_0^{[2]} \\ Q_1 &= Q_1^{[3]}. \end{aligned}$$

The equation for Q_1 is

$$q\frac{d}{dq}Q_1^{[3]} = \theta^{(1),[3]} = \left(\begin{array}{c|c|c|c} & & & a_3 q \\ \hline & & & \\ \hline & & & \\ \hline & & & \end{array}\right),$$

which gives $Q_1^{[3]} = \theta^{(1),[3]}$. The equation for Q_0 is

$$\begin{aligned} q\frac{d}{dq}Q_0^{[2]} &= \theta^{(0),[2]} + [Q_1^{[3]}, \omega^{[-1]}] \\ &= \left(\begin{array}{c|c|c|c} & & & \\ \hline & & & a_2 q \\ \hline & & & \\ \hline & & & \end{array}\right) + \left(\begin{array}{c|c|c|c} 1 & & a_3 q & \\ \hline & 1 & & -a_3 q \\ \hline & & 1 & \\ \hline & & & 1 \end{array}\right), \end{aligned}$$

which gives

$$Q_0 = \left(\begin{array}{c|c|c|c} 1 & & a_3 q & \\ \hline & 1 & & (a_2 - a_3) q \\ \hline & & 1 & \\ \hline & & & 1 \end{array}\right),$$

as stated in Example 6.24. □

So far we have assumed that $|q_i| \geq 4$. If $|q_i| = 2$ for some q_i, the algorithm works equally well, but we may have $\hat{c}_i \neq b_i$ when $1 \leq i \leq r$. We shall discuss this phenomenon in the next section. The cases $|q_i| = 3$ or $|q_i| = 1$ do not arise naturally in our situation.

6.7 Beyond Fano type examples

When $|q_i| > 0$, the grading simplifies both the theory and the calculations. In this section we consider very briefly, mainly through examples, what happens when we relax this condition to $|q_i| \geq 0$. (The case where some $|q_i| < 0$ is also interesting, but more complicated still—see [73].)

Most of the discussion of sections 6.1–6.5 still applies when $|q_i| \geq 0$. The map L_+ is holomorphic at $q = 0$ and we may normalize it so that $L_+|_{q=0} = I$, although it is not necessarily a polynomial function in this case (and indeed it may not be easy to express L_+ explicitly, as we shall see). However, the main problem—as illustrated in Example 6.26 even when $|q| = 2$—is that we cannot predict that $\hat{c}_0 = 1$ and $\hat{c}_i = b_i$ for $1 \leq i \leq r$, so we cannot immediately produce an abstract quantum cohomology algebra by the method of 6.4.

The homogeneity property of abstract quantum cohomology (Proposition 6.30) still holds. Even though we are no longer dealing with polynomials in q, we have the following information on the 'shape' of L_+:

Proposition 6.37 *The map* $L_+ = Q_0(I + \mathrm{h}Q_1 + \mathrm{h}^2 Q_2 + \cdots)$ *satisfies*

(1) $Q_0 = \exp X$ *where* $X_{\alpha,\beta} = 0$ *for* $\alpha \geq \beta - 1$,
(2) for $i \geq 1$, $(Q_i)_{\alpha,\beta} = 0$ *for* $\alpha \geq \beta - i - 1$.

Proof By Proposition 6.30 we have $(\Omega)_{\alpha,\beta} = 0$ for $\alpha \geq \beta+2$. Hence Ω takes values in the Lie algebra consisting of loops of the form $\sum_{i\in\mathbb{Z}} \mathrm{h}^i A_i$ whose coefficients satisfy the following conditions: $(A_i)_{\alpha,\beta} = 0$ for $\alpha \geq \beta - i - 1$ when $i \geq 0$, and $(A_i)_{\alpha,\beta} = 0$ for $\alpha \geq \beta - i + 1$ when $i < 0$. The method of the proof of Proposition 5.10 shows that L_+ take values in the corresponding loop group. In particular $(L_+)^{-1} dL_+ = \sum_{i\geq 0} \mathrm{h}^i A_i$ where $(A_i)_{\alpha,\beta} = 0$ for $\alpha \geq \beta - i - 1$, from which the stated properties of the Q_i follow. □

Example 6.38 Let us begin with a simple example based on Example 6.26 itself. Although this is still of Fano type, since $|q| = 2$, it illustrates in isolation the main difficulty above. The D-module

$$\mathcal{M} = D^{\mathrm{h}}/((\mathrm{h}\partial)^2 - 4q\mathrm{h}\partial - 2\mathrm{h}q)$$

is the quantum D-module of the degree two hypersurface M_3^2 in $\mathbb{C}P^2$ (see Example 6.24). This is just $\mathbb{C}P^1$, embedded in $\mathbb{C}P^2$ as a quadric, but the above D-module reflects (Gromov–Witten invariants based on) the embedding and is not simply the quantum D-module of $\mathbb{C}P^1$ (which would be $D^{\mathrm{h}}/((\mathrm{h}\partial)^2 - q)$, with $|q| = 4$).

With respect to $1, \mathrm{h}\partial$ we obtain

$$\Omega = \frac{1}{\mathrm{h}} \begin{pmatrix} 0 & 0 \\ 1 & 4q \end{pmatrix} + \begin{pmatrix} 0 & 2q \\ 0 & 0 \end{pmatrix},$$

and we obtain immediately (by the method of the last section)

$$L_+ = \begin{pmatrix} 1 & 2q \\ 0 & 1 \end{pmatrix}.$$

The new basis is given by $\hat{c}_0 = 1$, $\hat{c}_1 = \mathrm{h}\partial - 2q$, and with respect to this we have

$$\hat{\Omega} = \frac{1}{\mathrm{h}} \begin{pmatrix} 2q & 4q^2 \\ 1 & 2q \end{pmatrix},$$

which is not a 'matrix of structure constants of an algebra with unit element' with respect to the basis given by $1, b$. However, there is an easy way to modify this: let us apply the scalar gauge transformation $L_- \mapsto L_- \,\mathrm{e}^{-2q/\mathrm{h}}$. This gives a new connection matrix

$$\tilde{\Omega} = \frac{1}{\mathrm{h}} \begin{pmatrix} 0 & 4q^2 \\ 1 & 0 \end{pmatrix},$$

which is a matrix of structure constants. It 'de-quantizes' to the algebra $\mathbb{C}[b,q]/(b^2 - 4q^2)$, and this is essentially the same as the quantum cohomology algebra of $\mathbb{C}P^1$.

This example is discussed prominently in section 10 of [49], along with the quantum cohomology of M_N^{N-1} in general, where the same phenomenon occurs: the scalar gauge transformation

$$L_- \longmapsto L_- \, e^{-(N-1)!q/\mathrm{h}}$$

produces the quantum cohomology algebra

$$\mathbb{C}[b,q]/\big((b + (N-1)!q)^{N-1} - (N-1)^{N-1} q (b + (N-1)!q)^{N-2}\big).$$

From the o.d.e. point of view, we replace the solution y of

$$\big((\mathrm{h}\partial)^{N-1} - (N-1)\mathrm{h}^{N-2} q((N-1)\partial + N-2) \dots ((N-1)\partial + 1)\big)y = 0$$

by $\hat{y} = e^{-(N-1)!q/\mathrm{h}} y$; the formula

$$\mathrm{h}\partial e^{(N-1)!q/\mathrm{h}} = e^{(N-1)!q/\mathrm{h}} \, (\mathrm{h}\partial + (N-1)!q)$$

shows how the factor $b + (N-1)!q$ arises. From the D-module point of view we are using the isomorphism $D/(T) \to D/(Te^{(N-1)!q/\mathrm{h}})$ (cf. appendix 4.6). □

The question of why this procedure (and indeed, the simpler procedure for M_N^k when $1 \leq k \leq N-2$) gives the 'correct answer' for the quantum cohomology is a difficult one, which we do not address. The fact itself was established (in greater generality) in [49] and [91]–[93], and is known as the Mirror Theorem; evidently it requires making a connection with the rigorous definition of quantum cohomology. In this section our focus is on a weaker (but wider) question: How is it possible to extract an abstract quantum cohomology algebra from a D-module which is not of Fano type?

Example 6.39 The D-module

$$\mathcal{M} = D^{\mathrm{h}}/(\,(\mathrm{h}\partial)^2 - 3q\mathrm{h}^2(3\partial + 2)(3\partial + 1)\,)$$

has the same defect as the previous example, but we shall deal with it in a different way. This particular example—the quantum differential equation of a cubic hypersurface M_3^3 in $\mathbb{C}P^2$, or elliptic curve—will be of importance in mirror symmetry, which we shall discuss in Chapter 10.

With respect to $1, \mathrm{h}\partial$ we obtain

$$\Omega = \frac{1}{\mathrm{h}}\begin{pmatrix} 0 & 0 \\ 1 & 0 \end{pmatrix} + \begin{pmatrix} 0 & 0 \\ 0 & v \end{pmatrix} + \mathrm{h}\begin{pmatrix} 0 & u \\ 0 & 0 \end{pmatrix},$$

where $u = 27q/(1-27q)$ and $v = 6q/(1-27q)$. The proposition above implies that L_+ is of the form

$$L_+ = \begin{pmatrix} \alpha_0 & \beta \\ 0 & \alpha_1 \end{pmatrix}\left(\begin{pmatrix} 1 & 0 \\ 0 & 1 \end{pmatrix} + \mathrm{h}\begin{pmatrix} 0 & \gamma \\ 0 & 0 \end{pmatrix}\right).$$

In this example we have $|q| = 0$, so any function of q must have degree zero. Thus, by the homogeneity condition (see section 6.5), we must have $\beta = 0$. We have seen in section 6.6 that, when $|q| > 0$, we can find L_+ by 'quadrature' from the differential equation $\frac{1}{\mathrm{h}}Q_0\omega Q_0^{-1}L_+ = L_+\Omega - dL_+$. Here, substitution gives the following system for the functions $\alpha_0, \alpha_1, \gamma$:

$$\begin{aligned} \alpha_0' &= \alpha_0\gamma \\ \alpha_1' &= \alpha_1 v - \alpha_1\gamma \\ \gamma' &= u + \gamma v - \gamma^2 \end{aligned}$$

subject to $\alpha_0(0) = \alpha_1(0) = 1, \gamma(0) = 0$. This cannot be solved by quadrature.

The Ricatti equation for γ turns out to be the original o.d.e. in disguise; indeed, the latter is obtained by substituting $\gamma = \alpha_0'/\alpha_0$. Thus, if we find γ or α_0, we obtain the remaining functions from the relations

$$(\log\alpha_0)' = \gamma,\ (\log\alpha_1)' = v - \gamma,\ \alpha_0\alpha_1 = \exp(\textstyle\int v) = 1/(1-27q).$$

The easiest way to express L_+ is to solve the original equation

$$((\mathrm{h}\partial)^2 - 3q\mathrm{h}^2(3\partial + 2)(3\partial + 1))y = 0$$

by the Frobenius method. At the regular singular point $q = 0$ (where the indicial equation is $s^2 = 0$), we have a canonical 'Frobenius basis' of solutions u_0, u_1 of the form

$$u_0 = f_0, \quad u_1 = f_0 \log q + f_1$$

where f_0, f_1 are holomorphic at $q = 0$ and satisfy $f_0(0) = 1, f_1(0) = 0$. The Taylor series of f_0, f_1 can be computed explicitly. (See appendix 5.4 for a similar example; we shall also discuss this further in Example 10.11.)

Now, let us consider the Frobenius symbolic factors v_0, v_1 of appendix 4.6, that is, we factorize the o.d.e. as

$$(\mathrm{h}\partial)^2 - 3q\mathrm{h}^2(3\partial + 2)(3\partial + 1) = (1 - 27q)v_0v_1\mathrm{h}\partial\frac{1}{v_1}\mathrm{h}\partial\frac{1}{v_0}.$$

(There are several minor differences from the situation of appendix 4.6. First of all, the parameter h appears here, but it factors out of the equations and causes no problems. Next, the current o.d.e. is not monic, so we need the factor $(1 - 27q)$. Finally, the argument in appendix 4.6 was at a regular point, but this does not affect the normalization of the functions v_0, v_1; here we impose initial conditions on f_0, f_1 instead of on u_0, u_1.) We obtain

$$v_0 = u_0 = f_0, \quad v_1 = (u_1/u_0)' = 1 + (f_1/f_0)'$$

and we deduce that $\alpha_0 = v_0, \alpha_1 = v_0v_1$.

This allows us to express L_+ in terms of v_0, v_1, and it turns out that

$$\hat{P}_0 = \frac{1}{v_0}, \quad \hat{P}_1 = \frac{1}{v_1}\mathrm{h}\partial\frac{1}{v_0}.$$

Although this can be verified by computing L_+ from the formulae above, it follows directly from the form of $\hat{\Omega} = Q_0\Omega Q_0^{-1}$, which must be

$$\hat{\Omega} = \begin{pmatrix} \alpha_0 & \\ & \alpha_1 \end{pmatrix}^{-1} \begin{pmatrix} 0 & 0 \\ 1 & 0 \end{pmatrix} \begin{pmatrix} \alpha_0 & \\ & \alpha_1 \end{pmatrix} = \begin{pmatrix} 0 & 0 \\ \alpha_1/\alpha_0 & 0 \end{pmatrix} = \begin{pmatrix} 0 & 0 \\ v_1 & 0 \end{pmatrix}.$$

Namely, the gauge transformation L_+ converts the matrix system into

$$\mathrm{h}\partial \begin{pmatrix} \hat{P}_0 y \\ \hat{P}_1 y \end{pmatrix} = \begin{pmatrix} 0 & v_1 \\ 0 & 0 \end{pmatrix} \begin{pmatrix} \hat{P}_0 y \\ \hat{P}_1 y \end{pmatrix},$$

and this is equivalent to the o.d.e. $\mathrm{h}\partial(1/v_1)\mathrm{h}\partial(1/v_0)y = 0$ if $\hat{P}_0 = 1/v_0$ and $\hat{P}_1 = (1/v_1)\mathrm{h}\partial(1/v_0)$.

We are now in a position to extract an abstract quantum cohomology algebra from this D-module. Motivated by mirror symmetry (see Chapter 10), we expect to end up with the relation $\hat{b}^2 = 0$ in the (quantum or ordinary) cohomology algebra of an elliptic curve (two-dimensional torus). There is a straightforward way to obtain this: replace $(1/v_1)\partial$ by $\hat{\partial}$, that is, introduce a new variable $\hat{t}$ (and $\hat{q}$) by

$$\frac{\partial \hat{t}}{\partial t} = v_1 = \frac{\partial(u_1/u_0)}{\partial t}.$$

This can be achieved by defining $\hat{t} = u_1/u_0 = f_1/f_0 + \log q$, that is, $\hat{q} = qe^{f_1/f_0}$. The abstract quantum D-module is $\hat{\mathcal{M}} = \hat{\mathrm{D}}^{\mathrm{h}}/((\mathrm{h}\hat{\partial})^2(1/v_0)) \cong \hat{\mathrm{D}}^{\mathrm{h}}/((\mathrm{h}\hat{\partial})^2)$. □

Example 6.40 We have already met the D-module

$$\mathcal{M} = \mathrm{D}^{\mathrm{h}}/(\mathrm{h}^2\partial_1^2 - q_1\mathrm{h}^2(\partial_2 - 2\partial_1)(\partial_2 - 2\partial_1 - 1), \mathrm{h}^2\partial_2(\partial_2 - 2\partial_1) - q_2)$$

in Example 6.10. In fact this has a geometrical origin—it is related to the quantum D-module of the Hirzebruch surface Σ_2. The ordinary cohomology algebra of Σ_2 (cf. Examples 2.8 and 2.15) is

$$H^*\Sigma_2 = \mathbb{C}[b_1, b_2]/(b_1^2, b_2(b_2 - 2b_1)).$$

Naive consideration of rational curves in Σ_2 (see [12]) leads to the following candidate for the quantum cohomology algebra, where $|q_1| = 0$, $|q_2| = 4$:

$$QA = \mathbb{C}[b_1, b_2, q_1, q_2]/(b_1^2 - q_1(b_2 - 2b_1)^2, b_2(b_2 - 2b_1) - q_2).$$

However, this is not the 'correct answer'.

Remarkably, the correct answer may be salvaged by choosing the above quantization of QA, then extracting an abstract quantum cohomology algebra by the method of this section. This fact—which extends to more general toric manifolds and complete intersections in such manifolds—is the main assertion of the Mirror Theorem mentioned in the previous example (which was a special case). The D-modules which provide such quantizations have an independent existence; they were introduced before the mirror symmetry phenomenon was discovered, by I. Gelfand, M. Kapranov, and A. Zelevinsky, and are known as GKZ systems (see [26], section 5.5).

We shall just present the extraction of the abstract quantum cohomology algebra, as in Example 5.4 of [62]. First, with respect to the basis of the

D-module given by $1, h\partial_2, h\partial_1, h^2\partial_2^2$, we obtain

$$\Omega_1 = \frac{1}{h}\left(\begin{array}{c|cc|c} 0 & -\frac{q_2}{2} & \frac{-2q_1q_2}{4q_1-1} & 0 \\ \hline 0 & 0 & 0 & 2q_1q_2 \\ 1 & 0 & 0 & -q_2(4q_1-1) \\ \hline 0 & \frac{1}{2} & \frac{q_1}{4q_1-1} & 0 \end{array}\right) + \left(\begin{array}{c|cc|c} 0 & 0 & 0 & 0 \\ \hline 0 & 0 & \frac{q_1}{4q_1-1} & 0 \\ 0 & 0 & \frac{-2q_1}{4q_1-1} & 0 \\ \hline 0 & 0 & 0 & 0 \end{array}\right),$$

and

$$\Omega_2 = \frac{1}{h}\left(\begin{array}{c|cc|c} 0 & 0 & \frac{-q_2}{2} & 0 \\ \hline 1 & 0 & 0 & q_2(4q_1+1) \\ 0 & 0 & 0 & -q_2(4q_1-1) \\ \hline 0 & 1 & \frac{1}{2} & 0 \end{array}\right) + \left(\begin{array}{c|cc|c} 0 & 0 & 0 & q_2 \\ \hline 0 & 0 & 0 & 0 \\ 0 & 0 & 0 & 0 \\ \hline 0 & 0 & 0 & 0 \end{array}\right).$$

The connection form $\hat{\Omega}$ (obtained from the gauge transformation $L_+ = Q_0$) is defective for the same reason as the previous example. Indeed,

$$Q_0^{-1} = \left(\begin{array}{c|cc|c} 1 & 0 & 0 & -q_2 \\ \hline 0 & 1 & \frac{1}{2}(1-\sqrt{1-4q_1}) & 0 \\ 0 & 0 & \sqrt{1-4q_1} & 0 \\ \hline 0 & 0 & 0 & 1 \end{array}\right)$$

and we have $\hat{P}_1 = h\partial_2$ but $\hat{P}_2 = \frac{1}{2}(1-\sqrt{1-4q_1})h\partial_2 + \sqrt{1-4q_1}h\partial_1 \neq h\partial_1$.

We may remedy this problem by introducing new coordinates such that

$$\hat{\partial}_2 = \partial_2, \quad \hat{\partial}_1 = \frac{1}{2}(1-\sqrt{1-4q_1})\partial_2 + \sqrt{1-4q_1}\partial_1.$$

It may be verified that $q_1 = \hat{q}_1/(1+\hat{q}_1)^2$, $q_2 = \hat{q}_2(1+\hat{q}_1)$ achieves this.

This transformation converts $\mathcal{M}$ to

$$\hat{\mathcal{M}} = \hat{D}^h/(h^2\hat{\partial}_1^2 - \hat{q}_1\hat{q}_2, h^2\hat{\partial}_2(\hat{\partial}_2 - 2\hat{\partial}_1) - \hat{q}_2(1-\hat{q}_1)),$$

which de-quantizes to the algebra

$$\mathbb{C}[\hat{b}_1, \hat{b}_2, \hat{q}_1, \hat{q}_2]/(\hat{b}_1^2 - \hat{q}_1\hat{q}_2, \hat{b}_2(\hat{b}_2 - 2\hat{b}_1) - \hat{q}_2(1-\hat{q}_1)).$$

This will turn out to be the quantum cohomology algebra of Σ_2.

To verify this, one may compute the abstract quantum product, then use the fact (Chapter 11 of [26]) that the quantum products of Σ_2 can be deduced from those of $\Sigma_0 = \mathbb{C}P^1 \times \mathbb{C}P^1$, if one uses the symplectic invariance of Gromov–Witten invariants.

The abstract quantum products may be computed from $\hat{\mathcal{M}}$ by our usual method. With respect to $1, \mathrm{h}\hat{\partial}_2, \mathrm{h}\hat{\partial}_1, \mathrm{h}^2\hat{\partial}_2^2$, the matrices of the action of $\hat{\partial}_1, \hat{\partial}_2$ are

$$\hat{\Omega}_1 = \frac{1}{\mathrm{h}}\left(\begin{array}{c|cc|c} 0 & -\frac{\hat{q}_2(1-\hat{q}_1)}{2} & \hat{q}_1\hat{q}_2 & 0 \\ \hline 0 & 0 & 0 & 2\hat{q}_1\hat{q}_2 \\ 1 & 0 & 0 & \hat{q}_2(1-\hat{q}_1) \\ \hline 0 & \frac{1}{2} & 0 & 0 \end{array}\right) + \left(\begin{array}{c|cc|c} 0 & 0 & 0 & \hat{q}_1\hat{q}_2 \\ \hline 0 & 0 & 0 & 0 \\ 0 & 0 & 0 & 0 \\ \hline 0 & 0 & 0 & 0 \end{array}\right)$$

and

$$\hat{\Omega}_2 = \frac{1}{\mathrm{h}}\left(\begin{array}{c|cc|c} 0 & 0 & -\frac{\hat{q}_2(1-\hat{q}_1)}{2} & 0 \\ \hline 1 & 0 & 0 & \hat{q}_2(1+3\hat{q}_1) \\ 0 & 0 & 0 & 2\hat{q}_2(1-\hat{q}_1) \\ \hline 0 & 1 & \frac{1}{2} & 0 \end{array}\right) + \left(\begin{array}{c|cc|c} 0 & 0 & 0 & \hat{q}_2(1+\hat{q}_1) \\ \hline 0 & 0 & 0 & 0 \\ 0 & 0 & 0 & 0 \\ \hline 0 & 0 & 0 & 0 \end{array}\right).$$

This time we obtain

$$\hat{L}_+^{-1} = \hat{Q}_0^{-1} = \left(\begin{array}{c|cc|c} 1 & 0 & 0 & -\hat{q}_2(1+\hat{q}_1) \\ \hline 0 & 1 & 0 & 0 \\ 0 & 0 & 1 & 0 \\ \hline 0 & 0 & 0 & 1 \end{array}\right).$$

Conjugation by $\hat{Q}_0$ gives the following matrices of structure constants:

$$\left(\begin{array}{c|cc|c} 0 & \hat{q}_1\hat{q}_2 & \hat{q}_1\hat{q}_2 & 0 \\ \hline 0 & 0 & 0 & 2\hat{q}_1\hat{q}_2 \\ 1 & 0 & 0 & -2\hat{q}_1\hat{q}_2 \\ \hline 0 & \frac{1}{2} & 0 & 0 \end{array}\right), \left(\begin{array}{c|cc|c} 0 & (1+\hat{q}_1)\hat{q}_2 & \hat{q}_1\hat{q}_2 & 0 \\ \hline 1 & 0 & 0 & 2\hat{q}_1\hat{q}_2 \\ 0 & 0 & 0 & -2\hat{q}_2(-1+\hat{q}_1) \\ \hline 0 & 1 & \frac{1}{2} & 0 \end{array}\right).$$

This gives the abstract quantum products

$$\hat{b}_1 \circ \hat{b}_1 = \hat{q}_1\hat{q}_2,\ \hat{b}_1 \circ \hat{b}_2 = \hat{b}_1\hat{b}_2 + \hat{q}_1\hat{q}_2,\ \hat{b}_2 \circ \hat{b}_2 = \hat{b}_2^2 + \hat{q}_2(1+\hat{q}_1)$$
$$\hat{b}_1 \circ \hat{b}_2^2 = 2\hat{q}_1\hat{q}_2\hat{b}_2 - 2\hat{q}_1\hat{q}_2\hat{b}_1,\ \hat{b}_2 \circ \hat{b}_2^2 = 2\hat{q}_1\hat{q}_2\hat{b}_2 - 2\hat{q}_2(-1+\hat{q}_1)\hat{b}_1$$

which are in agreement with the computations of the quantum products for Σ_2 at the end of Chapter 11 of [26]. □

The ad hoc methods employed in these three examples may be unified by allowing coordinate changes which involve a new 'artificial' variable t_0. The simplest way to introduce this variable is to extend the original D-module—in the sense of section 4.4—by adjoining a new relation $\mathrm{h}\partial_0 - 1$.

In Example 6.38, let us make the coordinate change corresponding to

$$\begin{aligned}\mathrm{h}\hat{\partial}_0 &= \mathrm{h}\partial_0\\ \mathrm{h}\hat{\partial}_1 &= -2q\mathrm{h}\partial_0 + \mathrm{h}\partial_1,\end{aligned}$$

which is

$$\begin{aligned}\hat{t}_0 &= t_0 + 2e^{t_1}\\ \hat{t}_1 &= t_1.\end{aligned}$$

This has the *same effect* as the scalar gauge transformation. Namely, the t_0-extended version of L_- is easily seen to be

$$L_-(t_0, t_1, \mathrm{h}) = e^{t_0/\mathrm{h}} L_-(t_1, \mathrm{h}),$$

and this is equal to

$$e^{(\hat{t}_0 - 2q)/\mathrm{h}} L_-(t_1, \mathrm{h}) = e^{\hat{t}_0/\mathrm{h}} L_-(\hat{t}_1, \mathrm{h}) e^{-2q/\mathrm{h}}.$$

It can be verified directly that the transformed D-module is

$$\hat{D}^{\mathrm{h}}/(\,(\mathrm{h}\hat{\partial}_1)^2 - 4q^2, \mathrm{h}\hat{\partial}_0 - 1\,)$$

(where $\hat{D}^{\mathrm{h}}$ means $D^{\mathrm{h}}_{\hat{t}_0,\hat{t}_1}$).

In Example 6.39, working in the extended D-module

$$D^{\mathrm{h}}/(\,(\mathrm{h}\partial_1)^2 - 3q_1\mathrm{h}^2(3\partial_1 + 2)(3\partial_1 + 1), \mathrm{h}\partial_0 - 1\,),$$

a suitable change of variable would be given by

$$\begin{aligned}\mathrm{h}\hat{\partial}_0 &= \mathrm{h}\partial_0 = v_0 \quad (= 1)\\ \mathrm{h}\hat{\partial}_1 &= -\mathrm{h}\tfrac{\gamma}{v_1}\mathrm{h}\partial_0 + \tfrac{1}{v_1}\mathrm{h}\partial_1 \quad (= \tfrac{v_0}{v_1}\mathrm{h}\partial_1\tfrac{1}{v_0}).\end{aligned}$$

This converts the D-module directly to $\hat{D}^{\mathrm{h}}/(\,(\mathrm{h}\hat{\partial}_1)^2, \mathrm{h}\hat{\partial}_0 - 1\,)$. In both these examples, the extraneous scalar gauge transformation has been avoided (by incorporating it into the change of variable).

In all three examples, the change of variable is obtained in exactly the same way: the Jacobian matrix is obtained by normalizing the first $(r+1) \times (r+1)$ block of L_+^{-1} so that its first column is $(1, 0, \ldots, 0)^t$. A general approach has been given by H. Iritani in [72].

It is natural to ask whether it makes sense to extend further beyond $t_0, t_1, \ldots, t_r$ to $t_0, t_1, \ldots, t_s$. For the quantum cohomology D-module this is possible, and the extension is neither trivial nor artificial—it gives the so-called large quantum D-module, which we shall discuss in Chapter 9.

6.8 Towards integrable systems

A quantum cohomology (or abstract quantum cohomology) D-module is a particular flat connection with various special properties. As we shall see in the next chapter, integrable systems often arise from demanding that a connection has a special form, so it may be natural to expect that a quantum cohomology D-module is a solution of some integrable system.

In Chapter 9 we will meet the WDVV equations, an 'integrable p.d.e.' with exactly the above property. In Chapters 8 and 10 we will see that the quantum cohomology D-module is a particular kind of solution of the equations for pluriharmonic maps. (In both cases the equations admit many other solutions unrelated to quantum cohomology, however.)

It is natural to ask whether a quantum cohomology D-module can be regarded as an extension of a simpler D-module, or whether it extends to a more complicated one. Certainly quantum cohomology can be regarded as an extension of ordinary cohomology, though the corresponding D-modules are not related very closely; this is not an extension of the kind considered in section 4.4. On the other hand, (see the remarks at the end of that section) one expects to be able to write any D-module as an extension of that kind, 'in general'. The question is whether there is natural way of writing such an extension. For example, the quantum cohomology D-module of the flag manifold F_3 (Examples 3.4 and 6.6) has rank 6, and it may be expressed as a t-extension of a rank 6 D-module of the form $D_x/(T)$. However, this presentation is not very natural. The most natural presentation is the one we have seen, where there are two relations T_1, T_2 which have the special property that $[T_1, T_2] = 0$. In general, for a D-module $D/(T_1, \ldots, T_u)$, there is no reason to expect that the generators T_i can be chosen so that $[T_i, T_j] = 0$. The existence of such generators is related to the concept of quantum integrable systems (in which the Poisson-commuting conserved quantities of a classical system are 'quantized' to commuting differential operators).

In Chapter 9 we shall also see that there is a 'larger' quantum cohomology D-module in which the variable is $t \in H^*M$ (rather than $t \in H^2M$), and similar comments apply to that.

7 Integrable systems

In Chapter 4 we met the KdV equation, one of the most famous integrable systems or 'soliton equations' ever studied. To some extent it overshadows the family to which it belongs: since there is no universally accepted definition of integrable system, as a first approximation one could say that an integrable system is a partial differential equation which is analogous to the KdV equation. In this chapter we shall look at the KdV equation and some other well known integrable systems. There are hundreds of treatments of the KdV equation in the pure and applied mathematical literature, but it is an equation with many facets, and our exposition will emphasize D-modules and analogies with quantum cohomology. In particular, our point of view is that *the KdV equation arises from a certain universal construction with D-modules,* which suggests why the equation appears in so many different guises.

In Chapter 4 we described the KdV equation in two ways: it is the condition that a certain matrix-valued 1-form Ω satisfies the zero curvature condition, or the condition that a certain D-module D/I has rank 2. The form Ω and the ideal I depend on an auxiliary function u here, and the condition translates into a p.d.e. for u, which is the KdV equation. Less abstractly, these conditions say (respectively) that a certain 2×2 matrix system of p.d.e. is consistent, or that the corresponding scalar system of p.d.e. has solution space of dimension 2.

In this chapter we shall derive various concrete equations in order to provide the appropriate background for subsequent chapters. All these examples are well known. The only novel aspect of our exposition, as already mentioned, is that we shall emphasize the D-module point of view.

The reader will have no difficulty finding articles and books containing more information on the KdV equation. We just mention the book [27] and the survey articles [5], [37], [98], and [125]. For wider surveys of KdV-like equations, with many examples, we recommend [20], [42], [96], and [99].

7.1 The KdV equation

An equation of the form

$$u_t = \alpha u u_x + \beta u_{xxx}$$

(with α, β non-zero constants) may be brought into the form $u_t = uu_x + u_{xxx}$ by rescaling x and t, so we shall refer to any such equation as 'the KdV equation'.

This equation has a celebrated interpretation: it describes the motion of shallow water waves, an example being the 'solitary wave' observed by J. S. Russell. If the initial profile of the wave is given by the function $x \mapsto u(x, 0)$, the profile at time t is given by $x \mapsto u(x, t)$. Analytically speaking, the equation is an evolution equation. However, we shall be concerned with another aspect of the KdV equation, related to properties of families of second-order ordinary differential equations.

A second-order o.d.e.

$$(\partial_x^2 + a_1 \partial_x + a_0)y = 0$$

may be reduced to the case where $a_1 = 0$ by the well known device of replacing y by $ye^{\frac{1}{2}\int a_1}$. Without loss of generality, therefore, we consider operators of the form $\partial_x^2 + u$.

Consider next a t-family of operators $L = \partial_x^2 + u(x, t) \in D_x$, and the t-family of rank 2 D-modules $D_x/(L)$. Consider also a t-family of elements P of D_x.

Theorem 7.1 *Given t-families of operators $L = \partial_x^2 + u(x, t)$ and P, we have:*

(1) The D-module $D_{x,t}/(L, \partial_t - P)$ has rank 2 if and only if $[\partial_t - P, L] = 0$ mod L.
(2) Let $P = \frac{1}{2}u_x - u\partial_x$ (we shall explain below why this is the simplest non-trivial choice of P). Then the condition $[\partial_t - P, L] = 0$ mod L is equivalent to the KdV equation $u_t = 3uu_x + \frac{1}{2}u_{xxx}$.

Remark The condition

$$[\partial_t - P, L] = 0 \bmod L$$

can be written

$$L_t = [P, L] \bmod L$$

(recall that $[\partial_t, L] = L_t$, the result of differentiating the coefficients of L with respect to t). The condition is, therefore, a generalization of the Lax equation $L_t = [P, L]$.

The proof of the theorem is straightforward: part (1) has already been proved in section 4.4, and part (2) is an easy computation which we shall give in detail shortly. However, we regard it as fundamental. It shows that the KdV equation arises as the simplest non-trivial 't-extension' of the D-module $D_x/(\partial_x^2 + u(x))$. *Other choices of L and P, and more general extensions of more general D-modules, give many more 'integrable systems' which generalize the KdV equation.*

It should be noted that $P = \frac{1}{2}u_x - u\partial_x$ is *not* the usual choice. The usual choice (as in Examples 4.11 and 4.15) is $P = \partial_x^3 + \frac{3}{2}u\partial_x + \frac{3}{4}u_x$. In this section, we shall consider the significance of this and other choices of P.

We begin with $P = p_i(x,t)\partial_x^i + \cdots + p_0(x,t)$. We shall write $a = -u$ in order to simplify signs. As explained in section 4.4, we may choose a basis of the D-module and use the flatness of the connection form $\Omega_1 dx + \Omega_2 dt$ instead of the condition $[\partial_t - P, L] = 0 \bmod L$ (this is convenient for computations). We recall that $[T]$ denotes the equivalence class of an operator T in $D_x/(L)$. Let $[P] = f[1] + g[\partial_x]$ be the expression of P relative to the basis $[1], [\partial_x]$. Sometimes we write this as $P = f + g\partial_x \bmod L$, or just $P \equiv f + g\partial_x$.

From

$$\partial_x[1] = [\partial_x] = 0[1] + 1[\partial_x]$$
$$\partial_x[\partial_x] = [\partial_x^2] = [a] = a[1] + 0[\partial_x]$$

we obtain

$$\Omega_1 = \begin{pmatrix} 0 & a \\ 1 & 0 \end{pmatrix}.$$

From

$$\partial_t[1] = [\partial_t] = [P] = f[1] + g[\partial_x]$$
$$\partial_t[\partial_x] = [\partial_x P] = [\partial_x(f + g\partial_x)] = (f_x + ag)[1] + (f + g_x)[\partial_x]$$

(see section 4.4 for the definition of the action of ∂_t) we obtain

$$\Omega_2 = \begin{pmatrix} f & f_x + ag \\ g & f + g_x \end{pmatrix}.$$

The zero curvature condition $d\Omega + \Omega \wedge \Omega = 0$ reduces to

$$a_t = f_{xx} + a_x g + 2ag_x$$
$$0 = 2f_x + g_{xx}.$$

The second equation here gives $f(x,t) = -\frac{1}{2}g_x(x,t) + h(t)$, but $h(t)$ disappears on substituting f into the first equation, which becomes

$$a_t = -\tfrac{1}{2}g_{xxx} + ga_x + 2g_x a.$$

Thus, a specific choice of P amounts to a specific choice of g, and it is natural to choose an expression for g involving a and its x-derivatives in order to obtain an equation for a single function. The simplest choice $g = a = -u$ (giving $P = \frac{1}{2}u_x - u\partial_x$) produces the KdV equation in the form $u_t = 3uu_x + \frac{1}{2}u_{xxx}$, as in Theorem 7.1.

If $P = p_i\partial_x^i + \cdots + p_0$, where each p_j is a differential polynomial in a, then $P \equiv f + g\partial_x$ where f, g are also differential polynomials in a. In view of this, the case $i = 1$ is enough. However, a supply of natural examples can be obtained by starting with a *monic homogeneous* P, that is, which satisfies the following condition:

Assumption $P = \partial_x^i + p_{i-1}\partial_x^{i-1} + \cdots + p_0$ where each term of $\partial_t - P$ has weighted degree i, the weights being assigned as follows:

$$|\partial_x| = 1, \quad |\partial_t| = i, \quad |a| = 2.$$

(A weaker assumption would be that we start with an operator $P = f + g\partial_x$ which is homogeneous in the sense that $|\partial_t| = |f| = j, |g| = j - 1$ for some j.)

Example 7.2 The case $i = 1$: $P = f + \partial_x$.

Here $g = 1$, so $f = 0$, and the p.d.e. is just $a_t = a_x$. The original D-module has been extended by adding the trivial relation $\partial_t - \partial_x$. □

Example 7.3 The case $i = 2$: $P = f + g\partial_x + \partial_x^2 \equiv (f + a) + g\partial_x$.

There are no differential polynomials in a of degree 1, so the homogeneity condition forces $g = 0$ and $f + a = 0$. We obtain $a_t = 0$ and the t-family is constant. Note that $P = \partial_x^2 - a = L$ here, so the original D-module has been extended by adding the (even more) trivial relation ∂_t. □

Example 7.4 The case $i = 3$: $P = f + g\partial_x + h\partial_x^2 + \partial_x^3 \equiv (f + ha + a_x) + (g + a)\partial_x$.

As usual f is determined by g. If g, h are differential polynomials of degree 2, 1 in a, the homogeneity condition forces $g = \alpha a$ (for some constant α) and $h = 0$. Writing $\alpha = -(2\beta + 3)$ for convenience, the p.d.e. becomes

$$a_t = (\beta + 1)a_{xxx} - 6(\beta + 1)aa_x$$

with $P = \partial_x^3 - (2\beta + 3)a\partial_x + \beta a_x$. This is the KdV equation.

If we impose the stronger condition $[\partial_t - P, L] = 0$, we obtain the additional condition $3a_x + 2g_x = 0$, hence $\beta = -\frac{3}{4}$. This gives the usual Lax equation $L_t = [P, L]$, with $P = \partial_x^3 - \frac{3}{2}a\partial_x - \frac{3}{4}a_x$.

In contrast, the weaker assumption that $P = f + g\partial_x$ with $|f| = j = |g| + 1$ leads to more possibilities. For example, if $j = 5$, we can take $g = a^2$ or $g = a_{xx}$ or a combination of the two. The first choice, $g = a^2$, gives the p.d.e.

$$a_t = -aa_{xxx} - 3a_x a_{xx} + 5a^2 a_x,$$

while $g = a_{xx}$ gives

$$a_t = -\tfrac{1}{2}a_{xxxxx} + 2a_{xxx}a + a_x a_{xx}.$$

□

Example 7.5 The case $i = 4$: $P = f + g\partial_x + h\partial_x^2 + \partial_x^4$.

We have omitted the ∂_x^3 term, as this must be zero under the homogeneity assumption. Reducing mod L we obtain

$$\begin{aligned} P &= \partial_x^4 + h\partial_x^2 + g\partial_x + f \\ &\equiv \partial_x(a\partial_x + a_x) + ha + g\partial_x + f \\ &\equiv a^2 + 2a_x\partial_x + a_{xx} + ha + g\partial_x + f \\ &\equiv (a^2 + a_{xx} + ah + f) + (2a_x + g)\partial_x. \end{aligned}$$

The subsidiary equation '$2f + g_x = 0$' gives

$$2(a^2 + a_{xx} + ah + f) + (2a_{xx} + g_x) = 0.$$

The main equation becomes

$$a_t = -\tfrac{1}{2}(2a_{xxxx} + g_{xxx}) + 2a(2a_{xx} + g_x) + a_x(2a_x + g),$$

and we have $P \equiv -\frac{1}{2}(g_x + 2a_{xx}) + (g + 2a_x)\partial_x$. The homogeneity condition gives $g = \alpha a_x$ (for some constant α), so we obtain

$$a_t = -\tfrac{1}{2}(2+\alpha)a_{xxxx} + 2(2+\alpha)aa_{xx} + (2+\alpha)a_x^2,$$

with $P \equiv -\frac{1}{2}(2+\alpha)a_{xx} + (2+\alpha)a_x\partial_x$.

The case $\alpha = -2$ has special significance. In fact this corresponds to $P = L^2$, so the relation $\partial_t - P$ gives $\partial_t \equiv 0$, and the t-family is constant as in the case $i = 2$ above. For $\alpha \neq -2$, the p.d.e. is non-trivial. It bears some resemblance[1] to the Kuramoto–Sivashinsky equation $u_t + \nabla^4 u + \nabla^2 u + \frac{1}{2}|\nabla u|^2 = 0$. These equations fail the Painlevé test for integrability and are, therefore, 'less integrable' than (for example) the KdV equation. □

[1] The author is grateful to Kenichi Maruno for this observation.

Evidently many more equations arise as the order of P increases. We just mention one example, the case where the stronger condition $[\partial_t - P, L] = 0$ holds. It is known that there is precisely one such P of the form $P_i = \partial_x^i + \cdots$ for each i, and when i is even, this is just $L^{i/2}$. We shall give the formula for P_i in section 8.3 (see the comments after Lemma 8.10). Writing $\partial_i = \partial/\partial t_i$, it follows from the formula that $[\partial_i - P_i, \partial_j - P_j] = 0$ for all i, j. We are therefore in the situation of Corollary 4.21, so we can construct an extended D-module

$$\frac{D_{x,t_3,t_5,\ldots,t_{2k+1}}}{(L, \partial_3 - P_3, \partial_5 - P_5, \ \ldots, \ \partial_{2k+1} - P_{2k+1})}$$

for any k. For each k we have the D-module $D_{x,t_{2k+1}}/(L, \partial_{2k+1} - P_{2k+1})$, and the corresponding p.d.e. is called the k-th equation of the KdV hierarchy.

If the second-order operator L is replaced by an operator of order n, the possibilities multiply further. The special case $[\partial_t - P, L] = 0$ gives rise to a distinguished sequence of operators P_i, and the resulting partial differential equations constitute the n-KdV hierarchy.

We shall not discuss the equally famous KP hierarchy here, as this involves a D-module of infinite rank. However, a similar approach can be given in such cases. In particular, we should mention the Novikov–Veselov hierarchy, where the condition $[\partial_t - P, L] = 0 \bmod L$ appears naturally.

So far it has not been necessary to introduce the spectral parameter λ. To obtain the form of the KdV equation with spectral parameter given in Example 4.15, we can proceed in a similar way, using the D-module $D_x/(L - \lambda)$, but now we must allow P to depend on λ as well. If $P = f + g\partial_x$ then the D-module $D_{x,t}/(L - \lambda, \partial_t - P)$ has rank 2 if and only if $[\partial_t - P, L - \lambda] = 0 \bmod L - \lambda$. Replacing a by $a + \lambda$ in the computation at the beginning of this section, we see that this condition is equivalent to

$$\begin{aligned} a_t &= f_{xx} + a_x g + 2(a + \lambda) g_x \\ 0 &= 2f_x + g_{xx}, \end{aligned}$$

hence

$$a_t = -\tfrac{1}{2} g_{xxx} + g a_x + 2(a + \lambda) g_x.$$

Let us now impose the homogeneity conditions $|f| = 3$, $|g| = |\lambda| = 2$. Then, we must have $g = \alpha a + \beta\lambda$ for some constants α, β, and the p.d.e. becomes

$$\begin{aligned} a_t &= -\tfrac{1}{2}\alpha a_{xxx} + (\alpha a + \beta\lambda) a_x + 2\alpha(a + \lambda) a_x \\ &= -\tfrac{1}{2}\alpha a_{xxx} + \alpha a a_x + 2\alpha a a_x + \lambda(\beta + 2\alpha) a_x. \end{aligned}$$

This holds for all values of λ if and only if $\beta+2\alpha = 0$, and we are left with the KdV equation. The particular version given in Example 4.15 corresponds to $\alpha = -\frac{1}{2}$. With respect to the basis $[1], [\partial_x]$ we obtain the matrix form of the zero curvature condition stated there.

The spectral parameter can be incorporated into the ring of differential operators in this case: we obtain the D-module $D_{x,t} \otimes \mathbb{C}[\lambda]/(L-\lambda, \partial_t - P)$, which has rank 2 over the ring $\mathcal{H}_{x,t} \otimes \mathbb{C}[\lambda]$.

We conclude with a brief comment on the homogeneity assumption. The basic point is that if P is weighted homogeneous then the p.d.e. admits a scaling property. For the case of the KdV equation, this means that if $u(x,t)$ is a solution then so is $\epsilon^2 u(\epsilon x, \epsilon^3 t)$. This gives rise to a special class of solutions such that $u(x,t) = \epsilon^2 u(\epsilon x, \epsilon^3 t)$, the scaling invariant solutions. For such functions the KdV equation reduces to an o.d.e., which can in fact be transformed to the second Painlevé equation.

We shall discuss some solutions of the KdV equation in section 8.5. A more detailed treatment of the homogeneity property will be given in section 9.3.

7.2 The mKdV equation

Any equation of the form

$$q_t = \pm q_{xxx} \pm q^2 q_x$$

for $q = q(x,t)$ is known as the mKdV (modified KdV) equation. If q is a solution of the mKdV equation $q_t = q_{xxx} - q^2 q_x$, it can be verified that $u = q_x - q^2$ is a solution of the KdV equation. This is the Miura transformation.

The mKdV equation arises naturally if one writes

$$\partial_x^2 + u = (\partial_x - q)(\partial_x + q)$$

so we shall begin with a brief remark on such factorizations (more information can be found in appendix 4.6). First of all, starting with u, if one succeeds in finding q, then (for each t) one can solve the o.d.e. $(\partial_x^2 + u)y = 0$ by quadrature, as one solution is given by solving $(\partial_x + q)y = 0$. But q can be found only by solving the equation $\partial_x^2 + u = (\partial_x - q)(\partial_x + q)$, that is $u = q_x - q^2$, and this is just as difficult as solving the original equation. So the factorizaton is a formal matter; it cannot be found 'explicitly' in general.

From the D-module point of view, a factorization $L = L_1 L_2$ gives an exact sequence

$$0 \to D/(L_1) \xrightarrow{\alpha} D/(L_1 L_2) \xrightarrow{\beta} D/(L_2) \to 0$$

where $\alpha(X+DL_1) = XL_2+DL_1L_2$ and $\beta(X+DL_1L_2) = X+DL_2$. Thus, a choice of factorization corresponds to a choice of sub-D-module.

The mKdV equation arises from this situation just as the KdV equation arises from the D-module $D_x/(\partial_x^2+u)$: we ask whether the t-family

$$D_x/(L) \xrightarrow{\beta} D_x/(\partial_x+q) \to 0$$

can be extended to

$$D_{x,t}/(L, \partial_t - P) \xrightarrow{\beta} D_{x,t}/(\partial_x+q, \partial_t - P) \to 0.$$

(Note that it does not make sense to extend the entire exact sequence by adjoining $\partial_t - P$, as the map α is not compatible with this.)

Theorem 7.6 *Given t-families of operators $L = \partial_x^2 + q_x - q^2 = L_1L_2$ and P, we have:*

(1) *The D-modules $D_{x,t}/(L, \partial_t - P)$, $D_{x,t}/(L_2, \partial_t - P)$ have ranks 2, 1 respectively if and only if $[\partial_t - P, L] = 0$ mod L and $[\partial_t - P, L_2] = 0$ mod $\partial_x + q$.*
(2) *Let $P = \frac{1}{2}u_x - u\partial_x$ where $u = q_x - q^2$. Then the condition of (1) holds if and only if q satisfies the mKdV equation $q_t = -\frac{1}{2}q_{xxx} + 3q^2q_x$ and u satisfies the KdV equation $u_t = 3uu_x + \frac{1}{2}u_{xxx}$.*

Proof Part (1) has already been proved in section 4.4. For part (2) we use the discussion after Theorem 7.1. First, if $P = f+g\partial_x$ mod L, then the condition rank $D_{x,t}/(L, \partial_t - P) = 2$ is equivalent to $a_t = -\frac{1}{2}g_{xxx} + 2g_xa + ga_x$ (with $f = -\frac{1}{2}g_x$ and $g = a = q^2 - q_x$) which is the KdV equation. Next, the condition rank $D_{x,t}/(\partial_x + q, \partial_t - P) = 1$ is equivalent to $(f - gq)_x = -q_t$ (note that $P = f - gq$ mod $\partial_x + q$, so this condition comes directly from Example 4.8). Substituting for f and g in $(f - gq)_x = -q_t$ gives the mKdV equation $q_t = -\frac{1}{2}q_{xxx} + 3q^2q_x$. □

As in the case of the KdV equation, a matrix version may be obtained by choosing a basis of the D-module $D_x/(L)$. In view of the above discussion, it is natural to use the basis $[1], [\partial_x + q]$.

We have

$$\partial_x[1] = [\partial_x] = -q[1] + 1[\partial_x + q]$$

$$\partial_x[\partial_x + q] = [q^2 - q_x + q\partial_x + q_x] = 0[1] + q[\partial_x + q]$$

which gives

$$\Omega_1 = \begin{pmatrix} -q & 0 \\ 1 & q \end{pmatrix}.$$

Next, using the definition of the action of ∂_t from section 4.4, we have $\partial_t[\partial_x] = [\partial_x P]$ and $\partial_t[q] = \partial_t q[1] = q\partial_t[1] + q_t[1] = q[P] + [q_t]$, hence

$$\partial_t[1] = [P] = (f - gq)[1] + g[\partial_x + q]$$
$$\partial_t[\partial_x + q] = [\partial_x P + qP + q_t] = (f_x - (gq)_x + q_t)[1] + (f + g_x + gq)[\partial_x + q],$$

which gives

$$\Omega_2 = \begin{pmatrix} f - gq & f_x - (gq)_x + q_t \\ g & f + g_x + gq \end{pmatrix}.$$

Imposing the zero curvature condition at this point simply gives the KdV equation for $a = q^2 - q_x$, of course. The mKdV equation for q arises from an additional condition, and the natural condition (in the matrix version) is that the matrices Ω_1 and Ω_2 are of the form

$$\begin{pmatrix} * & 0 \\ * & * \end{pmatrix}.$$

Thus, the additional condition is just $f_x - (gq)_x + q_t = 0$. This is exactly the condition that the rank of $D_{x,t}/(\partial_x + q, \partial_t - P)$ is 1, as in the scalar version, and the case $g = a$ gives the mKdV equation. The zero curvature form of the mKdV equation turns out to be $d\Omega + \Omega \wedge \Omega = 0$ where

$$\Omega = \begin{pmatrix} -q & 0 \\ 1 & q \end{pmatrix} dx + \begin{pmatrix} \frac{1}{2}q_{xx} - q^3 & 0 \\ q^2 - q_x & -\frac{1}{2}q_{xx} + q^3 \end{pmatrix} dt.$$

The 'additional condition' $f_x - (gq)_x + q_t = 0$ may be obtained in yet another way, without reference to flags of D-modules, by introducing a spectral parameter. It will be convenient to write $\lambda = \mu^2$. The operator $L - \mu^2$ does not inherit a factorization from $L = L_1 L_2$, so we proceed directly with the matrix version.

Let us choose the basis $[1], [\frac{1}{\mu}(\partial_x + q)]$ of $D_x/(L - \mu^2)$ (we regard μ as a fixed complex number). The usual calculation gives

$$\partial_x[1] = -q[1] + \mu[\tfrac{1}{\mu}(\partial_x + q)]$$
$$\partial_x[\tfrac{1}{\mu}(\partial_x + q)] = \mu[1] + q[\tfrac{1}{\mu}(\partial_x + q)]$$

hence

$$\Omega_1 = \begin{pmatrix} -q & \mu \\ \mu & q \end{pmatrix}.$$

Similarly

$$\partial_t[1] = (f - gq)[1] + \mu g[\tfrac{1}{\mu}(\partial_x + q)]$$
$$\partial_t[\tfrac{1}{\mu}(\partial_x + q)] = (\tfrac{1}{\mu}(f_x - (gq)_x + q_t) + \mu g)[1] + (f + g_x + gq)[\tfrac{1}{\mu}(\partial_x + q)]$$

hence

$$\Omega_2 = \begin{pmatrix} f - gq & \frac{1}{\mu}(f_x - (gq)_x + q_t) + \mu g \\ \mu g & f + g_x + gq \end{pmatrix}.$$

Taking the degree of μ to be 1, the homogeneity condition forces $g = \alpha a + \beta\mu^2$.

The zero curvature condition reduces to $f = -\frac{1}{2}g_x = -\frac{1}{2}\alpha a_x$ together with the KdV equation for $a = q^2 - q_x$, providing we take $\beta = -2\alpha$. Let us now impose the condition that Ω_1 and Ω_2 are polynomial in μ, that is, no negative powers of μ occur. This means that, as μ varies, the extended D-modules $D_{x,t}/(L-\mu^2, \partial_t - P)$ constitute a module over the ring $D_{x,t}\otimes\mathbb{C}[\mu]$ which is free over $\mathcal{H}_{x,t}\otimes\mathbb{C}[\mu]$ of rank 2. Only the top right hand entry of Ω_2 needs examination; the negative power of μ disappears if and only if

$$f_x - (gq)_x + q_t = 0,$$

which is the usual 'additional condition'.

The mKdV equation is now seen to be equivalent to $d\Omega + \Omega\wedge\Omega = 0$ where

$$\Omega = \begin{pmatrix} -q & \mu \\ \mu & q \end{pmatrix} dx + \alpha \begin{pmatrix} \frac{1}{2}q_{xx} - q^3 + 2\mu^2 q & \mu(q^2 + q_x) - 2\mu^3 \\ \mu(q^2 - q_x) - 2\mu^3 & -\frac{1}{2}q_{xx} + q^3 - 2\mu^2 q \end{pmatrix} dt.$$

Since we are taking $g = \alpha a$ here, we obtain the mKdV equation in the form $q_t = -\frac{\alpha}{2}q_{xxx} + 3\alpha q^2 q_x$, and $\alpha = 1$ gives the version above. If we take $\alpha = -\frac{1}{2}$, we obtain

$$\Omega_1 = \mu\begin{pmatrix} & 1 \\ 1 & \end{pmatrix} + \begin{pmatrix} -q & \\ & q \end{pmatrix}$$
$$\Omega_2 = \mu^3\begin{pmatrix} & 1 \\ 1 & \end{pmatrix} + \mu^2\begin{pmatrix} a & \\ & -a \end{pmatrix} + \mu\begin{pmatrix} & b \\ c & \end{pmatrix} + \begin{pmatrix} d & \\ & -d \end{pmatrix}$$

where $a = -q$, $b = -\frac{1}{2}(q^2 + q_x)$, $c = -\frac{1}{2}(q^2 - q_x)$, $d = \frac{1}{4}(2q^3 - q_{xx})$. This gives the mKdV equation in the form $q_t = \frac{1}{4}q_{xxx} - \frac{3}{2}q^2 q_x$.

This reveals another very special feature of the mKdV equation, which leads to a Lie-theoretic generalization. First, while Ω takes values a priori in the loop algebra

$$\Lambda\mathfrak{gl}_2\mathbb{C} = \{\gamma : S^1 \to \mathfrak{gl}_2\mathbb{C} \mid \gamma \text{ is smooth }\},$$

it actually lies in the smaller subalgebra

$$(\Lambda\mathfrak{sl}_2\mathbb{C})_\sigma = \{\gamma \in \Lambda\mathfrak{sl}_2\mathbb{C} \mid \sigma(\gamma(\mu)) = \gamma(-\mu)\},$$

where σ is the involution of $\mathfrak{sl}_2\mathbb{C}$ given by conjugation with the diagonal matrix diag$(1, -1)$. This is called the twisted loop algebra of $\mathfrak{sl}_2\mathbb{C}$. The twistedness condition

$$\begin{pmatrix} 1 & \\ & -1 \end{pmatrix} \mu^i \begin{pmatrix} A & B \\ C & D \end{pmatrix} \begin{pmatrix} 1 & \\ & -1 \end{pmatrix}^{-1} = (-\mu)^i \begin{pmatrix} A & B \\ C & D \end{pmatrix}$$

says that the coefficient of μ^i is diagonal if i is even, and off-diagonal if i is odd. (There is another condition, namely the physical assumption that all quantities are considered real, but we shall ignore this for the time being.) Next, *any* flat connection of the form

$$\Omega_1 = \mu \begin{pmatrix} & 1 \\ 1 & \end{pmatrix} + \begin{pmatrix} -q & \\ & q \end{pmatrix}$$

$$\Omega_2 = \mu^3 \begin{pmatrix} & 1 \\ 1 & \end{pmatrix} + \mu^2 \begin{pmatrix} * & \\ & * \end{pmatrix} + \mu \begin{pmatrix} & * \\ * & \end{pmatrix} + \begin{pmatrix} * & \\ & * \end{pmatrix}$$

must be that above, as one can verify easily by writing down the zero curvature condition. In other words, having specified Ω_1, the 't-extension' Ω_2 is determined completely from the zero curvature condition and its leading term

$$\mu^3 \begin{pmatrix} & 1 \\ 1 & \end{pmatrix}.$$

Twisted loop algebras (and twisted loop groups, which are defined in the same way) appear naturally in the theory of affine Kac–Moody Lie algebras: an affine Kac–Moody Lie algebra has to be, essentially, either a loop algebra or a twisted loop algebra. In this sense, the mKdV equation has a purely Lie-theoretic description.

7.3 Harmonic maps into Lie groups

We recall (from Example 4.16) that the equation for harmonic maps $\mathbb{R}^2 \to G$ is

$$\partial(\phi^{-1}\bar{\partial}\phi) + \bar{\partial}(\phi^{-1}\partial\phi) = 0,$$

and that this can be regarded as the zero curvature equation

$$d\alpha^\lambda + \alpha^\lambda \wedge \alpha^\lambda = 0$$

where

$$\begin{aligned}\alpha^\lambda &= \tfrac{1}{2}(1 - \tfrac{1}{\lambda})\alpha_1 dz + \tfrac{1}{2}(1-\lambda)\alpha_2 d\bar{z}\\ &= \tfrac{1}{2}(1 - \tfrac{1}{\lambda})\phi^{-1}\partial\phi\, dz + \tfrac{1}{2}(1-\lambda)\phi^{-1}\bar{\partial}\phi\, d\bar{z}\end{aligned}$$

and λ is a parameter in $\mathbb{C}^*$. We have $\alpha_2 = c(\alpha_1)$, where $C : G^{\mathbb{C}} \to G^{\mathbb{C}}$ is conjugation with respect to a (compact or non-compact) real form G of $G^{\mathbb{C}}$, and $c : \mathfrak{g}^{\mathbb{C}} \to \mathfrak{g}^{\mathbb{C}}$ is its derivative. From the fact that $c(\alpha^\lambda) = \alpha^{1/\bar{\lambda}}$, it follows that there exists a fundamental solution matrix H which satisfies the reality condition

$$C(H(\lambda)) = H(1/\bar{\lambda}).$$

We have

$$\phi = H^t|_{\lambda=-1},$$

with $C(\phi) = \phi$.

Similar remarks apply to the case of harmonic maps $\mathbb{R}^{1,1} \to G$ (see Example 4.16).

This formulation of the harmonic map equation was introduced in integrable systems theory (see [86], [101], [133], and many papers on the principal chiral model in physics), but was first studied systematically in differential geometry by K. Uhlenbeck in [126]. In the context of harmonic maps, following [126], the fundamental solution H (or H^t) is called an *extended solution* or *extended harmonic map*.

If we restrict λ to the unit circle $S^1 = \{\lambda \in \mathbb{C} \mid |\lambda| = 1\}$, H can be regarded as a map into the loop group $\Lambda \mathrm{GL}_n\mathbb{C}$. The condition $C(H(\lambda)) = H(1/\bar{\lambda})$ means that (the restriction to S^1 of) H actually takes values in the real loop group $\Lambda \mathrm{U}_n$.

The following discussion applies also to the more general equation

$$\partial_1(\phi^{-1}\partial_2\phi) + \partial_2(\phi^{-1}\partial_1\phi) = 0,$$

which is the zero curvature equation for

$$\alpha^\lambda = \tfrac{1}{2}(1 - \tfrac{1}{\lambda})\alpha_1 dz_1 + \tfrac{1}{2}(1-\lambda)\alpha_2 dz_2$$

(where α_2 is not necessarily $c(\alpha_1)$). One might call this the complexified harmonic map equation. In fact it will be easier to explain the method in this case, then impose the reality conditions at the end, since these conditions plays no role in the argument. For explicitness we shall assume that $G = U_n$, $G^{\mathbb{C}} = GL_n\mathbb{C}$ (thus $c(\alpha) = -\alpha^*$), although the argument applies for any G.

The basic idea was introduced in various forms by I. M. Krichever in [86], by J. Dorfmeister, F. Pedit, and H. Wu in [32], by G. B. Segal in [111], and in Chapter 8 we shall give yet another formulation, based on [16] and [60]. It uses the Birkhoff factorization (see Chapter 6): almost every element γ of the loop group $\Lambda GL_n\mathbb{C}$ may be factorized as $\gamma = \gamma_-\gamma_+$ where $\gamma_- \in \Lambda_- GL_n\mathbb{C}$, $\gamma_+ \in \Lambda_+ GL_n\mathbb{C}$. The factorization is unique if we insist that the constant term in the Fourier expansion of γ_- is I, in other words $\gamma_-(\infty) = I$. Evidently there is an analogous factorization in the opposite order, and we shall denote this by $\gamma = \check{\gamma}_+\check{\gamma}_-$, with $\check{\gamma}_+(0) = I$. However, we regard the first factorization as the primary one; for example, if we write $(\gamma_1\gamma_2)_-$ (as in the proposition below) we mean the first factor of the right hand side of $\gamma_1\gamma_2 = (\gamma_1\gamma_2)_-(\gamma_1\gamma_2)_+$.

The main result is the following proposition and its corollary. The proof uses only the fact that $\frac{1}{2}(1 - \frac{1}{\lambda})\alpha_1$ is holomorphic for $\lambda \in \mathbb{C}^* \cup \{\infty\}$ and $\frac{1}{2}(1 - \lambda)\alpha_2$ is holomorphic for $\lambda \in \mathbb{C}$ (as stated in [86], it uses only the fact that $\frac{1}{2}(1 - \frac{1}{\lambda})\alpha_1$ and $\frac{1}{2}(1 - \lambda)\alpha_2$ have no common poles). For simplicity we state the result for $GL_n\mathbb{C}$, but it holds for any group (real or complex) for which the Birkhoff factorization is available.

Proposition 7.7 *Let* $\alpha_1, \alpha_2 : \mathbb{C}^2 \to M_n\mathbb{C}$ *and*

$$\alpha^\lambda = \tfrac{1}{2}(1 - \tfrac{1}{\lambda})\alpha_1 dz_1 + \tfrac{1}{2}(1 - \lambda)\alpha_2 dz_2.$$

Assume that F *is a map such that* $F^{-1}dF = \alpha^\lambda$. *Let*

$$F = F_-F_+ = \check{F}_+\check{F}_-$$

be the Birkhoff factorizations[2] *of* F. *Then,*

(1) F_- *depends only on* z_1,
(2) $\check{F}_+$ *depends only on* z_2,

and F *can be written in terms of* F_- *and* $\check{F}_+$ *as* $F = \check{F}_+(\check{F}_+^{-1}F_-)_-$.

Proof From $F_-F_+ = \check{F}_+\check{F}_-$ we obtain $\check{F}_- = \check{F}_+^{-1}F_-F_+ = (\check{F}_+^{-1}F_-F_+)_- = (\check{F}_+^{-1}F_-)_-$, hence $F = \check{F}_+\check{F}_- = \check{F}_+(\check{F}_+^{-1}F_-)_-$. The main task is to prove that

[2] We may assume (without loss of generality) that F is the identity loop $I \in \Lambda GL_n\mathbb{C}$ when $(z_1, z_2) = (a_1, a_2)$. Since I belongs to the big cell of $\Lambda GL_n\mathbb{C}$, the Birkhoff factorizations exist in a neighbourhood of the point (a_1, a_2).

$\partial_1 \check{F}_+ = \partial_2 F_- = 0$. We have

$$\begin{aligned}\check{F}_+^{-1}\partial_1\check{F}_+ &= (F\check{F}_-^{-1})^{-1}\partial_1(F\check{F}_-^{-1})\\ &= \check{F}_- F^{-1}\partial_1 F\check{F}_-^{-1} + \check{F}_-\partial_1\check{F}_-^{-1}\\ &= \tfrac{1}{2}(1-\tfrac{1}{\lambda})\check{F}_-\alpha_1\check{F}_-^{-1} + \check{F}_-\partial_1\check{F}_-^{-1}.\end{aligned}$$

Since $\check{F}_+ = I + \lambda * + \cdots$ the Fourier expansions of the left and right hand sides of the equation are of the form

$$\lambda * + \lambda^2 * + \cdots \quad = \quad * + \lambda^{-1} * + \lambda^{-2} * + \cdots$$

hence both sides must be zero. In particular $\partial_1 \check{F}_+ = 0$.

Similarly, from

$$\begin{aligned}F_-^{-1}\partial_2 F_- &= (FF_+^{-1})^{-1}\partial_2(FF_+^{-1})\\ &= F_+F^{-1}\partial_2 FF_+^{-1} + F_+\partial_2 F_+^{-1}\\ &= \tfrac{1}{2}(1-\lambda)F_+\alpha_2 F_+^{-1} + F_+\partial_2 F_+^{-1}\end{aligned}$$

we conclude that $\partial_2 F_- = 0$. □

Corollary 7.8 *Given F as above, we have*

$$F_-^{-1}\frac{dF_-}{dz_1} = \tfrac{1}{\lambda}\omega_1(z_1), \qquad \check{F}_+^{-1}\frac{d\check{F}_+}{dz_2} = \lambda\omega_2(z_2)$$

for some $\mathfrak{g}^{\mathbb{C}}$-valued functions $\omega_1 = \omega_1(z_1), \omega_2 = \omega_2(z_2)$. Conversely, given any functions $\omega_1 = \omega_1(z_1)$, $\omega_2 = \omega_2(z_2)$, we may obtain $F_-, \check{F}_+$ (and hence an extended solution F of the complexified harmonic map equation) by solving each o.d.e. separately and putting $F = \check{F}_+(\check{F}_+^{-1}F_-)_-$.

Proof In the case of F_-, we have $F_-^{-1}\partial_1 F_- = \frac{1}{2}(1-\frac{1}{\lambda})F_+\alpha_1 F_+^{-1} + F_+\partial_1 F_+^{-1}$, and since the Fourier expansion of the right hand side is of the form

$$\lambda^{-1} * + * + \lambda * + \lambda^2 * + \ldots$$

while that of the left hand side has no positive powers of λ, we must have

$$F_-^{-1}\partial_1 F_- = -\tfrac{1}{2}\tfrac{1}{\lambda}A\alpha_1 A^{-1}$$

where $F_+ = A + O(\lambda)$. This is of the required form. Similar consideration of $\check{F}_+^{-1}\partial_2\check{F}_+$ gives

$$\check{F}_+^{-1}\partial_2\check{F}_+ = -\tfrac{1}{2}\lambda B\alpha_2 B^{-1},$$

where $\check{F}_- = B + O(\frac{1}{\lambda})$. □

Thus, $F = F(z_1, z_2, \lambda)$ may be reconstructed from the 'data' $\omega_1 = \omega_1(z_1)$ and $\omega_2 = \omega_2(z_2)$. There is no restriction on ω_1, ω_2 here. The original p.d.e. has been transformed into two systems of $n \times n$ matrix o.d.e.

Krichever regards this result as a generalization of the D'Alembert formula for the solution to the scalar wave equation $\phi_{z_1 z_2} = 0$ (to which it reduces when $n = 1$ and ϕ, z_1, z_2 are real), and he uses it to solve the sine-Gordon equation $\phi_{xy} = \sin\phi$. We present this example here as a simple application of Proposition 7.7, before discussing the case of harmonic[3] maps.

Example 7.9 We begin by considering a connection form

$$\alpha^\lambda = \left(\begin{array}{cc|c} 0 & c_1 & \frac{1}{\lambda}a_1 \\ -c_1 & 0 & \frac{1}{\lambda}b_1 \\ \hline -\frac{1}{\lambda}a_1 & -\frac{1}{\lambda}b_1 & 0 \end{array}\right) dx + \left(\begin{array}{cc|c} 0 & c_2 & \lambda a_2 \\ -c_2 & 0 & \lambda b_2 \\ \hline -\lambda a_2 & -\lambda b_2 & 0 \end{array}\right) dy$$

where a_i, b_i, c_i are real functions of x and y. Thus we take real variables $z_1 = x, z_2 = y$ in Proposition 7.7, and impose the condition that $\alpha = \alpha^\lambda|_{\lambda=1}$ is the complexification of a 1-form with values in $\mathfrak{so}_3$. The λ-dependence differs slightly from that in Proposition 7.7, but exactly the same proof applies.

The condition $d\alpha^\lambda + \alpha^\lambda \wedge \alpha^\lambda = 0$ breaks up into the following equations:

$$\begin{array}{ll} (1) & -(c_1)_y + (c_2)_x - a_1 b_2 + a_2 b_1 = 0 \\ (2) & \quad (a_2)_x + b_2 c_1 = 0 \\ (3) & -(a_1)_y - b_1 c_2 = 0 \\ (4) & \quad (b_2)_x - a_2 c_1 = 0 \\ (5) & -(b_1)_y + a_1 c_2 = 0. \end{array}$$

Equations (2) and (4) imply that $(a_2^2 + b_2^2)_x = 0$, while equations (3) and (5) imply $(a_1^2 + b_1^2)_y = 0$. Subject to these conditions, equations (2)–(5) also determine c_1, c_2 in terms of a_1, a_2, b_1, b_2.

Let us impose the stronger conditions $a_1^2 + b_1^2 = a_2^2 + b_2^2 = 1$. A gauge transformation by an SO_2-valued map permits us to assume that $a_1 = 0$,

[3] In fact this example is closely related to a version of the harmonic map equation, as we shall see in the next section (Example 7.14). It also has a well known geometric interpretation, namely that solutions of the sine-Gordon equation correspond to pseudo-spherical surfaces in $\mathbb{R}^3$, that is, surfaces of constant negative Gauss curvature. As such, it is one of the original 'soliton equations' discovered (ahead of its time) in classical differential geometry.

$b_1 = 1$. We may also write $a_2 = -\sin\phi$, $b_2 = -\cos\phi$, for some real-valued function $\phi = \phi(x, y)$. Then equations (2)–(5) imply that $c_1 = -\phi_x$ and $c_2 = 0$, and equation (1) becomes

$$\phi_{xy} = \sin\phi$$

which is the sine-Gordon equation. Thus, the sine-Gordon equation is equivalent to $d\alpha^\lambda + \alpha^\lambda \wedge \alpha^\lambda = 0$ where

$$\alpha^\lambda = \left(\begin{array}{cc|c} 0 & -\phi_x & 0 \\ \phi_x & 0 & \frac{1}{\lambda} \\ \hline 0 & -\frac{1}{\lambda} & 0 \end{array}\right) dx + \left(\begin{array}{cc|c} 0 & 0 & -\lambda\sin\phi \\ 0 & 0 & -\lambda\cos\phi \\ \hline \lambda\sin\phi & \lambda\cos\phi & 0 \end{array}\right) dy.$$

The factorization method leads to the explicit formulae

$$\omega_i = \left(\begin{array}{cc|c} 0 & 0 & s_i \\ 0 & 0 & c_i \\ \hline -s_i & -c_i & 0 \end{array}\right)$$

where

$$\begin{aligned} s_1 &= -\sin(\phi(0,0) - \phi(x,0)) \\ c_1 &= -\cos(\phi(0,0) - \phi(x,0)) \end{aligned}$$

and

$$\begin{aligned} s_2 &= \sin\phi(0,y) \\ c_2 &= \cos\phi(0,y), \end{aligned}$$

which amount to giving a solution to the sine-Gordon equation from the data $\phi(x, 0), \phi(0, y)$. For the details of the calculation of ω_1 and ω_2, see [28] and [124]. □

Let us return to the case of harmonic maps $\mathbb{C} \to G$, where G is a real Lie group. Here we impose the reality condition $c(\alpha_1) = \alpha_2$ and restrict to $z_1 = z, z_2 = \bar{z}$.

Corollary 7.10 *If* $c(\alpha_1) = \alpha_2$, *then* $C(F_-(\lambda)) = \check{F}_+(1/\bar{\lambda})$ *and* $C(F_+(\lambda)) = \check{F}_-(1/\bar{\lambda})$.

Proof As remarked earlier, the hypothesis $c(\alpha_1) = \alpha_2$ means $C(H(\lambda)) = H(1/\bar{\lambda})$. On applying C to $F = F_-F_+ = \check{F}_+\check{F}_-$, the conclusion follows from the uniqueness of the Birkhoff factorization. □

This leads to a very simple parametrization of (local) solutions to the harmonic map equation, as follows. Let us use the standard notation

$$\alpha_1 = \alpha', \quad \alpha_2 = \alpha''$$

and write

$$\omega_1 = \omega.$$

Then an extended solution is any map F which satisfies

$$F^{-1}dF = \alpha^\lambda = \tfrac{1}{2}(1 - \tfrac{1}{\lambda})\alpha' dz + \tfrac{1}{2}(1 - \lambda)\alpha'' d\bar{z}$$

for some $\alpha', \alpha'' : \mathbb{R}^2 \to \mathfrak{g}^{\mathbb{C}}$ with $c(\alpha') = \alpha''$. The correspondence

$$F \longleftrightarrow (F_-, \check{F}_+) \longleftrightarrow (\omega_1, \omega_2)$$

reduces to a correspondence

$$F \longleftrightarrow F_- \longleftrightarrow \omega = \omega_1$$

because the reality condition relates F_- and $\check{F}_+$. In other words, there is a correspondence between extended harmonic maps and holomorphic $\mathfrak{g}^{\mathbb{C}}$-valued functions ω. We summarize this discussion in the following theorem, which we take as the principal formulation of the equations for a harmonic map into a (compact or non-compact) Lie group G:

Theorem 7.11 *There is an essentially one to one correspondence*

$$\phi \longleftrightarrow H^t = F \longleftrightarrow F_- \longleftrightarrow \omega$$

between (1) G-valued harmonic maps $\phi(z, \bar{z})$ and (2) holomorphic $\mathfrak{g}^{\mathbb{C}}$-valued functions $\omega(z)$, where $F(z, \bar{z})$ is the extended solution corresponding to $\phi(z, \bar{z})$, $F_-(z)$ is the first Birkhoff factor of $F(z, \bar{z})$, and $\frac{1}{\lambda}\omega = F_-^{-1} dF_-/dz$. The map F takes values in ΛG, while F_- takes values in $\Lambda_- G^{\mathbb{C}}$.

'Essentially one to one' refers to the ambiguity in solving the ordinary differential equation $\frac{1}{\lambda}\omega = F_-^{-1} dF_-/dz$ (an element of $\Lambda G^{\mathbb{C}}$) and in performing the Birkhoff factorization $F = F_- F_+$ (an element of $G^{\mathbb{C}}$).

Dorfmeister, Pedit, and Wu regard this result as a generalization of the Weierstrass formula for the solution of the minimal surface equation (to which it reduces when $n = 2$ and ω is nilpotent). Evidently some care is needed with the domain of definition of the various maps appearing here, as the correspondence involves solving ordinary differential equations and performing Birkhoff factorizations. However, as a local result, it gives a very satisfying correspondence between harmonic maps and 'holomorphic data'.

We conclude with some comments on the practical effectiveness of the correspondence. The passage from ϕ to ω is not entirely explicit as there is in general no simple algorithm which produces F from ϕ, nor F_- from F. In the reverse direction it is necessary to start by solving an ordinary differential equation to produce F_-, then perform a factorization of the form

$$F_- = F(F_+)^{-1}$$

to produce F, neither of which can be done explicitly in general. However, the reverse procedure can be carried out numerically, leading to computer-generated images (of surfaces corresponding to harmonic maps), as in the software CMCLab by N. Schmitt.

The factorization $F_- = F(F_+)^{-1}$ arises (locally) from the Lie algebra decomposition

$$\Lambda\mathfrak{g}^{\mathbb{C}} = \Lambda\mathfrak{g} + \Lambda_+\mathfrak{g}^{\mathbb{C}}$$

and is called the Iwasawa factorization (whereas the Birkhoff factorization arises from $\Lambda\mathfrak{g}^{\mathbb{C}} = \Lambda_-\mathfrak{g}^{\mathbb{C}} + \Lambda_+\mathfrak{g}^{\mathbb{C}}$). When G is compact, the Iwasawa factorization is an infinite-dimensional version of the Gram–Schmidt orthogonalization procedure, and it may be carried out for any loop (i.e. the complement of the Iwasawa big cell in the loop group is empty). When G is non-compact, there are generally one or more (connected) big cells for the Iwasawa factorization. The Iwasawa factorization will reappear in section 8.1 and Chapter 10.

7.4 Harmonic maps into symmetric spaces

Harmonic maps into symmetric spaces appear often in applications. The Cartan immersion allows us to regard a symmetric space as a totally geodesic submanifold of a Lie group; on the other hand, any Lie group is itself an example of a symmetric space. Thus, the theory of harmonic maps into symmetric spaces is equivalent to the theory of harmonic maps into Lie groups. However, for symmetric spaces, the formalism presented in this section (from [18] and [32]) is more convenient than the Lie group formalism because it incorporates explicitly the involution[4]

$$\sigma : G \to G$$

which defines the particular symmetric space.

[4] σ is an automorphism such that $\sigma \circ \sigma$ is the identity map.

Let G/K be a symmetric space with respect to σ, that is,

$$(G_\sigma)_o \subseteq K \subseteq G_\sigma$$

where $G_\sigma = \{g \in G \mid \sigma(g) = g\}$ and $(G_\sigma)_o$ is the identity component of G_σ. The Cartan immersion is the map $i : G/K \to G, \quad [g] \mapsto \sigma(g)g^{-1}$.

A map $f : \mathbb{R}^2 \to G/K$ can be represented (non-uniquely) by a 'framing' $\psi : \mathbb{R}^2 \to G$, that is, $f = [\psi]$. As in section 7.3 we write

$$\psi^{-1}d\psi = \alpha = \alpha' dz + \alpha'' d\bar{z}$$

but we may decompose this further into the eigenspaces of the Lie algebra involution $\sigma : \mathfrak{g} \to \mathfrak{g}$ (or rather its $\mathbb{C}$-linear extension $\sigma : \mathfrak{g}^{\mathbb{C}} \to \mathfrak{g}^{\mathbb{C}}$) as follows:

$$\alpha = \alpha_{\mathfrak{k}} \oplus \alpha_{\mathfrak{m}} = (\alpha'_{\mathfrak{k}} + \alpha'_{\mathfrak{m}})dz + (\alpha''_{\mathfrak{k}} + \alpha''_{\mathfrak{m}})d\bar{z}$$

where $\mathfrak{k} = \{X \in \mathfrak{g} \mid \sigma(X) = X\}$ and $\mathfrak{m} = \{X \in \mathfrak{g} \mid \sigma(X) = -X\}$.

The map f is harmonic if and only if the composition $i \circ f$ is harmonic, and an easy calculation shows this condition is

$$(\alpha'_{\mathfrak{m}})_{\bar{z}} + (\alpha''_{\mathfrak{m}})_z - [\alpha'_{\mathfrak{m}}, \alpha''_{\mathfrak{k}}] - [\alpha''_{\mathfrak{m}}, \alpha'_{\mathfrak{k}}] = 0.$$

(Whether ψ itself is a harmonic map to G is irrelevant here.)

As in the previous section, it is possible to write this harmonic map equation as a zero curvature condition if we introduce

$$\alpha^\lambda = (\alpha'_{\mathfrak{k}} + \frac{1}{\lambda}\alpha'_{\mathfrak{m}})dz + (\alpha''_{\mathfrak{k}} + \lambda\alpha''_{\mathfrak{m}})d\bar{z}.$$

The coefficients of $\lambda^{-1}, \lambda^0, \lambda^1$ in the zero curvature equation $d\alpha^\lambda + \alpha^\lambda \wedge \alpha^\lambda = 0$ are

$$\begin{array}{ll} (1) & -(\alpha'_{\mathfrak{m}})_{\bar{z}} + [\alpha'_{\mathfrak{m}}, \alpha''_{\mathfrak{k}}] \\ (2) & (\alpha''_{\mathfrak{k}})_z - (\alpha'_{\mathfrak{k}})_{\bar{z}} + [\alpha'_{\mathfrak{m}}, \alpha''_{\mathfrak{m}}] + [\alpha'_{\mathfrak{k}}, \alpha''_{\mathfrak{k}}] \\ (3) & (\alpha''_{\mathfrak{m}})_z - [\alpha''_{\mathfrak{m}}, \alpha'_{\mathfrak{k}}] \end{array}$$

so the harmonic map equation is (1) – (3) = 0. Starting from a map ψ, the equation (1) + (2) + (3) = 0 is simply the identity $d\alpha + \alpha \wedge \alpha = 0$, which is equivalent to (1) + (3) = 0 = (2) (since (1) + (3) $\in \mathfrak{k}$ and (2) $\in \mathfrak{m}$). Thus, the zero curvature condition (1) = (2) = (3) = 0 is equivalent to (1) – (3) = 0, the harmonic map equation. Conversely, starting from any $\mathfrak{g}^{\mathbb{C}}$-valued 1-form α on $\mathbb{R}^2$ such that $c(\alpha') = \alpha''$, the zero curvature condition (1) = (2) = (3) = 0 implies that (1) + (2) + (3) = 0, hence $\alpha = \psi^{-1}d\psi$ for some $\psi : \mathbb{R}^2 \to G$, and (1) – (3) = 0 says that $f = [\psi]$ must be harmonic.

As in the previous section, the zero curvature equation $d\alpha^\lambda + \alpha^\lambda \wedge \alpha^\lambda = 0$ says that there exists a map $F : \mathbb{R}^2 \to \Lambda G$ such that $\alpha^\lambda = F^{-1}dF$. This map is called an extended framing, and it is related to the original harmonic map by the simple formula

$$f = [F|_{\lambda=1}].$$

The discussion so far is essentially equivalent to section 7.3: we have used the harmonic map equation for G of section 7.3 to obtain the harmonic map equation for G/K; conversely, if the Lie group G is realized as the symmetric space $(G \times G)/G$, we recover the formulae of section 7.3.

The new feature of this section is that the 1-form α^λ can be regarded as a 1-form with values in the *twisted* loop algebra

$$(\Lambda\mathfrak{g}^{\mathbb{C}})_\sigma = \{\gamma \in \Lambda\mathfrak{g}^{\mathbb{C}} \mid \sigma(\gamma(\lambda)) = \gamma(-\lambda)\},$$

in fact in the subalgebra $(\Lambda\mathfrak{g})_\sigma$ consisting of loops which satisfy the reality condition $c(\gamma(\lambda)) = \gamma(1/\bar{\lambda})$. Equivalently, F takes values in the twisted loop group

$$(\Lambda G)_\sigma = \{\gamma \in \Lambda G \mid \sigma(\gamma(\lambda)) = \gamma(-\lambda)\}.$$

We have already encountered twisted loop groups in connection with quantum cohomology in section 5.3 and the mKdV equation in section 7.2.

The Birkhoff and Iwasawa factorizations generalize to the case of twisted maps (see [8]), as do Proposition 7.7 and its corollaries. We obtain the following analogue of Theorem 7.11.

Theorem 7.12 *There is an essentially one to one correspondence*

$$f \longleftrightarrow F \longleftrightarrow F_- \longleftrightarrow \omega$$

between (1) G/K-valued harmonic maps $f(z,\bar{z})$ and (2) holomorphic $\mathfrak{m}^{\mathbb{C}}$-valued functions $\omega(z)$, where $F(z,\bar{z})$ is the extended framing corresponding to $f(z,\bar{z})$, $F_-(z)$ is the first Birkhoff factor of $F(z,\bar{z})$, and $\frac{1}{\lambda}\omega = F_-^{-1}dF_-/dz$.

Here ω is $\mathfrak{m}^{\mathbb{C}}$-valued rather than $\mathfrak{g}^{\mathbb{C}}$-valued because the twisting condition

$$\sigma(\tfrac{1}{\lambda}\omega) = (\tfrac{1}{\lambda}\omega)|_{\lambda=-1}$$

is equivalent to $\sigma(\omega) = -\omega$.

Example 7.13 We shall prove the well known fact that harmonic maps into the symmetric space SO_3/SO_2 are precisely the Gauss maps of CMC surfaces, that is, surfaces in $\mathbb{R}^3$ of constant mean curvature. It is known (the Sym–Bobenko formula, see [30]) that a surface with *non-zero* constant mean curvature is determined by its Gauss map, hence Theorem 7.12 gives a 'generalized Weierstrass representation' for such surfaces. In other words, Theorem 7.12 solves the Gauss–Codazzi equations. This justifies the assertion made in appendix 4.5 that the 'λ-enhanced frame' is the appropriate geometrical object for the Cartan/Klein approach to solving the Gauss–Codazzi equations for CMC surfaces.

Recall (part 3 of appendix 4.5) that we have the frame

$$\psi = \begin{pmatrix} | & | & | \\ e^{-\frac{u}{2}}p_x & e^{-\frac{u}{2}}p_y & \nu \\ | & | & | \end{pmatrix}$$

of a surface p, and $\psi^{-1}d\psi = \alpha' dz + \alpha'' d\bar{z}$, $\alpha'' = \overline{\alpha'}$, with

$$\alpha' = \left(\begin{array}{cc|c} 0 & \frac{i}{2}u_z & -(Q+\frac{1}{2}e^uH)e^{-\frac{u}{2}} \\ -\frac{i}{2}u_z & 0 & -i(Q-\frac{1}{2}e^uH)e^{-\frac{u}{2}} \\ \hline (Q+\frac{1}{2}e^uH)e^{-\frac{u}{2}} & i(Q-\frac{1}{2}e^uH)e^{-\frac{u}{2}} & 0 \end{array}\right).$$

The Gauss map is

$$\nu = [\psi] : \mathbb{C} \to SO_3/SO_2,$$

where the involution defining the symmetric space SO_3/SO_2 is given by conjugation by the diagonal matrix $\mathrm{diag}(-1,-1,1)$. The horizontal and vertical lines in the above matrix α' indicate the decomposition $\mathfrak{g}^{\mathbb{C}} = \mathfrak{k}^{\mathbb{C}} \oplus \mathfrak{m}^{\mathbb{C}}$. Guided by this decomposition, we introduce the 1-form $\alpha^\lambda = (\alpha'_{\mathfrak{k}} + \frac{1}{\lambda}\alpha'_{\mathfrak{m}})dz + (\alpha''_{\mathfrak{k}} + \lambda\alpha''_{\mathfrak{m}})d\bar{z}$, with $\alpha'_{\mathfrak{k}} + \frac{1}{\lambda}\alpha'_{\mathfrak{m}} =$

$$\left(\begin{array}{cc|c} 0 & \frac{i}{2}u_z & -\frac{1}{\lambda}(Q+\frac{1}{2}e^uH)e^{-u/2} \\ -\frac{i}{2}u_z & 0 & -\frac{1}{\lambda}i(Q-\frac{1}{2}e^uH)e^{-u/2} \\ \hline \frac{1}{\lambda}(Q+\frac{1}{2}e^uH)e^{-u/2} & \frac{1}{\lambda}i(Q-\frac{1}{2}e^uH)e^{-u/2} & 0 \end{array}\right).$$

The zero curvature condition for α^λ is equivalent to the existence of a map ψ^λ such that $\alpha^\lambda = (\psi^\lambda)^{-1}d\psi^\lambda$, and we have $\psi^\lambda|_{\lambda=1} = \psi$. This is the λ-enhanced frame mentioned above. We are using ψ, ψ^λ here instead of the F, F^λ in appendix 4.5; we shall write F for ψ^λ from now on, in accordance with the notation of the present chapter.

To apply the general theory, we must identify the components (1), (2), (3) of the curvature $d\alpha^\lambda + \alpha^\lambda \wedge \alpha^\lambda$. It may be verified that (2) $= 0$ is the Gauss equation, and (1)+(3) $= 0$ is the Codazzi equation. These are, respectively, the $\mathfrak{k}^{\mathbb{C}}$ and $\mathfrak{m}^{\mathbb{C}}$ components of the usual equation $(1) + (2) + (3) = d\alpha + \alpha \wedge \alpha = 0$. The harmonic map equation $(1) - (3) = 0$ turns out to be

$$Q_{\bar{z}} + \tfrac{1}{2} e^u H_z = 0.$$

Comparison with the Codazzi equation shows that ν is harmonic if and only if $H_z = 0$, that is, (since H is real) if and only if H is constant. This establishes the fact that the Gauss map of a surface is harmonic if and only if the mean curvature of the surface is constant, as asserted.

We conclude this example with some remarks on the conclusions to be drawn from Theorem 7.12. First of all it follows that the Gauss map of a CMC surface is specified (locally, and up to Euclidean isometries) by giving a holomorphic 1-form

$$\frac{1}{\lambda}\left(\begin{array}{cc|c} & & p \\ & & q \\ \hline -p & -q & \end{array}\right),$$

that is, a pair of holomorphic functions p, q, together with the value of the mean curvature.

Finally, it should be noted that a harmonic map into SO_3/SO_2 is associated to any 1-form

$$\left(\begin{array}{cc|c} 0 & c & \frac{1}{\lambda}a \\ -c & 0 & \frac{1}{\lambda}b \\ \hline -\frac{1}{\lambda}a & -\frac{1}{\lambda}b & 0 \end{array}\right) dz + \left(\begin{array}{cc|c} 0 & \bar{c} & \lambda\bar{a} \\ -\bar{c} & 0 & \lambda\bar{b} \\ \hline -\lambda\bar{a} & -\lambda\bar{b} & 0 \end{array}\right) d\bar{z}$$

where a, b, c are complex functions. The condition $d\alpha^\lambda + \alpha^\lambda \wedge \alpha^\lambda = 0$ implies that $c = -\bar{a}_{\bar{z}}/\bar{b}$ (as well as the harmonic map equation for $[F]$), that is, c is determined by a, b. Comparison with the special form of α^λ which arises from a surface p shows that the functions a, b are equivalent to the functions Q, u (assuming that H is a given constant). The apparently special feature that H is real, or equivalently that $ia - b$ is real, can be achieved by making an $SO_2\mathbb{C}$-valued gauge transformation.

Thus, there is in fact nothing special about the case of a surface, and we conclude that the theory of (Gauss maps of) CMC surfaces is equivalent to the theory of harmonic maps into SO_3/SO_2. (Degenerate surfaces are permitted here, i.e. we do not assume that the map p is an immersion into $\mathbb{R}^3$.) □

The SU_2-version of the Gauss–Codazzi equations (see appendix 4.5) gives a correspondence between Gauss maps of CMC surfaces and holomorphic

1-forms

$$\frac{1}{\lambda}\begin{pmatrix} 0 & u \\ v & 0 \end{pmatrix}.$$

This (or the SO_3 version above) is the appropriate generalization of the Weierstrass representation for minimal surfaces mentioned at the end of appendix 4.5.

It is of interest to note how the minimal surface case (which we discussed at length in appendix 4.5) fits into the current loop group-theoretic discussion. In fact it is exactly the case where $u = 0$. We have

$$F_- = \psi^\lambda = \begin{pmatrix} 1 & \\ & \lambda \end{pmatrix}^{-1} \begin{pmatrix} 1 & 0 \\ \delta & 1 \end{pmatrix} \begin{pmatrix} 1 & \\ & \lambda \end{pmatrix}, \quad \psi = \begin{pmatrix} 1 & 0 \\ \delta & 1 \end{pmatrix},$$

which shows essentially how the loop parameter λ factors out. Of course, this is the 'non-enhanced' situation where we ignore $e^\theta = q/|q|$ and consider only δ. Similarly, the totally umbilic case is given by $v = 0$. The minimal surface case has been studied from the loop group point of view in [31]. The relation with our appendix 4.5 is that the functions r, s of [31] are given by $r = -\delta q = g\sqrt{f}, s = q = \sqrt{f}$.

Example 7.14 The sine-Gordon equation, which we discussed in a somewhat ad hoc fashion in Example 7.9, is also a case of the harmonic map equation. Only the 'reality conditions' are different. In terms of the general theory of this section, we recognise the zero curvature condition for the connection form

$$\alpha^\lambda = \left(\begin{array}{cc|c} 0 & c_1 & \frac{1}{\lambda}a_1 \\ -c_1 & 0 & \frac{1}{\lambda}b_1 \\ \hline -\frac{1}{\lambda}a_1 & -\frac{1}{\lambda}b_1 & 0 \end{array}\right) dx + \left(\begin{array}{cc|c} 0 & c_2 & \lambda a_2 \\ -c_2 & 0 & \lambda b_2 \\ \hline -\lambda a_2 & -\lambda b_2 & 0 \end{array}\right) dy$$

as the equation for harmonic maps $\mathbb{R}^{1,1} \to SO_3/SO_2$ (cf. Example 4.16). If we impose the conditions

$$a_1^2 + b_1^2 = a_2^2 + b_2^2 = 1$$

then we obtain the sine-Gordon equation as in Example 7.9. □

7.5 Pluriharmonic maps (and quantum cohomology)

The harmonic map equation is a good example of a 'zero curvature equation', and it is important in differential geometry, but the main reason for its appearance in this book is its relation with quantum cohomology. That is, if $\dim H^2M = 1$, then the map L_- of Chapter 5 satisfies

$$L_-^{-1}\frac{\partial L_-}{\partial t} = \tfrac{1}{\mathrm{h}}\,\hat{\omega}.$$

It plays the role of F_- in Theorems 7.11 and 7.12 (and h plays that of λ). Thus, the quantum cohomology of a manifold M (with $\dim H^2M = 1$) gives rise to a harmonic map.

The holomorphic data for this harmonic map consists of the structure constants of the quantum cohomology algebra. This means that the two theories have somewhat different directions: the starting point for harmonic map theory is the end point for quantum cohomology theory. There is also a fundamental difference in the treatment of singularities: we have always assumed that the harmonic map (and associated extended solution) is regular, whereas the 'extended solution' associated to quantum cohomology is necessarily singular at $q = 0$. We shall go into the relation between harmonic maps and quantum cohomology in more detail in Chapter 10. For the moment we just mention briefly some general aspects.

First, if $\dim H^2M > 1$, the quantum cohomology of M gives rise to a *pluriharmonic* map. The theory of pluriharmonic maps is not as well developed as the theory of harmonic maps, but the fundamental zero curvature formulation is still valid (see [29] and [97]). For the linear system associated to the harmonic map equation, Proposition 7.7 and its corollary express the fundamental solution $H(z,\bar{z})$ in terms of $F_-(z)$, and hence in terms of 'free holomorphic data' $\omega(z)$. In the case of pluriharmonic maps, the same argument expresses the fundamental solution $H(z_1,\bar{z}_1,\dots,z_r,\bar{z}_r)$ in terms of $F_-(z_1,\dots,z_r)$; it halves the number of variables. But when $r > 1$, F_- satisfies a partial differential equation, so we cannot expect a reduction to free holomorphic data in general. We just have a reduction to a system with half the number of variables.

Evidently any pluriharmonic map obtained from quantum cohomology is very special, but some general properties can be described within the framework of the present chapter. For example, the fundamental solution matrix of the quantum differential equations satisfies

$$\sigma(L(t,\mathrm{h})) = L(t,-\mathrm{h}),$$

where the involution $\sigma : \mathrm{GL}_{s+1}\mathbb{C} \to \mathrm{GL}_{s+1}\mathbb{C}$ is given by $\sigma(X) = {X^{(*)}}^{-1}$ (Proposition 5.10). Our discussion of quantum cohomology so far did not involve a particular real form of the twisted loop group, but on choosing any real form the theory of section 7.4 produces a particular symmetric space as the target space of the harmonic map. We shall discuss this matter in detail in Chapter 10.

7.6 Summary: zero curvature equations

It should be clear from the examples in this chapter that the structure of an integrable system involves more data than a flat connection or family of flat connections. The 'shape' of the connection matrix is crucial, and it is desirable to specify it precisely. Unfortunately there is no systematic way of doing this, at present. We shall summarize here the examples typified by the KdV and harmonic map equations, in order to set the scene for the rest of the book.

The best known systematic families of examples are the ZS–AKNS (Zakharov–Shabat and Ablowitz–Kaup–Newell–Segur) family of integrable systems (see [1] and [134]), and the Lie-theoretic generalizations of the KdV and mKdV equations introduced and studied by V. G. Drinfeld and V. V. Sokolov ([33]), and by G. Wilson ([128]). C.-L. Terng and K. Uhlenbeck (see [119]–[123]) have considered modifications of these families, in which reality conditions and symmetry conditions are imposed; many important examples in differential geometry arise this way.

All of these equations can be represented as zero curvature equations of a very special type: the flat connection has the form $\Omega = \Omega_1 dz_1 + \Omega_2 dz_2$ where

$$\Omega_1 = \lambda C_1 + U, \quad \Omega_2 = \lambda^k C_2 + \lambda^{k-1} V_{k-1} + \cdots + V_0,$$

C_1 and C_2 being constant elements of a Lie algebra, of a type which ensures that the zero curvature condition determines $V_0, \ldots, V_{k-1}$ in terms of U.

With appropriate technical assumptions, these systems can be regarded as infinite-dimensional generalizations of the completely integrable systems of classical mechanics (i.e. a system of n Poisson-commuting flows on a symplectic manifold of dimension $2n$). In particular, they have infinitely many conserved quantities, and this leads eventually to 'explicit' solutions. Many of the characteristic features of integrable systems appeared during the development of this theory: establishment of the evolutionary p.d.e. in question as the first of an infinite series (or hierarchy) of commuting flows, derivation of solutions by means of the inverse scattering method and the Riemann–Hilbert problem, existence of soliton solutions and solutions expressable in terms of theta functions for algebraic curves, and so on. Although it was far from clear in the early days of the subject, most of this structure was eventually traced back to an infinite-dimensional Lie group of symmetries of the system, leading to the above Lie-theoretic formulation.

This brief discussion is not meant to be a historical review of 'generalized KdV and mKdV equations' (the survey article [98] has a comprehensive account of this fascinating story). We just wish to make the point that, admirable though these equations are, they are extremely special.

The harmonic map equations represent another systematic family of integrable equations. In general, they are quite different from the previous family, as the flat connection

$$\Omega_1 = \lambda^{-1} U_1 + U_0, \quad \Omega_2 = \lambda V_1 + V_0,$$

contains negative as well as positive powers of λ, and the leading coefficients are in general not constant. Of course, they are also Lie-theoretic, in the sense that we consider harmonic maps into a Lie group or symmetric space. An infinite-dimensional Lie group of symmetries plays a key role here too. The most important property is that these equations can in principle be *solved* by Lie-theoretic methods. However, they are of much less specific form than the previous family, and are perhaps 'too general' to be regarded as true integrable systems. Geometrically interesting harmonic maps (and pluriharmonic maps, which include those arising from quantum cohomology) are to be found by considering harmonic maps into specific Lie groups or symmetric spaces and by imposing additional restrictions. We have seen this already in the case of CMC surfaces (which includes the sinh-Gordon equation) and the sine-Gordon equation.

These two families—which may informally be described as generalized KdV/mKdV equations and specialized harmonic/pluriharmonic map equations—offer some guidance for the construction of more general integrable systems, by the following procedure:

Step 1: Write down a connection form $\alpha = \alpha_1 dz_1 + \alpha_2 dz_2$, where the coefficient matrices α_1, α_2 are expressed in terms of arbitrary functions (and their derivatives). For example, each component of α_1, α_2 may be regarded as an independent function, as in the case of the harmonic map equation. At the other extreme, each component may be written in terms of a single function, as in the KdV equation.

Step 2: Insert λ (the 'spectral parameter') into this connection matrix, at various strategic points, to give a λ-dependent connection form $\Omega = \alpha^\lambda$, in such a way that the original connection form is given by taking a specific value of λ.

Step 3: Write down the zero curvature condition $d\Omega + \Omega \wedge \Omega = 0$. The zero curvature condition for α is the condition for the existence of a map ψ such that $\psi^{-1}d\psi = \alpha$; the zero curvature condition for $\Omega = \alpha^\lambda$ imposes conditions on this map, which is by definition the integrable p.d.e. (if α, and hence ψ, is expressed in terms of other functions, then we obtain a p.d.e. for those functions).

This simple procedure is certainly effective—integrable partial differential equations in the above sense can be generated effortlessly. However, to produce a p.d.e. worthy of the name 'integrable system', the procedure should be initiated in some way by simple general principles (which should in turn lead to general results, in particular concerning the solutions). That is, the shape of α, and the λ-dependence of α^λ, should not be completely random; they should be consequences of natural geometrical or algebraic requirements.

For example, the mKdV equation can be regarded as the zero curvature condition for

$$\alpha = \Omega = \begin{pmatrix} -q & 0 \\ 1 & q \end{pmatrix} dx + \begin{pmatrix} \frac{1}{2}q_{xx} - q^3 & 0 \\ q^2 - q_x & -\frac{1}{2}q_{xx} + q^3 \end{pmatrix} dt$$

(i.e. with trivial λ-dependence). However, the form

$$\Omega = \begin{pmatrix} -q & \mu^2 \\ 1 & q \end{pmatrix} dx + \begin{pmatrix} \frac{1}{2}q_{xx} - q^3 + 2\mu^2 q & \mu^2(q^2 + q_x) - 2\mu^4 \\ q^2 - q_x - 2\mu^2 & -\frac{1}{2}q_{xx} + q^3 - 2\mu^2 q \end{pmatrix} dt$$

with $\alpha = \Omega|_{\mu=0}$ is more natural, as, after conjugation by the diagonal matrix $\operatorname{diag}(1, \mu)$, the apparently complicated q-dependence follows merely by specifying that $\Omega = \Omega_1 dx + \Omega_2 dt$ be of the form

$$\Omega_1 = \mu \begin{pmatrix} & 1 \\ 1 & \end{pmatrix} + \begin{pmatrix} -q & \\ & q \end{pmatrix}$$

$$\Omega_2 = \mu^3 \begin{pmatrix} & 1 \\ 1 & \end{pmatrix} + \mu^2 \begin{pmatrix} a & \\ & -a \end{pmatrix} + \mu \begin{pmatrix} & b \\ c & \end{pmatrix} + \begin{pmatrix} d & \\ & -d \end{pmatrix}.$$

This illustrates how a 'strong shape' can follow from a simple general principle, in this case a Lie-theoretic one.

A broader general principle, and the underlying philosophy of this book, is that the D-module approach suggests another way to specify the shape of $\Omega = \alpha^\lambda$, for example, by specifying generators of a certain form for the left ideal which annihilates a cyclic element. This provides an alternative formulation of the KdV and mKdV equations, as we have seen in this chapter, but it also accommodates quite different examples such as quantum cohomology and certain classes of harmonic maps. In the case of quantum cohomology the idea of 'quantization of a given commutative algebra' (Chapter 6) provides the extra structure.

Natural conditions imposed on connection matrices, such as reality conditions, can equally well be imposed on D-modules. Some conditions, such as the form of the λ-dependence of the connection matrices, are more natural at the level of D-modules, as we shall see in the next chapter.

8 Solving integrable systems

Our objective in this chapter is to review how the 'Grassmannian model of the loop group', usually referred to as the Sato or Segal–Wilson Grassmannian, provides a conceptual framework for solving integrable systems with spectral parameter. While it is certainly possible to solve integrable systems without mentioning the Grassmannian model, such methods always involve complicated ad hoc calculations. The Grassmannian model does not eliminate the complicated calculations, but it organises them efficiently and provides an intelligent compromise between the concrete and the abstract. It also reveals the infinite-dimensional symmetry groups behind such systems. We shall summarize the three main approaches which use the Grassmannian model, and then show how they are unified by the D-module point of view.

Infinite-dimensional Grassmannians first appeared in the work of M. Sato and his collaborators (see [108]–[110]). These Grassmannians were algebraic objects, based on infinite-rank D-module structures observed in the KdV and KP hierarchies. They provided a natural setting for the approach of I. M. Gelfand and L. A. Dickii (see [27] and [45]) using fractional powers of differential operators, which was based on the Lax form of the KdV equation. A geometrical approach based on infinite-dimensional Lie groups and homogeneous spaces was given by G. Segal and G. Wilson in [112].

In [111], Segal showed how the same Grassmannian model sheds light on the harmonic map equation. A more detailed discussion with further developments can be found in [60]. We shall review this theory briefly in section 8.6. As in the case of the KdV equation, the advantage of the Grassmannian model is that it provides a geometrical framework, suppressing to some extent the complicated formulae which appear in explicit solutions.

We shall not discuss quantum cohomology in this chapter, as this will be the main focus of Chapters 9 and 10.

8.1 The Grassmannian model

In Chapter 4 we studied equivalences

$$T_j y = 0, \quad 1 \le j \le u \iff \partial_i Y = A_i Y, \quad 1 \le i \le r$$

of scalar systems and matrix systems. The scalar system should be viewed as a D-module

$$D/(T_1, \ldots, T_u)$$

which is free and has finite rank over the ring $\mathcal{H}$ of coefficient functions. A choice of basis gives rise to a connection form

$$\Omega = \sum_{i=1}^{r} \Omega_i dz_i$$

such that $d\Omega + \Omega \wedge \Omega = 0$, that is, a flat connection. Putting $A_i = \Omega_i^t$, we obtain the matrix system, and the dimension of the solution space of this system is equal to the rank of the D-module. Conversely, given a matrix system $\partial_i Y = A_i Y$, the dimension of the solution space is either zero or n (where each A_i is an $n \times n$ matrix). It has dimension n, that is, the system is 'compatible', if and only if the connection $d - A = d - \sum_{i=1}^{r} A_i dz_i$ is flat. In this case it corresponds to a cyclic D-module of rank n.

Let $y_{(1)}, \ldots, y_{(n)}$ be any basis of solutions of the scalar system and let

$$J = (y_{(1)}, \ldots, y_{(n)}).$$

A fundamental solution[1] H of the matrix system (a solution of $dHH^{-1} = A$) may be written as

$$H = \begin{pmatrix} | & & | \\ Y_{(1)} & \cdots & Y_{(n)} \\ | & & | \end{pmatrix} = \begin{pmatrix} - & P_1 J & - \\ & \vdots & \\ - & P_n J & - \end{pmatrix}$$

where $[P_1], \ldots, [P_n]$ is a basis of the D-module.

Similar statements apply when the system of differential equations depends on a spectral parameter λ. The usefulness of λ is clear in the case of the harmonic map equation (Example 4.16 of Chapter 4); it would not be possible to express the equation as a zero curvature condition without λ.

[1] Recall that $H^t = F = L$. We shall avoid L in discussions of the KdV equation as it conflicts with the standard notation for the differential operator $L = \partial_x^2 + u$.

In the case of the KdV equation, λ appears naturally as an eigenvalue of the Schrödinger operator. But a purely computational argument in favour of λ is hardly satisfying. We shall therefore present a more substantial justification for studying 'differential equations with spectral parameter'. This may look complicated at first, but we shall soon see that it is a productive point of view.

First of all, since H depends on λ as well as $z_1, \dots, z_r$, we may regard it as a function of $z_1, \dots, z_r$ which takes values in the loop group

$$\Lambda \mathrm{GL}_n\mathbb{C} = \{\gamma : S^1 \to \mathrm{GL}_n\mathbb{C} \mid \gamma \text{ is smooth}\}.$$

As in Chapter 6 we shall generally write $H(z_1, \dots, z_r, \lambda)$ for the matrix and $H(z_1, \dots, z_r)$ for the loop.

The loop group is an infinite-dimensional Lie group. Its Lie algebra is the loop algebra

$$\Lambda \mathfrak{gl}_n\mathbb{C} = \{f : S^1 \to \mathfrak{gl}_n\mathbb{C} \mid f \text{ is smooth}\}.$$

The theory of loop groups and loop algebras is particularly well behaved; in particular, it shares many common features with finite-dimensional Lie theory.

An even more powerful version of this theory can be obtained by regarding the loop group as a group of geometrical transformations of an infinite-dimensional homogeneous space, the 'Grassmannian model' $\mathrm{Gr}^{(n)}$. This generalizes the well known action of the finite-dimensional Lie group $\mathrm{GL}_n\mathbb{C}$ on the finite-dimensional Grassmannian $\mathrm{Gr}_k(\mathbb{C}^n)$, and much of the intuition carries over to the infinite-dimensional case. We shall review the definition and basic properties of $\mathrm{Gr}^{(n)}$ next.

We construct a vector space which has

$$\dots, \lambda^{-2}, \lambda^{-1}, 1, \lambda^1, \lambda^2, \dots$$

as basis vectors. To be precise, this is the Hilbert space

$$H^{(1)} = L^2(S^1, \mathbb{C}) = \mathrm{Span}\{\lambda^i \mid i \in \mathbb{Z}\}$$

of square-integrable complex functions. Similarly,

$$H^{(n)} = L^2(S^1, \mathbb{C}^n) = \mathrm{Span}\{\lambda^i e_j \mid i \in \mathbb{Z}, 1 \leq j \leq n\}$$

where $e_1, \dots, e_n$ is the standard basis of $\mathbb{C}^n$. We use 'Span' to indicate the closed linear subspace spanned by the indicated vectors. Next, let

$$H_+^{(n)} = \text{Span}\{\lambda^i e_j \mid i \geq 0, 1 \leq j \leq n\}.$$

The Sato or Segal–Wilson Grassmannian $\text{Gr}^{(n)}$ is the set of all closed linear subspaces V of $H^{(n)}$ which

(1) are 'commensurate' with $H_+^{(n)}$,
(2) consist of smooth elements, and which
(3) are preserved by the operation of multiplication by λ (i.e. $\lambda V \subseteq V$)

(see [102] for the precise definition). It is proved in [102] that $\text{Gr}^{(n)}$ has the structure of an infinite-dimensional Kähler manifold.

There is a natural action of the loop group $\Lambda\text{GL}_n\mathbb{C}$ on $\text{Gr}^{(n)}$, given by $\gamma \cdot V = \gamma V$. It is proved in [102] that this action is transitive, with isotropy subgroup at $V = H_+^{(n)}$ given by $\Lambda_+\text{GL}_n\mathbb{C}$, the subgroup of $\Lambda\text{GL}_n\mathbb{C}$ consisting of maps which extend holomorphically (as maps into $\text{GL}_n\mathbb{C}$) to the unit disk. Thus, we have an identification

$$\text{Gr}^{(n)} \cong \Lambda\text{GL}_n\mathbb{C}/\Lambda_+\text{GL}_n\mathbb{C}, \quad \gamma H_+^{(n)} \leftrightarrow [\gamma].$$

This is a diffeomorphism of complex manifolds. Replacing the smooth loops by (for example) analytic loops, rational loops, or finite Laurent series loops, we obtain loop groups[2] $\Lambda_{\text{an}}\text{GL}_n\mathbb{C}$, $\Lambda_{\text{rat}}\text{GL}_n\mathbb{C}$, or $\Lambda_{\text{alg}}\text{GL}_n\mathbb{C}$, and corresponding Grassmannians $\text{Gr}_{\text{an}}^{(n)}$, $\text{Gr}_{\text{rat}}^{(n)}$, or $\text{Gr}_{\text{alg}}^{(n)}$.

There is another identification, which strengthens further the analogy with finite-dimensional Grassmannians. Namely, $\text{Gr}^{(n)}$ can be identified with the 'unitary based loop group'

$$\Omega\text{U}_n = \{\gamma : S^1 \to \text{U}_n \mid \gamma(1) = I \text{ and } \gamma \text{ is smooth}\}.$$

This arises from the fact that the action on $\text{Gr}^{(n)}$ of the unitary loop group

$$\Lambda\text{U}_n = \{\gamma : S^1 \to \text{U}_n \mid \gamma \text{ is smooth }\}$$

is also transitive, with isotropy subgroup U_n (the constant loops) at $H_+^{(n)}$. Thus, we have another identification

$$\text{Gr}^{(n)} \cong \Lambda\text{U}_n/\text{U}_n \cong \Omega\text{U}_n, \quad \gamma H_+^{(n)} \leftrightarrow [\gamma] \leftrightarrow \gamma(\lambda)\gamma(1)^{-1}.$$

[2] Since the reciprocal of the determinant of a finite Laurent series loop is not necessarily a finite Laurent series, it is necessary to specify that γ is an element of $\Lambda_{\text{alg}}\text{GL}_n\mathbb{C}$ if and only if both γ and γ^{-1} are finite Laurent series.

The transitivity of the action of ΛU_n, like the transitivity of the action of U_n on $Gr_k(\mathbb{C}^n)$, can be proved using a Gram–Schmidt orthogonalization procedure. This gives a factorization

$$\Lambda GL_n\mathbb{C} = \Lambda U_n \, \Lambda_+ GL_n\mathbb{C}$$

which (in analogy with the case of finite-dimensional Lie groups) is called the Iwasawa factorization of the loop group. We have $\Lambda U_n \cap \Lambda_+ GL_n\mathbb{C} = U_n$ so the factorization $\gamma = \gamma_u \gamma_+$ of an element of $\Lambda GL_n\mathbb{C}$ is unique up to multiplication by an element of U_n.

Using analytic loops, rational loops, or finite Laurent series loops, we obtain loop groups denoted by $\Omega_{\mathrm{an}} U_n$, $\Omega_{\mathrm{rat}} U_n$, or $\Omega_{\mathrm{alg}} U_n$, which may be identified with the corresponding Grassmannians $Gr^{(n)}_{\mathrm{an}}$, $Gr^{(n)}_{\mathrm{rat}}$, or $Gr^{(n)}_{\mathrm{alg}}$.

We shall also need the 'opposite' Grassmannian, obtained by using λ^{-1} instead of λ. That is, we define

$$\check{H}^{(n)}_+ = \mathrm{Span}\{\lambda^i e_j \mid i \leq 0, 1 \leq j \leq n\},$$

then $\check{Gr}^{(n)}$ is the set of all closed linear subspaces V of $H^{(n)}$ which

(1) are 'commensurate' with $\check{H}^{(n)}_+$,
(2) consist of smooth elements, and which
(3) are preserved by the operation of multiplication by λ^{-1} (i.e. $\lambda^{-1} V \subseteq V$).

We obtain homogeneous space representations

$$\check{Gr}^{(n)} \cong \Lambda GL_n\mathbb{C}/\Lambda_- GL_n\mathbb{C} \cong \Lambda U_n/U_n \cong \Omega U_n,$$

together with their analogues for the other types of loop groups.

8.2 The fundamental construction

Now let us consider a system of partial differential equations with spectral parameter λ. We assume that we have equivalent scalar and $n \times n$ matrix systems, that the matrix equations are compatible, and that the scalar system corresponds to a D-module which is free and of rank n over an appropriate ring of functions. (We shall be more precise about the D-module structure later.)

In the matrix system

$$\partial_i Y = A_i^\lambda Y, \quad 1 \leq i \leq r,$$

it is the λ-dependence of A^λ which determines the nature of the integrable system. (To emphasize this dependence we write $A^\lambda, \Omega^\lambda$ instead of A, Ω from now on.) Let us make the following assumption.

Assumption There exist non-negative integers k, l such that, for $i = 1, \ldots, r$, each map A_i^λ is of the form

$$A_i^\lambda(z,\lambda) = \sum_{j=-k}^{l} A_{i,j}^\lambda(z)\lambda^j.$$

(More generally we could allow a different k_i, l_i for each i.) It follows from this assumption that the fundamental solution matrix H is holomorphic in λ except possibly for singularities at $\lambda = 0, \infty$.

The effect of these singularities may be studied geometrically by introducing the following maps $W, \check{W}$. We regard this as the 'fundamental construction'.

Definition 8.1 $W(z) = H^t(z)H_+^{(n)}$, $\check{W}(z) = H^t(z)\check{H}_+^{(n)}$.

It is easy to show that $W, \check{W}$ take values in $\mathrm{Gr}^{(n)}, \check{\mathrm{Gr}}^{(n)}$ respectively.

Proposition 8.2 *W and $\check{W}$ determine H.*

Proof Let $H^t = F = F_-F_+$ (Birkhoff factorization). We have $W = H^tH_+^{(n)} = F_-H_+^{(n)}$, and this determines F_-. Similarly, $\check{W}$ determines $\check{F}_+$. As in the proof of Proposition 7.7, F_- and $\check{F}_+$ determine H by $H^t = \check{F}_+(\check{F}_+^{-1}F_-)_-$. □

From now on, therefore, we can study the geometrical objects W and $\check{W}$ instead of H.

Proposition 8.3 *We have:*

(1) $\lambda^k\partial_i W \subseteq W$, $\quad i = 1, \ldots, r$.
(2) $\lambda^{-l}\partial_i \check{W} \subseteq \check{W}$, $\quad i = 1, \ldots, r$.

To be more precise, the statement $\lambda^k\partial_i W \subseteq W$ here means that the operator $\lambda^k\partial_i$ preserves the space[3] of local sections of W, that is,

$$\lambda^k(\partial_i s)(z) \in W(z)$$

[3] This could be expressed more abstractly by saying that the operator $\lambda^k\partial_i$ acts on sections of the (trivial) pullback bundle $W^*\mathcal{T}$, where $\mathcal{T}$ is the tautologous bundle on $\mathrm{Gr}^{(n)}$ whose fibre over the point V is the Hilbert space V.

for any locally defined map s such that $s(z) \in W(z)$. We shall abbreviate this to '$\lambda^k \partial_i$ preserves the space of sections ΓW of W', or just '$\lambda^k \partial_i W \subseteq W$' as in the proposition.

Proof From $(H^t)^{-1} dH^t = (A^\lambda)^t$ we have

$$\begin{aligned}\lambda^k \partial_i W &= \lambda^k \partial_i H^t H_+^{(n)} \\ &= \lambda^k H^t (A_i^\lambda)^t H_+^{(n)} \\ &= H^t \lambda^k (A_i^\lambda)^t H_+^{(n)} \\ &\subseteq H^t H_+^{(n)} = W.\end{aligned}$$

The proof of (2) is similar. □

This says that (the spaces of sections of) W and $\check{W}$ acquire D-module structures. We shall discuss this in detail for W; the case of $\check{W}$ is similar.

Definition 8.4 *Let k be an integer. Let $D_z^{\lambda^k}$ be the ring of differential operators with coefficients in $\mathcal{H}_z \otimes \mathcal{H}_\lambda^+$ which is generated by $\lambda^k \partial_1, \dots, \lambda^k \partial_r$.*

We recall that $\mathcal{H}_\lambda^+$ denotes the space of functions of λ which are holomorphic in a neighbourhood of the origin.

In the general case we use the notation $D_{z_1,\dots,z_r}^{\lambda^{k_1},\dots,\lambda^{k_r}}$, with D^A as a convenient abbreviation for any of these rings when no confusion is likely (the suffix A is meant to indicate that the λ-dependence of the D-module depends on the form of the matrix system).

The analogues of the D-module isomorphisms $\mathcal{M} \cong \mathcal{H}^{s+1}$ of section 4.2 are particularly useful in the current situation. Let us assume that $\mathcal{M}$ is a free module over $\mathcal{H}_z \otimes \mathcal{H}_\lambda^+$ of rank n. Then these isomorphisms take the following form:

$$D^A/(T_1, \dots, T_u) \quad \cong \quad \mathcal{H}_z^n \otimes \mathcal{H}_\lambda^+ \quad \cong \quad \Gamma W \ \ (\subseteq \mathcal{H}_z^n \otimes \mathcal{H}_\lambda)$$

$$[P] = [\textstyle\sum_{i=1}^n f_i P_i] \longleftrightarrow P \cdot \begin{pmatrix} 1 \\ 0 \\ \vdots \\ 0 \end{pmatrix} = \begin{pmatrix} f_1 \\ f_2 \\ \vdots \\ f_n \end{pmatrix} \longleftrightarrow PJ.$$

These are isomorphisms of modules over D^A, if the action of D^A in each case is defined as in section 4.2. The only new feature is the appearance of ΓW (rather than $\mathcal{H}_z^n \otimes \mathcal{H}_\lambda^+$ or $\mathcal{H}_z^n \otimes \mathcal{H}_\lambda$) on the right hand side. To see that

ΓW is the image of the map $[P] \mapsto PJ$, observe that

$$W = H^t H_+^{(n)}$$

$$= \begin{pmatrix} | & & | \\ P_1 J & \cdots & P_n J \\ | & & | \end{pmatrix} \mathrm{Span}\{e_1, \ldots, e_n, \lambda e_1, \ldots, \lambda e_n, \ldots\}$$

$$= \mathrm{Span}\{P_1 J, \ldots, P_n J, \lambda P_1 J, \ldots, \lambda P_n J, \ldots\}.$$

Hence the space of sections of W can be represented as

$$\Gamma W = \mathrm{Span}\{XJ \mid X \in D^A\} \cong D^A/(T_1, \ldots, T_u).$$

This diagram provides the crucial link between the D-module version of an integrable system and the Grassmannian model. The abstract D-module is more intrinsic, but the space of sections of W is a geometrical realization of it, which is very useful for calculations.

Before giving some examples, we should emphasize that the 'semi-infinite' status of W is related to the assumption above on the nature of the singularity of A^λ at $\lambda = 0$. If λ were not present (or fixed) then the appropriate ring of differential operators would be D_z, and we would have $D/(T_1, \ldots, T_u) \cong \mathrm{Map}(N_z, \mathbb{C}^n)$ as in section 4.2. If the λ-dependence were unrestricted, then the appropriate ring of differential operators would be $D_z \otimes \mathcal{H}_\lambda$, and we would have $D_z \otimes \mathcal{H}_\lambda \,/\, (T_1, \ldots, T_u) \cong \mathrm{Map}(N_z, H^{(n)}) = \mathcal{H}_z^n \otimes \mathcal{H}_\lambda$. Our situation is intermediate between these extremes.

Example 8.5 The form of the harmonic map equation (section 7.3) gives

$$\lambda \partial W \subseteq W, \quad \bar{\partial} W \subseteq W,$$

so the appropriate ring of differential operators for this problem is

$$D^A = D_{z,\bar{z}}^{\lambda^{k_1}, \lambda^{k_2}} = D_{z,\bar{z}}^{\lambda, \lambda^0}.$$

The condition $\bar{\partial} W \subseteq W$ says (as in the case of maps into a finite-dimensional Grassmannian) that W is *holomorphic*; this is consistent with Proposition 7.7 (applied to $H^t = F = F_- F_+$), which says that F_- depends only on z (and not on $\bar{z}$).

Since $\check{W}$ is equivalent to W in this example, there is no need to consider it. The Grassmannian-theoretic version of the harmonic map equation is simply the condition $\lambda \partial W \subseteq W$ for a holomorphic map W. □

Example 8.6 The usual matrix version of the KdV equation (section 4.3) gives

$$\partial_x W \subseteq W, \quad \partial_t W \subseteq W,$$

so the appropriate ring of differential operators is

$$D^A = D_{x,t}^{\lambda^0,\lambda^0} = D_{x,t} \otimes \mathcal{H}_\lambda^+.$$

As in the previous example, we conclude that W is independent of both x and t, that is, W is constant in this case.

For $\check{W}$ we have

$$\lambda^{-1}\partial_x \check{W} \subseteq \check{W}, \quad \lambda^{-2}\partial_t \check{W} \subseteq \check{W},$$

so the appropriate ring of differential operators here is $D^A = D_{x,t}^{\lambda^{-1},\lambda^{-2}}$. □

Example 8.7 For the quantum differential equations we have

$$\mathrm{h}\partial_i W \subseteq W, \quad i = 1, \ldots, r,$$

so the appropriate ring of differential operators is

$$D^A = D_{t_1,\ldots,t_r}^{\mathrm{h},\ldots,\mathrm{h}},$$

which we shall abbreviate as D_t^{h}, or simply D^{h} as in Chapter 5.

The equations for $\check{W}$ are the trivial equations

$$\partial_i \check{W} \subseteq \check{W}, \quad i = 1, \ldots, r$$

and $\check{W}$ is constant. □

The infinite-dimensional Grassmannian can be used to construct and describe solutions of integrable systems, and this will be our main focus in the rest of this chapter. To conclude this section, we make some brief remarks on the infinite-dimensional symmetry groups which underlie these constructions.

The basic fact, which is obvious from the Grassmannian point of view, is that the condition

$$D^A\, W \subseteq W$$

is preserved by the natural action of the loop group $\Lambda GL_n\mathbb{C}$ on $\mathrm{Gr}^{(n)}$. That is, if W satisfies this condition, then so does γW for any $\gamma \in \Lambda GL_n\mathbb{C}$. In the case of the harmonic map equation, this gives immediately an action of $\Lambda GL_n\mathbb{C}$ on the space of harmonic maps. For other systems, we expect an action of an appropriate subgroup of $\Lambda GL_n\mathbb{C}$ on the space of solutions (namely the subgroup which preserves the conditions on W beyond $D^A\, W \subseteq W$). In the case of equations of KdV or mKdV type, this action is usually called the dressing action.

8.3 Solving the KdV equation: the Guiding Principle

In the next section we shall summarize three well known methods of producing families of solutions of the KdV equation, which derive from a single 'Guiding Principle' stated in this section. These methods are specially designed for the KdV equation (and related equations) and exploit its peculiar features. A central role is played by the infinite-dimensional Grassmannian which was introduced in section 8.1. In fact, all methods lead to the conclusion that these solutions of the KdV equation 'are in one to one correspondence with the points of a Grassmannian'. This is an unusual description (to say the least) from the point of view of p.d.e. theory.

We treat the KdV equation in some detail, as it is an important and well known example and because the D-module approach is particularly illuminating in this case. But we do not insist on studying integrable systems of exactly this type; we have in mind examples (such as those related to quantum cohomology) which share only certain features of the KdV equation.

As with any p.d.e., it is necessary to specify what kind of solutions one has in mind. This is a very important matter in analysis, but it is less important for us because we are concerned primarily with the algebraic or geometric process of constructing solutions, rather than the properties of the solutions themselves. It turns out that certain classes of solutions correspond to different types of Grassmannians, so the problem breaks naturally into two parts: first establish the general principle which relates solutions and points of Grassmannians, then adjust the Grassmannian according to whatever application one has in mind. We shall concentrate on the first part.

From the point of view of analysis, the KdV equation is an example of an evolution equation, that is, it governs the t-evolution

$$t \mapsto u(x,t)$$

of a wave whose initial shape is given at $t = 0$ by the graph of $u(x) = u(x,0)$. Under reasonable smoothness and boundary conditions, therefore, one expects that solutions of the KdV equation correspond to initial conditions of this kind. These initial conditions, in turn, should correspond to the points of the Grassmannian just referred to. While this is essentially correct, there is much more to be gained from the Grassmannian; it turns out that the t-evolution of the initial wave can be represented as a 'straight line' in the Grassmannian. In other words, the non-linear KdV equation is thereby linearized! (To be precise, the straight line is a line in the tangent space of the Grassmannian; its image under the exponential map is a curve in the Grassmannian itself which represents the solution of the KdV equation.)

In this section we shall give the standard explanation of how this linearization of the KdV equation arises from the Lax pair formulation. Unfortunately, however, this explanation is not rigorous. For this reason, we refer to it as the Guiding Principle. The following section presents three different ways of making it rigorous. In section 8.5, we shall explain how these three ways are related to each other, thanks to the fundamental construction of section 8.2 which relates the D-module to the Grassmannian model.

As we have seen in section 7.1, the KdV equation may be expressed in the form

$$L_t = [P, L]$$

where $L = \partial^2 + u$, $P = \partial^3 + \frac{3}{2}u\partial + \frac{3}{4}u_x$. This was discovered by P. Lax. The fundamental property of an equation of this type is that the t-evolution of the operator L is *isospectral*. That is, if we regard L, P as evolving in time from $L_0 = L|_{t=0}, P_0 = P|_{t=0}$, the evolution of each eigenvector λ of L is independent of time (although corresponding eigenfunctions may evolve non-trivially). The operators L, P may be regarded as evolving in some infinite-dimensional vector space, and this will lead eventually to the Grassmannian interpretation.

Isospectrality can be checked by showing directly that $\mathrm{d}\lambda/\mathrm{d}t = 0$, but there is a more conceptual reason. Namely, making the Ansatz $L = UL_0U^{-1}$ gives

$$\begin{aligned} L_t &= U_tL_0U^{-1} - UL_0U^{-1}U_tU^{-1} \\ &= U_tU^{-1}UL_0U^{-1} - UL_0U^{-1}U_tU^{-1} \\ &= [U_tU^{-1}, L], \end{aligned}$$

so o.d.e. theory suggests that the unique solution of the Lax equation

$$L_t = [P, L], \quad L|_{t=0} = L_0$$

should be $L = UL_0U^{-1}$ where U is the unique solution of the equation

$$U_tU^{-1} = P, \quad U|_{t=0} = I.$$

It follows that L has the same spectrum as L_0, as required.

The difficulty with this argument is that, if we regard the t-evolution of L as a curve in an infinite-dimensional Lie algebra of differential operators, it is not clear where U lives. It should live in a corresponding infinite-dimensional Lie group, but the existence of 'a corresponding Lie group' is problematical in infinite dimensions.

In the case of finite-dimensional Lie algebras, there is nothing to worry about, so let us first consider this case as a model.

Let $L, P : \mathbb{R} \to \mathfrak{g}$ be functions taking values in $\mathfrak{g}$, the Lie algebra of a finite-dimensional Lie group G. For notational convenience we shall assume that G is a matrix group, so that the adjoint action of G on $\mathfrak{g}$ is given by conjugation, and the Lie bracket is given by commutator product. Then a standard application of o.d.e. theory shows that the unique local solution L of the o.d.e.

$$L_t = [P, L], \quad L|_{t=0} = L_0$$

is

$$L = UL_0U^{-1}$$

where U is the unique local solution of the equation

$$U_tU^{-1} = P, \quad U|_{t=0} = I.$$

Here, L takes values in $\mathfrak{g}$ and U takes values in G.

Even better, for certain kinds of P, the group-valued map U can be found explicitly. Thus, we have not only the qualitative result that the eigenvalues of L are constant, but also an explicit formula for the eigenvectors (these are obtained from the eigenvectors of L_0 by applying U). In particular, we have an explicit formula for the solution L to the original problem.

Example 8.8 Let us examine the equation $L_t = [P, L]$ for various P.

(1) $P = \text{constant}$.

Obviously $U(t) = \exp tP$ satisfies the equation $U_tU^{-1} = P$ and the initial condition $U|_{t=0} = I$, so it is the unique solution. Thus, $L(t) = (\exp tP)L_0(\exp tP)^{-1}$.

(2) $P = \pi_2 L$ where $\pi_2 : \mathfrak{g} \to \mathfrak{g}_2$ is projection with respect to a Lie algebra decomposition $\mathfrak{g} = \mathfrak{g}_1 \oplus \mathfrak{g}_2$.

In this case $U(t) = (\exp tL_0)_2$, assuming that there is a corresponding Lie group decomposition $G = G_1G_2$ such that each G_i is a closed subgroup of G, $\mathfrak{g}_i$ is the Lie algebra of G_i, and $G_1 \cap G_2 = \{1\}$. (The group element $\exp tL_0$ admits a unique factorization $\exp tL_0 = (\exp tL_0)_1(\exp tL_0)_2$, under this assumption.) To prove that $U(t) = (\exp tL_0)_2$ satisfies $\dot{U}U^{-1} = \pi_2 L$, observe that

$$\begin{aligned} L_0 &= (\exp tL_0)^{-1}(\exp tL_0)^{\cdot} \\ &= (\exp tL_0)_2^{-1}(\exp tL_0)_1^{-1}(\exp tL_0)_1^{\cdot}(\exp tL_0)_2 + (\exp tL_0)_2^{-1}(\exp tL_0)_2^{\cdot} \end{aligned}$$

hence $L = UL_0U^{-1} = (\exp tL_0)_1^{-1}(\exp tL_0)_1^{\cdot} + \dot{U}U^{-1}$, from which we obtain $\dot{U}U^{-1} = \pi_2 L$.

We have obtained the explicit solution

$$L(t) = (\exp tL_0)_2 L_0 (\exp tL_0)_2^{-1}.$$

Since $(\exp tL_0)_2 = (\exp tL_0)_1^{-1} \exp tL_0$, the solution can also be written

$$L(t) = (\exp tL_0)_1^{-1} L_0 (\exp tL_0)_1,$$

since $\exp tL_0$ commutes with L_0.

(3) $P = \pi_2 L^n$ for $n = 1, 2, 3, \ldots$.

Exactly the same argument shows that $U(t) = (\exp tL_0^n)_2$. Thus, the explicit solution is

$$\begin{aligned} L(t) &= (\exp tL_0^n)_2 L_0 (\exp tL_0^n)_2^{-1} \\ &= (\exp tL_0^n)_1^{-1} L_0 (\exp tL_0^n)_1. \end{aligned}$$

(4) $P = \pi_1 L$.

This gives the same Lax equation as $P = \pi_2(-L)$. The above method gives

$$\begin{aligned} L(t) &= (\exp -tL_0)_2 L_0 (\exp -tL_0)_2^{-1} \\ &= (\exp -tL_0)_1^{-1} L_0 (\exp -tL_0)_1. \end{aligned}$$

To summarize, in each of these examples we have $P = \pi_2 F(L)$ and

$$\begin{aligned} L(t) &= (\exp tF(L_0))_2 L_0 (\exp tF(L_0))_2^{-1} \\ &= (\exp tF(L_0))_1^{-1} L_0 (\exp tF(L_0))_1. \end{aligned}$$ □

As a final remark on the finite-dimensional case, we point out that it is sometimes convenient to conjugate L_0 to a special form Λ (such as a diagonal matrix, or other normal form):

$$L_0 = K_0 \Lambda K_0^{-1}.$$

Then we have

$$L = K \Lambda K^{-1}, \quad \text{where } K = UK_0$$

and, writing $P = \pi_2 F(L)$, we have

$$K(t) = (\exp tK_0 F(\Lambda) K_0^{-1})_2 K_0 = (K_0 (\exp tF(\Lambda)) K_0^{-1})_2 K_0$$

or

$$K(t) = (\exp tK_0F(\Lambda)K_0^{-1})_1^{-1}K_0 = (K_0(\exp tF(\Lambda))K_0^{-1})_1^{-1}K_0.$$

A suitable context for the Lax equation of the KdV equation is provided by the Lie algebra of formal pseudo-differential operators:

$$\mathrm{Psd} = \left\{ \textstyle\sum_{i=-\infty}^{N} \alpha_i(x)\partial_x^i \mid \alpha_i \in \mathcal{H}, N \in \mathbb{Z} \right\}.$$

This is a Lie algebra under the commutator product $[P,Q] = PQ - QP$ (see [27] and [112]). The subalgebra of differential operators

$$\mathrm{Psd}_+ = D = \left\{ \textstyle\sum_{i=0}^{N} \alpha_i(x)\partial_x^i \mid \alpha_i \in \mathcal{H}, N = 0,1,2,\ldots \right\}$$

provides a home for the t-evolution of L and P, but we shall also use the complementary subalgebra

$$\mathrm{Psd}_- = \left\{ \textstyle\sum_{i=-\infty}^{-1} \alpha_i(x)\partial_x^i \mid \alpha_i \in \mathcal{H} \right\}.$$

For the application of this theory to the KdV equation, the first key observation (which goes back to I. Schur) is:

Lemma 8.9 *Let* $L = \partial_x^2 + u$. *Then there exists* $K \in \mathrm{Psd}$ *of the form* $K = 1 + \alpha_1\partial_x^{-1} + \alpha_2\partial_x^{-2} + \cdots$ *such that* $L = K\partial_x^2K^{-1}$. *If both* K_1 *and* K_2 *satisfy these conditions, then there exists a constant coefficient series* $C = 1 + c_1\partial_x^{-1} + c_2\partial_x^{-2} + \cdots$ *such that* $K_1 = K_2C$.

Proof The condition $LK = K\partial_x^2$ leads to a sequence of recursive formulae for $\alpha_1, \alpha_2, \ldots$. □

In view of this it is natural to define the square root of L to be

$$L^{\frac{1}{2}} = K\partial_x K^{-1}$$

and more generally $L^{r/2} = K\partial_x^r K^{-1}$.

The second key observation is:

Lemma 8.10 *With respect to the decomposition* $\mathrm{Psd} = \mathrm{Psd}_- \oplus \mathrm{Psd}_+$ *of vector spaces, we have* $P = (L^{3/2})_+$.

Proof Direct calculation. □

In isolation this seems mysterious, so now is an appropriate time to say that all the operators $P_n = (L^{n/2})_+$, and (essentially) only these operators, satisfy the Lax condition $[\partial_t - P, L] = 0$; these are the operators referred to in section 7.2. This very strong condition says that '$\partial_t - P$ and L are simultaneously diagonalizable'; in fact one has $\partial_t - P_n = K(\partial_t - \partial_x^n)K^{-1}$. We shall now explain briefly the formal calculation which leads to these conclusions (and to the proof of the lemma).

Let us assume that a differential operator P satisfies $[\partial_t - P, L] = 0$, where $L = K\partial_x^2 K^{-1}$ as usual. Then $L_t = [K_t K^{-1}, L]$, hence $[P - K_t K^{-1}, L] = 0$, or $[K^{-1}PK - K^{-1}K_t, \partial_x^2] = 0$. This suggests that $K^{-1}PK - K^{-1}K_t$ must be a constant coefficient operator $\sum c_i \partial_x^i$. Hence $P - K_t K^{-1} = \sum c_i L^{i/2}$. Since P is a differential operator and $K_t K^{-1}$ contains only negative powers of ∂_x, we obtain $P = \sum c_i (L^{i/2})_+$. If P is required to be homogeneous, then all c_i but one must vanish. Finally, taking $c_n = 1$ (and all others zero), it follows from $P_n - K_t K^{-1} = L^{n/2} = K\partial_x^n K^{-1}$ that

$$\partial_t - P_n = K(\partial_t - \partial_x^n)K^{-1}.$$

Thus, L and all the operators $\partial_t - P_n$ are simultaneously conjugated to constant coefficient operators by K.

The finite-dimensional theory above suggests the following guiding principle.

The Guiding Principle The solution of the Lax form $L_t = [(L^{3/2})_+, L]$ of the KdV equation is given by

$$\boxed{L = UL_0U^{-1}, \quad U = (\exp tL_0^{3/2})_-^{-1}}$$

where the (hypothetical) group element $\exp tL_0^{3/2}$ is factorized as

$$\exp tL_0^{3/2} = (\exp tL_0^{3/2})_-(\exp tL_0^{3/2})_+$$

with respect to a (hypothetical) group decomposition

$$\mathcal{G} = \mathcal{G}_-\mathcal{G}_+$$

of the group $\mathcal{G}$ of invertible elements of Psd.

Since $L_0 = K_0\partial_x^2 K_0^{-1}$ and $K_0 \in \mathcal{G}_-$, we can write

$$\begin{aligned} K = UK_0 &= (K_0(\exp t\partial_x^3)K_0^{-1})_-^{-1}K_0 \\ &= [K_0((\exp t\partial_x^3)K_0^{-1})_-]^{-1}K_0 \\ &= ((\exp t\partial_x^3)K_0^{-1})_-^{-1}. \end{aligned}$$

Hence the Guiding Principle can be expressed as follows: the solution of the Lax form $L_t = [(L^{3/2})_+, L]$ of the KdV equation is given by

$$\boxed{L = K\partial_x^2 K^{-1}, \quad K = ((\exp t\partial_x^3)K_0^{-1})_-^{-1}.}$$

This is the form that we shall use. In principle it tells us how K evolves from K_0, and this in turn tells us how $u(x,t)$ evolves from $u(x,0)$ since $\partial_x^2 + u(x,t) = L = K\partial_x^2 K^{-1}$.

In the next section we shall give three 'realizations' of the Guiding Principle, that is, three ways of expressing solutions of the KdV equation which do not rely on properties of the hypothetical group $\mathcal{G}$.

8.4 Solving the KdV equation

The Sato D-module

Definition 8.11 $\mathcal{J} = DK = \{PK \mid P \in D\}$.

For each t, this is an infinite-dimensional linear subspace of the infinite-dimensional vector space Psd. It is also an infinite-rank submodule of the infinite-rank D-module Psd.

The Guiding Principle says that $\mathcal{J}$ evolves from $\mathcal{J}_0 = \mathcal{J}|_{t=0}$ through the formula

$$\begin{aligned}
\mathcal{J} &= D((\exp t\partial_x^3)K_0^{-1})_-^{-1} \\
&= D((\exp t\partial_x^3)K_0^{-1})_+^{-1}((\exp t\partial_x^3)K_0^{-1})_-^{-1} \text{ as } DQ = D \text{ for any } Q \in \mathcal{G}_+ \\
&= D((\exp t\partial_x^3)K_0^{-1})^{-1} \\
&= DK_0(\exp t\partial_x^3)^{-1} \\
&= \mathcal{J}_0 \exp -t\partial_x^3.
\end{aligned}$$

Now, all this is still 'hypothetical'. But the last expression for the evolution of $\mathcal{J}$ is very simple, its only defect being that the exponential $\exp -t\partial_x^3$ (and its action on $\mathcal{J}_0$) has not been rigorously defined. The approach used in [117] is based on the infinitesimal version of this expression, which can be defined without difficulty. We shall not go in this direction, as we mention the Sato approach only as motivation for the methods of the next two parts.

The Segal–Wilson Grassmannian and the Baker function

Let V be a point of the Grassmannian $\mathrm{Gr}^{(2)}$. The construction of Segal and Wilson associates to V a solution $\psi_V(x)$ of the equation $(L_0 - \lambda)\psi_V = 0$, and, more generally, a solution $\psi_V(x,t)$ of the system $(L-\lambda)\psi_V = 0$, $(\partial_t - P)\psi_V = 0$. This ψ_V is called a Baker function.

The motivation for this construction is the formula $L = K\partial_x^2 K^{-1}$ above. Since $e^{\mu x}$ is a solution of $(\partial_x^2 - \mu^2)y = 0$, the formula implies that

$$Ke^{\mu x} = e^{\mu x}(1 + \alpha_1\mu^{-1} + \alpha_2\mu^{-2} + \cdots)$$
$$Ke^{-\mu x} = e^{-\mu x}(1 - \alpha_1\mu^{-1} + \alpha_2\mu^{-2} - \cdots)$$

are formal solutions of $(L-\mu^2)y = 0$:

$$(L-\mu^2)Ke^{\pm\mu x} = (K\partial_x^2 K^{-1} - \mu^2)Ke^{\pm\mu x} = K(\partial_x^2 - \mu^2)e^{\pm\mu x} = 0.$$

The Grassmannian model can be used to study the nature of the formal solutions arising this way.

Evidently, it is convenient to use the spectral parameter μ with $\mu^2 = \lambda$ here, so, instead of

$$L^1(S^1, \mathbb{C}^2) = H^{(2)} = \text{Span}\{\lambda^i e_j \mid i \in \mathbb{Z}, j = 1,2\},$$

we introduce

$$L^1(S^1, \mathbb{C}) = H^{\text{scalar}} = \text{Span}\{\mu^i \mid i \in \mathbb{Z}\}$$

and identify these Hilbert spaces by identifying specific ordered bases as follows:

$$\begin{array}{cccccccc} \cdots & \mu^{-2} & \mu^{-1} & 1 & \mu & \mu^2 & \mu^3 & \cdots \\ \cdots & \lambda^{-1}e_1 & \lambda^{-1}e_2 & e_1 & e_2 & \lambda e_1 & \lambda e_2 & \cdots \end{array}$$

This isomorphism converts $f(\mu) \in L^1(S^1, \mathbb{C})$ to

$$\tfrac{1}{2}(f(\mu) + f(-\mu))e_1 + \tfrac{1}{2\mu}(f(\mu) - f(-\mu))e_2.$$

Conversely, $f_1(\lambda)e_1 + f_2(\lambda)e_2 \in L^1(S^1, \mathbb{C}^2)$ is converted to $f_1(\mu^2) + \mu f_2(\mu^2) \in L^1(S^1, \mathbb{C})$.

We define the 'scalar Grassmannian' $\text{Gr}^{\text{scalar},(2)}$ by replacing the condition $\lambda V \subseteq V$ (in the definition of $\text{Gr}^{(2)}$) by $\mu^2 V \subseteq V$. The isomorphism $H^{\text{scalar}} \cong H^{(2)}$ identifies $\text{Gr}^{\text{scalar},(2)}$ with $\text{Gr}^{(2)}$. Like $\text{Gr}^{(2)}$, $\text{Gr}^{\text{scalar},(2)}$ has an open dense 'big cell'; this is explained in more detail in appendix 8.8.

The result of Segal and Wilson (see [112]) is as follows:

Theorem 8.12 *Let V be in the big cell of $Gr^{\text{scalar},(2)}$. Let $g(x,t) = \exp(x\mu + t\mu^3)$. Define $\psi_V(x,t)$ by $\psi_V = g(\pi_+|_{g^{-1}V})^{-1}(1)$, where π_+ denotes orthogonal projection on H_+^{scalar}. (This is well defined for (x,t) near the origin, since $g^{-1}V$ is in the big cell for (x,t) near the origin.) Then $(L-\mu^2)\psi_V = 0$, $(\partial_t - P)\psi_V = 0$.*

Observe that ψ_V is necessarily of the form

$$\psi_V(x,t) = \exp(x\mu + t\mu^3)(1 + \alpha_1(x,t)\mu^{-1} + \alpha_2(x,t)\mu^{-2} + \cdots)$$

so the theorem produces a formal solution of the type predicted by the formal calculation above. However, it is not just a formal solution, but a genuine (local) solution whose analytic status is inherited from that of the point V. Moreover, its singularities have a geometrical origin: they are the points (x,t) where $\exp(x\mu + t\mu^3)V$ leaves the big cell.

Dressing

A third approach to the KdV equation, in which solutions are obtained by applying 'dressing transformations' to the 'vacuum solution', depends on using the matrix version rather than the scalar version of the equation. We recall from Example 4.15 that the KdV equation is equivalent to the zero curvature condition $dA^\lambda - A^\lambda \wedge A^\lambda = 0$, where $A^\lambda = A_1^\lambda dx + A_2^\lambda dt$ and

$$A_1^\lambda = \lambda\begin{pmatrix}0 & 0\\ 1 & 0\end{pmatrix} + \begin{pmatrix}0 & 1\\ -u & 0\end{pmatrix}$$

$$A_2^\lambda = \lambda^2\begin{pmatrix}0 & 0\\ 1 & 0\end{pmatrix} + \lambda\begin{pmatrix}0 & 1\\ -u/2 & 0\end{pmatrix} + \begin{pmatrix}-u_x/4 & u/2\\ -u^2/2 - u_{xx}/4 & u_x/4\end{pmatrix}.$$

The vacuum solution means $u = 0$. This corresponds to the simple connection form $\Lambda = \Lambda_1 dx + \Lambda_2 dt$, where

$$\Lambda_1 = \lambda\begin{pmatrix}0 & 0\\ 1 & 0\end{pmatrix} + \begin{pmatrix}0 & 1\\ 0 & 0\end{pmatrix} = \begin{pmatrix}0 & 1\\ \lambda & 0\end{pmatrix}$$

$$\Lambda_2 = \lambda^2\begin{pmatrix}0 & 0\\ 1 & 0\end{pmatrix} + \lambda\begin{pmatrix}0 & 1\\ 0 & 0\end{pmatrix} = \begin{pmatrix}0 & \lambda\\ \lambda^2 & 0\end{pmatrix}.$$

This is exactly analogous to the relation between $L = \partial_x^2 + u - \lambda$ and $\partial_x^2 - \lambda$: Λ arises from the D-module $D/(\partial_x^2 - \lambda, \partial_t - \partial_x^3)$ in the same way that A^λ arises from $D/(L-\lambda, \partial_t - P)$. But, while pseudo-differential operators were needed to convert L to ∂_x^2, an uncontroversial gauge transformation converts A^λ to Λ. We shall now give an explicit formula for this gauge transformation.

We have $A^\lambda = dHH^{-1}$, where H is a fundamental solution matrix. It is obvious that $\Lambda = dEE^{-1}$, where $E = \exp(x\Lambda_1 + t\Lambda_2)$. The required gauge transformation G will be such that $H = GE$ (or $H = GE\delta$ where δ is a constant element of $\Lambda GL_2\mathbb{C}$). This is equivalent to $A^\lambda = dGG^{-1} + G\Lambda G^{-1}$ or $d - A^\lambda = G(d - \Lambda)G^{-1}$.

It follows that an explicit formula for G amounts to an explicit formula for H. Such a formula (see [112], [129] and [134]) arises as follows:

Theorem 8.13 *Let*

$$\gamma \in \Lambda_- GL_2\mathbb{C}\ \Lambda_+ GL_2\mathbb{C} \quad (\subseteq \Lambda GL_2\mathbb{C}).$$

Thus, $E\gamma E^{-1} \in \Lambda_- GL_2\mathbb{C}\ \Lambda_+ GL_2\mathbb{C}$ *for* (x,t) *in some neighbourhood of* $(0,0)$. *Define* $G = (E\gamma E^{-1})_-^{-1}$. *Then* $d - A^\lambda = G(d - \Lambda)G^{-1}$ *is a flat connection of the above form, so the function* u *(which appears in* A^λ*) is a solution of the KdV equation.*

Proof We claim that A_1^λ is linear in λ, and that A_2^λ is a polynomial in λ of degree 2. Writing $D = E\gamma$, we have $G = (E\gamma E^{-1})_-^{-1} = D_-^{-1} = D_+ D^{-1}$. Recall that $A = dHH^{-1} = d(GE)(GE)^{-1} = dGG^{-1} + G\Lambda G^{-1}$. Using $G = D_-^{-1}$, we see that A_1^λ, A_2^λ have, respectively, no higher powers of λ than Λ_1, Λ_2. Using $G = D_+ D^{-1}$ we obtain $A = dD_+ D_+^{-1}$, hence A_1^λ, A_2^λ have no negative powers of λ. This proves the assertion.

An argument similar to that in section 7.6 shows that the coefficients of A_1^λ, A_2^λ have exactly the right form (in terms of u); cf. the example in the next section. □

Since E takes values in $\Lambda_+ GL_2\mathbb{C}$, the definition of G simplifies to $G = (E\gamma)_-^{-1}$, as in the proof above. Thus, the 'explicit' formula for the fundamental solution matrix is

$$H = (E\gamma)_-^{-1} E.$$

We shall explain briefly in the next section how this can be used to obtain truly explicit solutions $u(x,t)$ of the KdV equation.

Let us consider this result from the viewpoint of Proposition 7.7. The factorizations

$$F = F_- F_+ = \check{F}_+ \check{F}_-$$

(of $F = H^t$) produce two matrix-valued functions $F_-, \check{F}_+$ which are equivalent to F, via $F = \check{F}_+ (\check{F}_+^{-1} F_-)_-$. The formula $H = (E\gamma)_-^{-1} E$ is of exactly this type: we have $F = H^t = E^t(E\gamma)_-^{t\,-1} = E^t(E^{t-1}\gamma^{t-1})_-$, that is,

$$F_- = \gamma^{t-1}, \quad \check{F}_+ = E^t.$$

Alternatively, in terms of the fundamental construction of section 8.2 we have

$$\begin{aligned} W &= H^t H_+^{(2)} \\ &= E^t (E\gamma)_-^{t\,-1} H_+^{(2)} \\ &= E^t (E\gamma)_-^{t\,-1} (E\gamma)_+^{t\,-1} H_+^{(2)} \\ &= E^t (E\gamma)^{t-1} H_+^{(2)} \\ &= \gamma^{t-1} H_+^{(2)}, \end{aligned}$$

and

$$\begin{aligned} \check{W} &= H^t \check{H}_+^{(2)} \\ &= E^t (E\gamma)_-^{t\,-1} \check{H}_+^{(2)} \\ &= E^t \check{H}_+^{(2)}. \end{aligned}$$

In the situation of Proposition 7.7 (harmonic maps), F_- and $\check{F}_+$ are functions of one variable, and are therefore much simpler than F itself. In the current situation (the dressing construction for KdV), F_- and $\check{F}_+$ are again much simpler than F, but in a different way: F_- is constant and $\check{F}_+$ is the exponential of a linear function.

The dressing construction is usually described by saying that $H = (E\gamma)_-^{-1} E$ is the result of dressing the vacuum solution E by the loop group element γ. The dressing 'action' is a group action, defined abstractly as follows. If $\mathcal{G} = \mathcal{G}_1 \mathcal{G}_2$ is a multiplicative decomposition of groups, with $\mathcal{G}_1 \cap \mathcal{G}_2 = \{e\}$, then the action of $\mathcal{G}$ on itself is given by

$$g \cdot h = (hgh^{-1})_1^{-1} h.$$

With this notation we have

$$H = \gamma \cdot E.$$

In terms of the Grassmannian model, the vacuum solution is given by

$$W_{\text{vac}} = H_+^{(2)}, \quad \check{W}_{\text{vac}} = E^t \check{H}_+^{(2)}$$

and the result of dressing by the loop γ is

$$W = \gamma^{t-1} H_+^{(2)}, \quad \check{W} = E^t \check{H}_+^{(2)}$$

(cf. the remarks on the 'natural action' at the end of section 8.2).

Another point of view would be to focus on the main ingredient $(E\gamma)_-$ in the formula for H. This corresponds to the $\mathrm{Gr}^{(2)}$-valued map $E\gamma H_+^{(2)}$, which can be regarded as a 'linear flow on the Grassmannian', with initial point $\gamma H_+^{(2)}$ at $(x,t) = (0,0)$.

8.5 Solving the KdV equation: summary

We begin by explaining how the three standard approaches are related to each other. In view of the D-module characterization of the KdV equation in section 7.1, it is perhaps not surprising that they all have the same D-module-theoretic origin.

The situation at $t = 0$ is depicted in the following commutative diagram, where $L_0 = L|_{t=0}$ etc. and all maps are isomorphisms of D-modules:

$$\begin{array}{ccc} D_x \otimes \mathcal{H}_\lambda^+ \;/\,(L_0 - \lambda) & \xrightarrow{[X]\mapsto XJ_0} & \Gamma W_0 \\ \downarrow & & \downarrow \\ \mathcal{J}_0 = D_x K_0 & \xrightarrow{XK_0 \mapsto XK_0 e^{x\mu}} & \Gamma W_0^{\mathrm{scalar}} \end{array}$$

The top row is the 'fundamental construction' of section 8.2. The left hand vertical map is given by $[X] \mapsto XK_0$. This is an isomorphism of $D_x \otimes \mathcal{H}_\lambda^+$-modules if we define the action of λ on $D_x K_0$ by $\lambda \cdot XK_0 = XK_0 \partial_x^2$. Observe that this gives $(L_0 - \lambda) \cdot K_0 = 0$.

The t-extension or 't-flow' gives a similar diagram of extended D-modules:

$$\begin{array}{ccc} D_{x,t} \otimes \mathcal{H}_\lambda^+ \;/\,(L - \lambda, \partial_t - P) & \xrightarrow{[X]\mapsto XJ} & \Gamma W \\ \downarrow & & \downarrow \\ \mathcal{J} = D_x K & \xrightarrow{XK \mapsto XK e^{x\mu + t\mu^3}} & \Gamma W^{\mathrm{scalar}} \end{array}$$

The left hand vertical map is given by $[X] \mapsto XK$, and this is an isomorphism of $D_{x,t} \otimes \mathcal{H}_\lambda^+$-modules if the action of ∂_t on $D_x K_0$ is the one determined by $\partial_t \cdot K = PK$. Observe that this gives $(\partial_t - P) \cdot K = 0$.

We just mention briefly how the diagram above should be modified for the mKdV equation. As explained in section 7.2, the mKdV equation arises from a 'flag' of D-modules (rather than a single D-module). The fundamental construction converts this to a map

$$W_1 \subseteq W_0$$

into an infinite-dimensional flag manifold $\mathrm{Fl}^{(2)}$. This flag manifold will be discussed later in Examples 8.18 and 10.15. It can be described as the homogeneous space for the twisted loop group $(\Lambda\mathrm{SL}_2\mathbb{C})_\sigma$ (see section 7.2) which is analogous to the homogeneous space $\mathrm{Gr}^{(2)}$ for the loop group $\Lambda\mathrm{SL}_2\mathbb{C}$.

To illustrate all this abstract theory, let us produce a concrete solution of the KdV equation, starting with a choice of loop γ and ending with a formula for $u(x,t)$. The main step will be the computation of $(E\gamma)_-$; then it is easy to compute $H = (E\gamma)_-^{-1}E$ and $A = dHH^{-1}$, from which u can be read off.

The simplest type of example occurs for a loop of 'finite uniton number', that is, a loop which is polynomial in λ and λ^{-1}. Such loops will play an important role for harmonic maps in section 8.6 and for quantum cohomology in Chapter 10. In the case of the KdV equation, they lead to solutions which are rational functions of x and t.

Somewhat perversely, we shall begin with a loop

$$\gamma = \begin{pmatrix} \lambda & \\ & \lambda^{-1} \end{pmatrix}$$

which is *not* in the big cell. However, this is a natural choice from the viewpoint of the geometry of $\mathrm{Gr}^{(2)}$ (as we shall see), and $E\gamma$ is in the big cell for (x,t) near $(0,0)$. Thus, the previous discussion will apply, away from the singular point $(0,0)$.

As a computational tool, we use the fact that the orbit $\Lambda_+\mathrm{SL}_2\mathbb{C}\,\gamma\, H_+^{(2)}$ has a dense open finite-dimensional 'Schubert cell' (in a sense that will be explained shortly), consisting of points in $\mathrm{Gr}^{(2)}$ of the form

$$\tilde{E}\begin{pmatrix} \lambda & \\ & \lambda^{-1} \end{pmatrix} H_+^{(2)}, \quad \tilde{E} = \exp\left(\begin{pmatrix} & a_0 \\ & \end{pmatrix} + \lambda\begin{pmatrix} & a_1 \\ & \end{pmatrix}\right) = \begin{pmatrix} 1 & a_0 + \lambda a_1 \\ 0 & 1 \end{pmatrix}$$

with $a_0, a_1 \in \mathbb{C}$. Since

$$\begin{pmatrix} \lambda & \\ & \lambda^{-1} \end{pmatrix} H_+^{(2)} = \mathrm{Span}\left\{\tfrac{1}{\lambda}\begin{pmatrix} 0 \\ 1 \end{pmatrix}, \begin{pmatrix} 0 \\ 1 \end{pmatrix}, \lambda H_+^{(2)}\right\},$$

we have

$$\tilde{E}\begin{pmatrix} \lambda & \\ & \lambda^{-1} \end{pmatrix} H_+^{(2)} = \mathrm{Span}\left\{\tfrac{1}{\lambda}\begin{pmatrix} a_0 \\ 1 \end{pmatrix} + \begin{pmatrix} a_1 \\ 0 \end{pmatrix}, \begin{pmatrix} a_0 \\ 1 \end{pmatrix}, \lambda H_+^{(2)}\right\}$$

and (from the expansion of E)

$$E\begin{pmatrix} \lambda & \\ & \lambda^{-1} \end{pmatrix} H_+^{(2)} = \mathrm{Span}\left\{\tfrac{1}{\lambda}\begin{pmatrix} x \\ 1 \end{pmatrix} + \begin{pmatrix} t + \frac{1}{6}x^3 \\ \frac{1}{2}x^2 \end{pmatrix}, \begin{pmatrix} x \\ 1 \end{pmatrix}, \lambda H_+^{(2)}\right\}.$$

Comparing these, we see that $E\gamma H_+^{(2)} = \tilde{E}\gamma H_+^{(2)}$ (and hence $(E\gamma)_- = (\tilde{E}\gamma)_-$) if we take $a_0 = x$, $a_1 = t - x^3/3$.

The Birkhoff factorization of $\tilde{E}\gamma$ must be of the form $(I + \frac{1}{\lambda}X)Y(I + \lambda Z)$, where X, Z are nilpotent matrices. Comparing $(I + \frac{1}{\lambda}X)H_+^{(2)}$ with $\tilde{E}\gamma H_+^{(2)}$, we obtain X and hence Y, Z as well. This gives

$$\tilde{E}\gamma = \left[I + \frac{1}{\lambda}\frac{1}{a_1}\begin{pmatrix} a_0 & -a_0^2 \\ 1 & -a_0 \end{pmatrix}\right] \frac{1}{a_1}\begin{pmatrix} -a_0 & a_1^2 \\ -1 & 0 \end{pmatrix} \left[I + \lambda\frac{1}{a_1}\begin{pmatrix} 0 & 0 \\ 1 & 0 \end{pmatrix}\right].$$

With $H = GE$ and $G = (\tilde{E}\gamma)_-^{-1}$, we obtain

$$H_x H^{-1} = G_x G^{-1} + G\begin{pmatrix} 0 & 1 \\ \lambda & 0 \end{pmatrix} G^{-1} = \frac{1}{a_1}\begin{pmatrix} a_0^2 & a_1 \\ 2a_0 & -a_0^2 \end{pmatrix} + \lambda\begin{pmatrix} 0 & 0 \\ 1 & 0 \end{pmatrix}.$$

Unfortunately, this is not of the expected form (because the diagonal entries should be zero). To remedy the problem, we use the ambiguity in the Birkhoff factorization — the central factor — and redefine $(\tilde{E}\gamma)_-$ by

$$(\tilde{E}\gamma)_- = \left(I + \frac{1}{\lambda}\frac{1}{a_1}\begin{pmatrix} a_0 & -a_0^2 \\ 1 & -a_0 \end{pmatrix}\right)\begin{pmatrix} a & 0 \\ c & a \end{pmatrix}$$

where a, c are functions of z. For any a, c we obtain $H_x H^{-1}$ with the same coefficient of λ, and if we choose $a = 1$, $c = -a_0^2/a_1$, we obtain

$$H_x H^{-1} = \begin{pmatrix} 0 & 1 \\ -u & 0 \end{pmatrix} + \lambda\begin{pmatrix} 0 & 0 \\ 1 & 0 \end{pmatrix},$$

where $-u = (a_0^2/a_1)_x + a_0^4/a_1^2 + 2a_0/a_1$. The formula for $H_t H^{-1}$ in terms of u follows automatically from that for $H_x H^{-1}$ and the zero curvature equation (cf. the discussion in section 7.6 for the mKdV equation), giving the KdV equation $4u_t = u_{xxx} + 6uu_x$, as in Example 4.15.

Substituting the values of a_0, a_1 computed earlier, we find the following explicit solution of this equation:

$$u(x,t) = -\frac{6x(x^3 + 6t)}{(3t - x^3)^2}.$$

This is singular at $(x,t) = (0,0)$, as expected.

While this is just one example, it turns out that the most general 'finite uniton number solutions' arise in essentially the same way, replacing $E\gamma$ by

$$\exp\left(x\begin{pmatrix} 0 & 1 \\ \lambda & 0 \end{pmatrix} + t\begin{pmatrix} 0 & \lambda \\ \lambda^2 & 0 \end{pmatrix} + t_3\begin{pmatrix} 0 & \lambda^2 \\ \lambda^3 & 0 \end{pmatrix} + \cdots\right)\begin{pmatrix} \lambda^k & \\ & \lambda^{-k} \end{pmatrix}.$$

This is because the identity component of $\mathrm{Gr}^{(2)}_{\mathrm{alg}}$, which can be identified with the loop group $\Omega_{\mathrm{alg}}\mathrm{SU}_2$, has a Morse-Bott decomposition given by the unstable manifolds of the energy functional (cf. Example 8.17), and these are precisely the orbits of the geodesics $\mathrm{diag}(\lambda^k,\lambda^{-k})$ under $\Lambda^{\mathrm{alg}}_+\mathrm{SL}_2\mathbb{C}$. For $\mathrm{diag}(\lambda^k,\lambda^{-k})$, the orbit is a complex vector bundle of rank $2k-1$ over $\mathbb{C}P^1$, and the part lying over a big cell of $\mathbb{C}P^1$ may be identified with the set of points of the form

$$\exp\left(\begin{pmatrix} & a_0 \\ & \end{pmatrix} + \lambda\begin{pmatrix} & a_1 \\ & \end{pmatrix} + \cdots + \lambda^{2k-1}\begin{pmatrix} & a_{2k-1} \\ & \end{pmatrix}\right)\begin{pmatrix} \lambda^k & \\ & \lambda^{-k} \end{pmatrix} H^{(2)}_+,$$

where $a_0, a_1, \ldots, a_{2k-1} \in \mathbb{C}$. This illustrates the principle mentioned in the Introduction, concerning the usefulness of natural coordinates on a Schubert cell of the Grassmannian (in this case, a finite-dimensional cell).

In this description we have ignored the complement of the big cell of the Schubert variety. To obtain *all* types of finite uniton number solutions, we must consider this set as well. For example, in the calculation above (where $k = 1$), the complement of the big cell consists of points of the form

$$\mathrm{Span}\left\{\tfrac{1}{\lambda}\begin{pmatrix}1\\0\end{pmatrix} + \begin{pmatrix}0\\b\end{pmatrix}, \begin{pmatrix}1\\0\end{pmatrix}, \lambda H^{(2)}_+\right\}$$

for any $b \in \mathbb{C}$, together with the single point $H^{(2)}_+$. (The Schubert variety has three cells, of the form $\mathbb{C}^2$, $\mathbb{C}^1$, $\mathbb{C}^0$.) The special point $H^{(2)}_+$ gives the trivial solution $u = 0$ of the KdV equation, as we have seen; this is just the case $k = 0$. As an example of the remaining type of solution, let us consider the 'centre' of the cell $\mathbb{C}$, given by $b = 0$; this corresponds to

$$\gamma = \begin{pmatrix}\lambda^{-1} & \\ & \lambda\end{pmatrix}.$$

Exactly as above, we obtain

$$E\gamma H^{(2)}_+ = \tilde{E}H^{(2)}_+, \quad \tilde{E} = \exp\tfrac{1}{\lambda}\begin{pmatrix} & \frac{1}{x} \\ & \end{pmatrix}.$$

Observe that $\tilde{E}_- = \tilde{E}$ in this case, and the variable t has disappeared. Taking $H = GE$, with $G = \tilde{E}^{-1}$, we obtain

$$H_xH^{-1} = \tfrac{1}{x}\begin{pmatrix}-1 & x\\ 0 & 1\end{pmatrix} + \lambda\begin{pmatrix}0 & 0\\ 1 & 0\end{pmatrix}.$$

To achieve the correct shape (zero diagonal), we multiply $\tilde{E}$ on the right by a λ-independent matrix as in the previous calculation, but this time with

$a = 1$ and $c = 1/x$. We obtain the explicit solution

$$u(x,t) = -\frac{2}{x^2}.$$

Because of the absence of t, this is called a stationary solution; it is a solution of $0 = u_{xxx} + 6uu_x$.

Similarly, from the big cell for general k, the variables $t_{4k-1}, t_{4k+1}, \ldots$ disappear, and we obtain rational solutions $u(x, t, t_3, t_5, \ldots, t_{4k-3})$ of the KdV hierarchy depending on the first $2k$ variables which are stationary with respect to the remaining variables. The awkwardness of dealing separately with an odd number of variables may be removed by passing to the scalar Grassmannian, as in [112].

The stationary solution property leads to a celebrated construction which generalizes the case of rational solutions, based on the following observation. In the simple example above, the operators L, P (of Example 4.15) satisfy $L_t = [P, L]$, and therefore $0 = [P, L]$, when $u = -2/x^2$. By a theorem of J. L. Burchnall and T. W. Chaundy, these commuting differential operators must satisfy an algebraic relation, which in this example is $P^2 = L^3$. Formally, this equation may be regarded as the equation of a rational curve C; the coordinate ring of the curve may be identified with the algebra A_C of polynomials in L and P, or the abstract algebra $\mathbb{C}[x, y]/(x^3 - y^2)$. In terms of the scalar Grassmannian, the rational solutions are then characterized by the condition $A_C W^{\text{scalar}} \subseteq W^{\text{scalar}}$. We shall meet this kind of condition again in section 10.6. It can be interpreted as saying that the D-module for this particular solution may be 'enhanced' by the algebra A_C.

More generally, a stationary solution with respect to *any* variable t_{2i-1} satisfies $0 = [P_{2i-1}, L]$, leading to an algebraic curve C and explicit formulae for such solutions in terms of theta functions. The famous soliton solutions are the next simplest class of solutions that arise this way, from curves of the form $y^2 = x(x - \alpha)^2$. A detailed introduction to this theory can be found in [5].

8.6 Solving the harmonic map equation

In contrast to the KdV equation, the harmonic map equation did not arise from a 'D-module extension problem'. Nevertheless, using the fundamental construction of section 8.2, the zero curvature formulation of the harmonic map equation leads to a correspondence between harmonic maps and maps into the infinite-dimensional Grassmannian. A fortiori these correspond to certain D-modules. We shall say later on (section 8.7) how these D-modules are relevant to solving the harmonic map equation. However, the main purpose of this section is to show how the infinite-dimensional Grassmannian can be used to find some families of solutions of the harmonic map equation.

Let us recall some notation and results from section 7.3. Harmonic maps

$$\phi : \mathbb{R}^2 = \mathbb{C} \to G$$

correspond to extended solutions

$$F : \mathbb{C} \to \Lambda G$$

and these are restrictions of extended solutions $F : \mathbb{C} \to \Lambda G^{\mathbb{C}}$ satisfying the reality condition $C(F(\lambda)) = F(1/\bar{\lambda}))$. Using the Birkhoff factorization

$$F = F_- F_+ = \check{F}_+ \check{F}_-$$

and Proposition 7.7, one sees that F corresponds to the pair $(F_-, \check{F}_+)$. Since $\check{F}_+(\lambda) = C(F_-(1/\bar{\lambda}))$ here, one can say that F corresponds to F_- alone. The advantage of F_- is that it is a (holomorphic) function of z, whereas the original F is a function of $z, \bar{z}$. Going one step further, $F_- = F_-(z)$ corresponds to a $\mathfrak{g}^{\mathbb{C}}$-valued function ω where

$$\tfrac{1}{\lambda}\omega = F_-^{-1} \partial F_- = F_-^{-1} F_-'.$$

This establishes the DPW (or generalized Weierstrass) correspondence of [32]: harmonic maps ϕ correspond, locally[4], to $\mathfrak{g}^{\mathbb{C}}$-valued holomorphic maps ω. Whereas the smooth map ϕ satisfies a non-linear p.d.e., in two real variables, the holomorphic map $\omega = \omega(z)$ is unconstrained.

We have just described the passage from ϕ to ω. The reverse procedure entails

(i) integration of the o.d.e. $\tfrac{1}{\lambda}\omega = F_-^{-1} F_-'$ to produce F_-,
(ii) Iwasawa factorization $F_- = F F_+^{-1}$ to produce F, then
(iii) substitution of $\lambda = -1$ in (the ΩG-valued) map F to produce ϕ.

Now we turn to the infinite-dimensional Grassmannian. The fundamental construction of section 8.2 produces a holomorphic map

$$W = F H_+^{(n)} = F_- H_+^{(n)}$$

[4] The global properties of ϕ are not easily read off from the global properties of ω, however. Even in the classical Weierstrass correspondence for minimal surfaces, this is a difficult problem.

(and the equivalent map $\check{W} = F\check{H}_+^{(n)} = \check{F}_+\check{H}_+^{(n)}$, which we can ignore). Here we are assuming that G has been realized as a group of $n\times n$ matrices. From Example 8.5 the harmonic map equation is equivalent to

$$\lambda \partial W \subseteq W.$$

This version of the harmonic map equation was introduced by G. B. Segal in [111]. It has the advantage that holomorphic maps into (finite-dimensional) Grassmannians are very familiar and provide a source of intuition.

Thinking of W as a holomorphic map into the homogeneous space

$$\mathrm{Gr}^{(n)} = \Lambda \mathrm{GL}_n\mathbb{C}/\Lambda_+\mathrm{GL}_n\mathbb{C}$$

suggests that we consider more general representations[5]

$$W = LH_+^{(n)},$$

where L is any holomorphic map of the form F_-X such that X takes values in $\Lambda_+\mathrm{GL}_n\mathbb{C}$. (Representing W by L is analogous to representing a holomorphic map into $\mathbb{C}P^{n-1}$ by a holomorphic map into $\mathrm{GL}_n\mathbb{C}$ or $\mathbb{C}^n - \{0\}$.)

It follows that

$$L^{-1}\partial L = \tfrac{1}{\lambda}\theta_{-1} + \theta_0 + \lambda\theta_1 + \cdots$$

whereas for $L = F_-$ we would have $\theta_{-1} = \omega$ and $\theta_i = 0$ for $i \geq 0$. We shall refer to L as a *complex extended solution,* and the particular choice $L = F_-$ as a *normalized complex extended solution,* following the usage in the harmonic maps literature, where $(\frac{1}{\lambda}\theta_{-1} + \theta_0 + \lambda\theta_1 + \cdots)dz$ is called a DPW potential and $\frac{1}{\lambda}\omega dz$ a normalized DPW potential.

From any complex extended solution L the Birkhoff factorization $L = L_-L_+$ produces $F_- = L_-$, and the Iwasawa factorization $L = L_{\mathbb{R}}L_+$ (not necessarily the same L_+) produces $F = L_{\mathbb{R}}$. Thus, no information is lost by using L to represent a harmonic map, and in fact this kind of representation arises naturally in examples, as we shall see. A specific choice of L reflects the specific nature of the example.

Before giving examples, we summarize the above discussion by incorporating W into a modified version of Theorem 7.11:

[5] This L plays the same role as the L in Chapters 5 and 6. In sections 8.3–8.5 of this chapter we used the notation L for the differential operator $\partial_x^2 + u$, as is traditional. No confusion is likely in the remainder of this chapter.

Theorem 8.14 *There is an essentially one to one correspondence*

$$\phi \longleftrightarrow F \longleftrightarrow W \longleftrightarrow L_- \longleftrightarrow \omega$$

between G-valued harmonic maps ϕ, ΛG*-valued extended solutions F,* $Gr^{(n)}$*-valued holomorphic maps W such that* $\lambda \partial W \subseteq W$, $\Lambda G^{\mathbb{C}}$*-valued normalised complex extended solutions* L_-*, and holomorphic* $\mathfrak{g}^{\mathbb{C}}$*-valued functions* ω*.*

With W as the central object, the Birkhoff factorization gives L_- and the Iwasawa factorization gives F. In particular, we have $L_- = F_-$ and $F = L_{\mathbb{R}}$.

Example 8.15 Holomorphic maps between Kähler manifolds are amongst the simplest examples of harmonic maps. In the present context these arise as follows. Let G/K be a Hermitian symmetric space immersed in G by means of the Cartan immersion

$$i : G/K \to G, \quad [g] \mapsto \sigma(g)g^{-1}$$

(see section 7.4). Then any holomorphic map $f : \mathbb{C} \to G/K$ is known to be harmonic. Moreover, since i is totally geodesic, it follows that $\phi = i \circ f$ is also harmonic. This class of examples includes the Gauss maps of minimal surfaces in $\mathbb{R}^3$, which are harmonic maps into the symmetric space $\mathbb{C}P^1 \cong \mathrm{U}_2/(\mathrm{U}_1 \times \mathrm{U}_1) \subseteq \mathrm{U}_2$.

Concrete expressions for all maps appearing in Theorem 8.14 are easily found for such harmonic maps. For example, in the case $G = \mathrm{U}_n$, the Hermitian symmetric spaces are the Grassmannians

$$\mathrm{Gr}_k(\mathbb{C}^n) \cong \frac{\mathrm{U}_n}{\mathrm{U}_k \times \mathrm{U}_{n-k}} \cong \frac{\mathrm{GL}_n\mathbb{C}}{G_k}$$

where G_k is the subgroup of $\mathrm{GL}_n\mathbb{C}$ consisting of $n \times n$ matrices of the form

$$\begin{pmatrix} * & * \\ 0 & * \end{pmatrix},$$

the bottom left corner being the $(n-k) \times k$ zero matrix. It is easily verified that

$$F = \pi_f + \lambda \pi_{f^\perp} : \mathbb{C} \to \Omega \mathrm{U}_n$$

is an extended solution corresponding to $f : \mathbb{C} \to \mathrm{Gr}_k(\mathbb{C}^n)$, where $\pi_{f(z)}$ denotes the orthogonal projection map $\mathbb{C}^n \to f(z)$ with respect to the Hermitian inner product of $\mathbb{C}^n$, and $\pi_{f(z)^\perp}$ the orthogonal projection map to $f(z)^\perp$.

The holomorphic map $f : \mathbb{C} \to \mathrm{Gr}_k(\mathbb{C}^n)$ may be represented (locally) as $f = [\psi]$, where ψ is a holomorphic $\mathrm{GL}_n\mathbb{C}$-valued map. This gives rise to a natural complex extended solution

$$L = \psi \operatorname{diag}(\underbrace{1, \ldots, 1}_{k}, \underbrace{\lambda, \ldots, \lambda}_{n-k}).$$

We shall abbreviate this as $L = \psi(P_k + \lambda P_k^\perp)$, where $P_k, P_k^\perp$ are the projection operators onto $\mathbb{C}^k, (\mathbb{C}^k)^\perp$.

We have

$$W = LH_+^{(n)} = f \oplus \lambda H_+^{(n)}.$$

This is just the original map f composed with the natural embedding

$$\mathrm{Gr}_k(\mathbb{C}^n) \to \mathrm{Gr}^{(n)}, \quad V \mapsto V \oplus \lambda H_+^{(n)}$$

of the finite-dimensional Grassmannian into the infinite-dimensional Grassmannian.

To find ω in this example, we must find $L_- = I + O(\frac{1}{\lambda})$ and compute $L_-^{-1}L_-'$. Without loss of generality we may choose ψ to be in the big cell of $\mathrm{GL}_n\mathbb{C}$ and perform a factorization

$$\psi = \begin{pmatrix} I & 0 \\ \alpha & I \end{pmatrix} \begin{pmatrix} * & * \\ 0 & * \end{pmatrix}$$

where α is holomorphic. In order to move L into the (big cell of) the identity component of the loop group, we pre-multiply it by $(P_k + \lambda P_k^\perp)^{-1}$. We have

$$\begin{aligned}(P_k + \lambda P_k^\perp)^{-1} L = &\begin{pmatrix} I & 0 \\ 0 & \lambda I \end{pmatrix}^{-1} \begin{pmatrix} I & 0 \\ \alpha & I \end{pmatrix} \begin{pmatrix} I & 0 \\ 0 & \lambda I \end{pmatrix} \\ &\times \begin{pmatrix} I & 0 \\ 0 & \lambda I \end{pmatrix}^{-1} \begin{pmatrix} * & * \\ 0 & * \end{pmatrix} \begin{pmatrix} I & 0 \\ 0 & \lambda I \end{pmatrix}\end{aligned}$$

which shows that

$$((P_k + \lambda P_k^\perp)^{-1} L)_- = \begin{pmatrix} I & 0 \\ \frac{1}{\lambda}\alpha & I \end{pmatrix}, \quad ((P_k + \lambda P_k^\perp)^{-1} L)_+ = \begin{pmatrix} * & \lambda * \\ 0 & * \end{pmatrix}.$$

We obtain

$$\begin{pmatrix} I & 0 \\ \frac{1}{\lambda}\alpha & I \end{pmatrix}^{-1} \begin{pmatrix} I & 0 \\ \frac{1}{\lambda}\alpha & I \end{pmatrix}' = \frac{1}{\lambda} \begin{pmatrix} 0 & 0 \\ \alpha' & 0 \end{pmatrix},$$

so the holomorphic data corresponding to the original holomorphic map f is just α'. This example is very simple, but it illustrates all the maps involved in the DPW correspondence. It has the special feature that the Iwasawa and Birkhoff factorizations reduce to their finite-dimensional counterparts. □

Example 8.16 A substantial generalization of the previous example is provided by the 'twistor construction' of F. E. Burstall and J. H. Rawnsley (see [19]), which generalizes earlier work of E. Calabi, J. Eells and J. Wood, R. Bryant, and others.

Let

$$G/H \cong G^{\mathbb{C}}/P$$

be a generalized flag manifold, where G is a compact real form of a complex semisimple Lie group $G^{\mathbb{C}}$, and P is a parabolic subgroup. Let

$$\pi : G/H \to G/K$$

be the associated canonical twistor fibration in the sense of [19]; G/K is a symmetric space canonically associated to $G^{\mathbb{C}}/P$. The main property of π is that it gives a method of constructing harmonic maps. Namely, if $h : \mathbb{C} \to G/H$ is a holomorphic map which is superhorizontal with respect to π, then the map $f = \pi \circ h : \mathbb{C} \to G/K$ is harmonic. The previous example is the trivial case $G/H = G/K$.

A typical example is the following construction of harmonic maps into $\mathbb{C}P^n$ from the 'osculating spaces' of a holomorphic map $f : \mathbb{C} \to \mathbb{C}P^n$, which is similar to the construction of the Frenet frame of a curve in Euclidean space. Let us represent f (locally) as $f = [\tilde{f}]$ where $\tilde{f} : \mathbb{C} \to \mathbb{C}^{n+1} - \{0\}$ is holomorphic. Then

$$f_{\langle i\rangle} = \mathrm{Span}\{\tilde{f}, \tilde{f}', \dots, \tilde{f}^{(i)}\} = \mathrm{Span}\{\tilde{f}, \partial\tilde{f}, \dots, \partial^i\tilde{f}\}$$

defines a holomorphic map into $\mathrm{Gr}_{i+1}(\mathbb{C}^{n+1})$. The domain of this map consists of those points z where $\tilde{f}(z), \tilde{f}'(z), \dots, \tilde{f}^{(i)}(z)$ are linearly independent. Without loss of generality we can assume that there is a non-empty open subset of $\mathbb{C}$ for which $\tilde{f}(z), \tilde{f}'(z), \dots, \tilde{f}^{(n)}(z)$ are linearly independent. On this domain

$$f_i = f_{\langle i-1\rangle}^{\perp} \cap f_{\langle i\rangle}$$

is a well defined smooth map taking values in $\mathbb{C}P^n$, for each $i = 0, 1, \dots, n$, and it turns out that f_i is harmonic. For $i = 0$ we have the original holomorphic map $f_0 = f$, and for $i = n$ we obtain an anti-holomorphic map, but for $0 < i < n$ we obtain harmonic maps which are neither holomorphic nor anti-holomorphic.

The fact that f_i is harmonic is a consequence of the fact that $(f_{\langle i-1\rangle}, f_{\langle i\rangle})$ is a holomorphic superhorizontal map into the total space of a twistor fibration

$$\pi : \frac{U_{n+1}}{U_i \times U_1 \times U_{n-i}} \to \frac{U_{n+1}}{U_1 \times U_n}.$$

A corresponding extended solution $F : \mathbb{C} \to \Omega U_n$ is

$$F = \pi_{i-1} + \lambda \pi_{i-1}^{\perp} \pi_i + \lambda^2 \pi_i^{\perp}$$

where $\pi_i(z) : \mathbb{C}^{n+1} \to f_{\langle i\rangle}(z)$ is the orthogonal projection map.

There is a natural choice of complex extended solution,

$$L = \begin{pmatrix} | & & | \\ \tilde{f} & \cdots & \tilde{f}^{(n)} \\ | & & | \end{pmatrix} \operatorname{diag}(\underbrace{1, \ldots, 1}_{i}, \lambda, \underbrace{\lambda^2, \ldots, \lambda^2}_{n-i}),$$

which we write as $\psi\,(P_{i-1} + \lambda P_{i-1}^{\perp} P_i + \lambda^2 P_i^{\perp})$.

We obtain

$$W = f_{\langle i-1\rangle} \oplus \lambda f_{\langle i\rangle} \oplus \lambda^2 H_+^{(n+1)} = \operatorname{Span}\{f_{\langle i-1\rangle}, \lambda f_{\langle i\rangle}, \lambda^2 H_+^{(n)}\},$$

which is the composition of $(f_{\langle i-1\rangle}, f_{\langle i\rangle})$ with a natural embedding of the flag manifold (the total space of the twistor fibration) into the infinite-dimensional Grassmannian.

Next we turn to L_-, which can be computed from a finite-dimensional factorization, exactly as in the previous example. Choosing the case $i = 1, n = 2$ for simplicity, let us (without loss of generality) make a factorization of the form

$$\psi = \begin{pmatrix} | & | & | \\ \tilde{f} & \tilde{f}' & \tilde{f}'' \\ | & | & | \end{pmatrix} = \begin{pmatrix} 1 & & \\ \alpha_1 & 1 & \\ \beta_1 & \alpha_2 & 1 \end{pmatrix} \begin{pmatrix} * & * & * \\ & * & * \\ & & * \end{pmatrix}.$$

Then we replace L by $\gamma^{-1} L$ where γ is the constant loop

$$\gamma(\lambda) = \begin{pmatrix} 1 & & \\ & \lambda & \\ & & \lambda^2 \end{pmatrix}.$$

From

$$\gamma^{-1}\begin{pmatrix} | & | & | \\ \tilde{f} & \tilde{f}' & \tilde{f}'' \\ | & | & | \end{pmatrix}\gamma = \gamma^{-1}\begin{pmatrix} 1 & & \\ \alpha_1 & 1 & \\ \beta_1 & \alpha_2 & 1 \end{pmatrix}\gamma\,\gamma^{-1}\begin{pmatrix} * & * & * \\ & * & * \\ & & * \end{pmatrix}\gamma$$

$$= \begin{pmatrix} 1 & & \\ \frac{1}{\lambda}\alpha_1 & 1 & \\ \frac{1}{\lambda^2}\beta_1 & \frac{1}{\lambda}\alpha_2 & 1 \end{pmatrix}\begin{pmatrix} * & \lambda * & \lambda^2 * \\ & * & \lambda * \\ & & * \end{pmatrix}$$

we obtain

$$(\gamma^{-1}L)_- = \begin{pmatrix} 1 & & \\ \frac{1}{\lambda}\alpha_1 & 1 & \\ \frac{1}{\lambda^2}\beta_1 & \frac{1}{\lambda}\alpha_2 & 1 \end{pmatrix} = \gamma^{-1}\begin{pmatrix} 1 & & \\ \alpha_1 & 1 & \\ \beta_1 & \alpha_2 & 1 \end{pmatrix}\gamma.$$

This normalized complex extended solution produces the holomorphic data

$$(\gamma^{-1}L)_-^{-1}(\gamma^{-1}L)_-' = \frac{1}{\lambda}\begin{pmatrix} 0 & 0 & 0 \\ \alpha_1' & 0 & 0 \\ 0 & \alpha_2' & 0 \end{pmatrix} + \frac{1}{\lambda^2}\begin{pmatrix} 0 & 0 & 0 \\ 0 & 0 & 0 \\ \beta_1' - \alpha_1'\alpha_2 & 0 & 0 \end{pmatrix}.$$

The second term must be zero, a fact which follows directly from the definition of $\alpha_1, \alpha_2, \beta_1$ or from the the general theory which says that $(\gamma^{-1}L)_-^{-1}(\gamma^{-1}L)_-'$ must be of the form $\frac{1}{\lambda}\omega$. Thus, the holomorphic data consists of α_1', α_2'.

So far this seems very similar to the previous example, but there is an important difference, which is indicated by the condition $\beta_1' = \alpha_1'\alpha_2$. To understand this it is instructive to reverse the construction and produce a harmonic map from arbitrary holomorphic data

$$\omega = \begin{pmatrix} 0 & 0 & 0 \\ v_1 & 0 & 0 \\ 0 & v_2 & 0 \end{pmatrix}.$$

First we solve the o.d.e. $L_-^{-1}L_-' = \frac{1}{\lambda}\omega$. Since the coefficient matrix is nilpotent, the o.d.e. can be solved by quadrature, and we obtain

$$L_- = \begin{pmatrix} 1 & & \\ & 1 & \\ & & 1 \end{pmatrix} + \frac{1}{\lambda}\begin{pmatrix} 0 & 0 & 0 \\ \int v_1 & 0 & 0 \\ 0 & \int v_2 & 0 \end{pmatrix} + \frac{1}{\lambda^2}\begin{pmatrix} 0 & 0 & 0 \\ 0 & 0 & 0 \\ \int(v_1\int v_2) & 0 & 0 \end{pmatrix}$$

$$= \begin{pmatrix} 1 & & \\ & \lambda & \\ & & \lambda^2 \end{pmatrix}^{-1}\begin{pmatrix} 1 & & \\ \int v_1 & 1 & \\ \int(v_1\int v_2) & \int v_2 & 1 \end{pmatrix}\begin{pmatrix} 1 & & \\ & \lambda & \\ & & \lambda^2 \end{pmatrix},$$

which recovers the map denoted $(\gamma^{-1}L)_-$ above. Then we perform the Iwasawa factorization. This amounts to applying the Gram–Schmidt orthogonalization procedure to the columns of the middle matrix in the last line, starting from the left. Substituting $\lambda = -1$ gives the harmonic map into U_3, and it is clear that this factors through $\mathbb{C}P^2$. Comparing this with the original notation, we have $v_1 = \alpha_1', v_2 = \alpha_2'$ and the additional relation $\beta_1 = \int(v_1 \int v_2)$, that is, $\beta_1' = \alpha_1'\alpha_2$.

We conclude that each harmonic map of this type corresponds to (a) the holomorphic data $\alpha_1, \alpha_2, \beta_1$ (i.e. L_-), constrained by the equation $\beta_1' = \alpha_1'\alpha_2$, or (b) the unconstrained holomorphic data α_1', α_2' (i.e. ω). In either case, the differential equation $\beta_1' = \alpha_1'\alpha_2$ arises, a feature which was not present in the previous example. □

In the above examples, the extended solutions were *polynomial* in λ, λ^{-1}. In general, harmonic maps with this property are called 'harmonic maps of finite uniton number' (the concept of uniton number, essentially the highest power of λ or λ^{-1} in an extended solution, was introduced by K. K. Uhlenbeck in [126]). Harmonic maps obtained from the twistor construction are particular examples, but there are many harmonic maps of finite uniton number which do not arise from the twistor construction—and the infinite-dimensional Grassmannian is the key to understanding these, despite their 'finite-dimensional' nature.

First of all, the finite uniton number condition means that W takes values in a finite-dimensional subvariety[6] of the form

$$\{V \in \mathrm{Gr}^{(n)} \mid \lambda^k H_+^{(n)} \subseteq V \subseteq \lambda^{-k} H_+^{(n)}\}$$

(see [111]). This is analogous to a Schubert variety in a finite-dimensional Grassmannian; in particular it can be described as the closure of $\exp \mathfrak{n}_k$ where $\mathfrak{n}_k$ is a nilpotent Lie subalgebra of the loop algebra $\Lambda\mathfrak{gl}_n\mathbb{C}$. It turns out that the harmonic map equation greatly simplifies if we use coordinates provided by $\mathfrak{n}_k$, and this leads to explicit formulae for all harmonic maps of this type. This point of view was developed in Chapter 22 of [60], and it leads to the general theory given in the next example.

Example 8.17 Let $\phi : \mathbb{C} \to G$ be a harmonic map of finite uniton number k. It was shown in [16] that ϕ admits a complex extended solution of the form

$$L = \exp C\gamma,$$

[6] In fact, k is bounded by an integer which depends on G (see [16]), so the problem still has a finite-dimensional nature. But the structure of these varieties is best understood in the context of the infinite-dimensional Grassmannian.

where $\gamma : S^1 \to G$ is a homomorphism and $C = C_0 + \lambda C_1 + \cdots + \lambda^{k-1} C_{k-1}$ is a holomorphic map with values in the (polynomial) loop algebra. This has the geometrical interpretation that $W = LH_+^{(n)}$ takes values in a Schubert subvariety of $\Omega G \subseteq \mathrm{Gr}^{(n)}$. The map $\gamma^{-1} L$ plays the role of L_- in the general theory above. Moreover,

$$\gamma^{-1} L = \exp(\gamma^{-1} C \gamma) = \exp(\tfrac{1}{\lambda} B_1 + \cdots + \tfrac{1}{\lambda^k} B_k),$$

where $\frac{1}{\lambda} B_1 + \cdots + \frac{1}{\lambda^k} B_k$ has values in a certain nilpotent Lie subalgebra of $\Lambda \mathfrak{g}^{\mathbb{C}}$.

Conversely, given any such holomorphic map $B = \frac{1}{\lambda} B_1 + \cdots + \frac{1}{\lambda^k} B_k$, $\exp B$ is a normalized complex extended solution for some harmonic map if and only if B satisfies a certain system of ordinary differential equations. This system has a very simple form—it is always integrable by quadrature. The holomorphic data is

$$(\exp B)^{-1} (\exp B)' = \tfrac{1}{\lambda} B_1'.$$

Thus, harmonic maps of finite uniton number fall into types given by homomorphisms γ. For each type there is a normalized extended solution given in terms of holomorphic maps $B_1, \ldots, B_k$ satisfying a system of o.d.e., or in terms of the unconstrained holomorphic data B_1'. Further details, and examples, can be found in [16], [60] and the update [61].

The twistor construction can be viewed in the following way. The Schubert variety above is in fact the closure of an unstable manifold of the energy functional $E : \Omega G \to \mathbb{R}$ on the loop group, and the corresponding critical submanifold is the conjugacy class of the geodesic γ. The harmonic maps obtained from the twistor construction are the ones for which W is contained in a such a critical manifold. Since the energy functional is the Hamiltonian function associated to the action of the circle group which acts on $\Omega G \cong \Lambda G / G$ by rotation, its critical points are the fixed points of this action. Thus, the harmonic maps obtained from the twistor construction are the harmonic maps into G whose extended solutions are fixed by this circle action. This is the case $C = C_0$ in the above description.

Let us mention a concrete example, in order to illustrate how harmonic maps of finite uniton number go beyond the twistor construction. The general theory says that all harmonic maps into $G = \mathrm{U}_3$ of finite uniton number corresponding to the homomorphism

$$\gamma = \begin{pmatrix} 1 & & \\ & \lambda & \\ & & \lambda^2 \end{pmatrix}$$

may be represented by normalized complex extended solutions of the form $\exp B$ with

$$B = \frac{1}{\lambda}\begin{pmatrix} 0 & 0 & 0 \\ v_1 & 0 & 0 \\ w & v_2 & 0 \end{pmatrix} + \frac{1}{\lambda^2}\begin{pmatrix} 0 & 0 & 0 \\ 0 & 0 & 0 \\ x & 0 & 0 \end{pmatrix}.$$

Those maps which arise from the twistor construction of Example 8.16 are characterized by $w = 0$. □

Example 8.18 As a continuation of the previous example, let us focus on harmonic maps of finite uniton number into a specific symmetric space G/K, with corresponding involution σ. The infinite-dimensional Grassmannian provides geometrical insight in this situation as well.

According to the general theory of section 7.4, it is necessary only to impose the restriction that all loops satisfy the twisting condition

$$\sigma(\gamma(\lambda)) = \gamma(-\lambda).$$

In particular, this applies to

$$B = \tfrac{1}{\lambda}B_1 + \cdots + \tfrac{1}{\lambda^k}B_k$$

and to the holomorphic data $\frac{1}{\lambda}B_1'$, and the twisting condition here is

$$\sigma(B_i) = (-1)^i B_i.$$

In the previous example, where

$$B = \frac{1}{\lambda}\begin{pmatrix} 0 & 0 & 0 \\ v_1 & 0 & 0 \\ w & v_2 & 0 \end{pmatrix} + \frac{1}{\lambda^2}\begin{pmatrix} 0 & 0 & 0 \\ 0 & 0 & 0 \\ x & 0 & 0 \end{pmatrix},$$

the twisting condition with respect to the involution

$$\sigma(X) = \operatorname{diag}(1, -1, 1)\ X\ \operatorname{diag}(1, -1, 1)$$

coincides with the twistor condition $w = 0$. For the involution

$$\sigma(X) = \operatorname{diag}(1, -1, -1)\ X\ \operatorname{diag}(1, -1, -1)$$

we obtain the condition $v_2 = 0$. In both cases we obtain harmonic maps into $\mathbb{C}P^2$.

There is a version of the infinite-dimensional Grassmannian for each symmetric space G/K, or rather for each involution σ. We shall just give the main example here, for $\mathrm{Gr}^{(n)}$ and an involution σ of the form

$$\sigma(X) = \operatorname{diag}(I_{n_1}, -I_{n_2}, I_{n_3})\ X\ \operatorname{diag}(I_{n_1}, -I_{n_2}, I_{n_3})$$

where $n_1 + n_2 + n_3 = n$. We define the twisted Grassmannian by

$$\mathrm{Gr}^{(n)}_\sigma = (\Lambda \mathrm{GL}_n\mathbb{C})_\sigma H^{(n)}_+ \subseteq \mathrm{Gr}^{(n)}.$$

This is the natural target space of W. The twisted Grassmannian can be identified with a certain infinite-dimensional flag manifold, bringing it closer to the original symmetric space, which in this case is the Grassmannian

$$G/K = \frac{U_n}{U_{n_1+n_3} \times U_{n_2}} \, .$$

The identification uses the isomorphism[7] $T : \Lambda \mathrm{GL}_n\mathbb{C} \to (\Lambda \mathrm{GL}_n\mathbb{C})_\sigma$ given by

$$T(\gamma)(\lambda) = \mathrm{diag}(I_{n_1}, \lambda I_{n_2}, I_{n_3})^{-1}\ \gamma(\lambda^2)\ \mathrm{diag}(I_{n_1}, \lambda I_{n_2}, I_{n_3}).$$

The inverse of T induces an identification

$$\mathrm{Gr}^{(n)}_\sigma \to \mathrm{Fl}^{(n)}_{n_1+n_3,n_2}$$

where the right hand side denotes the 'periodic flag manifold' consisting of flags of subspaces of $\mathrm{Gr}^{(n)}$ of the form

$$\lambda V_0 = V_2 \subseteq V_1 \subseteq V_0$$

with $\dim V_1/V_2 = n_2$. The harmonic map equation $\lambda \partial W \subseteq W$ takes the form $\partial W_i \subseteq W_{i-1}$, where

$$\lambda W_0 = W_2 \subseteq W_1 \subseteq W_0$$

denotes the map into the periodic flag manifold corresponding to W. All this is explained in more detail in Chapter 19 of [60].

As a concrete example, let us re-examine the twistor construction of harmonic maps into $\mathbb{C}P^n$ from Example 8.16. A twisted complex extended solution is given by

$$(P_{i-1} + \lambda P^\perp_{i-1}P_i + \lambda^2 P^\perp_i)^{-1} \psi (P_{i-1} + \lambda P^\perp_{i-1}P_i + \lambda^2 P^\perp_i)$$

but T^{-1} converts this to $(P_{i-1} + \lambda P^\perp_{i-1})^{-1}\psi(P_{i-1} + \lambda P^\perp_{i-1})$, so the corresponding map into the periodic flag manifold is the result of applying this to the 'standard' flag $\lambda H^{(n+1)}_+ \subseteq (\mathbb{C}^i)^\perp \cap \mathbb{C}^{i+1} \oplus \lambda H^{(n+1)}_+ \subseteq H^{(n+1)}_+$. Ignoring the constant factor $(P_{i-1} + \lambda P^\perp_{i-1})^{-1}$ on the left, we obtain

$$\lambda f_{\langle i-1\rangle} \oplus \lambda^2 H^{(n+1)}_+ \subseteq \lambda f_{\langle i\rangle} \oplus \lambda^2 H^{(n+1)}_+ \subseteq f_{\langle i-1\rangle} \oplus \lambda H^{(n+1)}_+$$

[7] More generally, it is known that $\Lambda G \cong (\Lambda G)_\sigma$ for any inner automorphism σ.

and the harmonic map condition is simply $\partial f_{\langle i-1\rangle} \subseteq f_{\langle i\rangle}$. Thus, the twisted Grassmannian is better adapted to the symmetric space. □

At this point, it is worth emphasizing the local nature of the Birkhoff and Iwasawa factorizations which play a key role in this chapter (and in the whole book). Each factorization arises from a situation where there is a Lie algebra decomposition $\mathrm{Lie}(\mathcal{G}) = \mathrm{Lie}(\mathcal{G}_1) \oplus \mathrm{Lie}(\mathcal{G}_2)$, but the product $\mathcal{G}_1\mathcal{G}_2$ is not in general equal to $\mathcal{G}$. In the case of the Birkhoff factorization, $\mathcal{G}_1\mathcal{G}_2$ is the open and dense 'big cell'; it is never equal to $\mathcal{G}$. In the case of the Iwasawa factorization, $\mathcal{G}_1\mathcal{G}_2$ is equal to $\mathcal{G}$ when $\mathcal{G}_1$ is the loop group (or twisted loop group) of a *compact* Lie group; in general $\mathcal{G}_1\mathcal{G}_2$ is open but not dense (as there may be more than one big cell). Since the extended solution condition is not affected by left translation in the loop group, an extended solution may always be translated so that it intersects the open Birkhoff cell or the open Iwasawa cell (or both) at some given basepoint. However, away from this point, the extended solution may leave one or both of these open sets, a phenomenon which can give rise to singularities in the associated harmonic map. For more detailed analysis, it is necessary to understand the structure of the orbits of $\mathcal{G}_1$ on the homogeneous space $\mathcal{G}/\mathcal{G}_2$.

In our examples we have emphasized the finite uniton number case because we shall need it later in Chapter 10, and because it is essentially the only case where a description of all harmonic maps is available. It was proved in [126] that all harmonic maps $\phi : S^2 \to G$ are of finite uniton number, when G is compact. However, in the context of general harmonic maps the finite uniton number condition is very strong. For example, in the context of CMC surfaces, the finite uniton number condition means minimal or totally umbilic.

8.7 D-module aspects

We conclude with some comments on the D-module aspects of the harmonic map equations. The main difference from what we have seen for the KdV/mKdV equations (and from what we will see for quantum cohomology) is the lack of a natural cyclic generator of the D-module corresponding to a general harmonic map. Of course, the harmonic maps arising in a particular differential geometric problem may give rise to such a cyclic generator, and indeed there are many such examples based on 'Frenet frame' or 'osculating space' constructions with (real or complex) curves.

We have seen this in Example 8.16, where the successive derivatives of a holomorphic map f are used to construct a harmonic map into complex projective space. The map f in this case can be regarded as a cyclic element of the D-module corresponding to that harmonic map. All harmonic maps of finite uniton number may be constructed this way (see [61]).

For all harmonic (or pluriharmonic) maps arising from quantum cohomology, the corresponding D-module admits a natural cyclic generator. Overall, however, it remains to be investigated how useful this D-module point of view is in differential geometry.

Perhaps the most important lesson to be drawn from the examples in this chapter is the effectiveness of the Grassmannian model in implementing the Birkhoff factorization method of solution (Proposition 7.7 and Theorem 8.14 for harmonic maps; sections 8.4 and 8.5 for the KdV equation). This suggests that there should be a more fundamental 'D-module decomposition theorem', whereby the original D-module $\mathcal{M}$ decomposes into two simpler D-modules $\mathcal{M}_-, \mathcal{M}_+$. In the case of harmonic maps, $\mathcal{M}_-$ and $\mathcal{M}_+$ are equivalent, and each may be regarded as a 1-variable reduction of the original 2-variable equation. For the KdV equation, $\mathcal{M}_-$ is constant and $\mathcal{M}_+$ corresponds to an exponential function. In both cases, the decomposition is the key to describing the relevant solutions.

8.8 Appendix: the Birkhoff and Iwasawa decompositions

The Birkhoff factorization (section 6.1, Chapters 7 and 8) is based on the existence of an open dense 'big cell'

$$\Lambda_- \mathrm{GL}_n\mathbb{C}\, H_+^{(n)} \subseteq \mathrm{Gr}^{(n)} \cong \frac{\Lambda \mathrm{GL}_n\mathbb{C}}{\Lambda_+ \mathrm{GL}_n\mathbb{C}}$$

in the infinite-dimensional Grassmannian. In fact this is just one piece of a decomposition of $\mathrm{Gr}^{(n)}$ into infinitely many 'cells', generalizing the Schubert cell decomposition of a finite-dimensional Grassmannian. The origin of this is a decomposition of the loop group

$$\Lambda \mathrm{GL}_n\mathbb{C} = \bigsqcup_k \Lambda_- \mathrm{GL}_n\mathbb{C}\; \gamma_k\; \Lambda_+ \mathrm{GL}_n\mathbb{C}$$

where $k = (k_1, \dots, k_n) \in \mathbb{Z}^n$ and

$$\gamma_k = \mathrm{diag}(\lambda^{k_1}, \dots, \lambda^{k_n}) = \begin{pmatrix} \lambda^{k_1} & & \\ & \ddots & \\ & & \lambda^{k_n} \end{pmatrix}.$$

This induces the Birkhoff decomposition

$$\mathrm{Gr}^{(n)} = \bigsqcup_k \Lambda_- \mathrm{GL}_n\mathbb{C}\, \gamma_k\, H_+^{(n)}.$$

For $n \geq 2$, the group $\Lambda \mathrm{SL}_n\mathbb{C}$ produces the same decomposition.

Similar remarks apply to the loop group $\Lambda G^{\mathbb{C}}$ of any complex semisimple group $G^{\mathbb{C}}$, which decomposes into pieces indexed by the affine Weyl group, and this induces a decomposition of any generalized flag manifold of $\Lambda G^{\mathbb{C}}$ (see [8] and [102]). This generalizes the Bruhat decomposition of $G^{\mathbb{C}}$ and the induced cell decompositions of the generalized flag manifolds $G^{\mathbb{C}}/P$.

The big cell admits the following geometric characterization:

$$\{W \in \mathrm{Gr}^{(n)} \mid \pi_+|_W : W \to H_+^{(n)} \text{ is an isomorphism}\}$$

where $\pi_+ : H^{(n)} \to H_+^{(n)}$ is orthogonal projection with respect to the Hermitian inner product

$$(f,g) \mapsto \int_{S^1} f^t \bar{g}\, \tfrac{d\lambda}{\lambda}$$

of $H^{(n)}$. If $W = \gamma H_+^{(n)}$ is in the big cell, that is, $\gamma = \gamma_- \gamma_+$, then γ_- is given explicitly by

$$\gamma_- = \begin{pmatrix} | & & | \\ (\pi_+|_W)^{-1}(e_1) & \cdots & (\pi_+|_W)^{-1}(e_n) \\ | & & | \end{pmatrix},$$

where $e_1, \ldots, e_n$ are the standard basis vectors of $\mathbb{C}^n$.

The orbits of $\Lambda_+\mathrm{GL}_n\mathbb{C}$ are quite different from the orbits of $\Lambda_-\mathrm{GL}_n\mathbb{C}$: the former have finite dimension, while the latter have finite codimension. Harmonic maps of finite uniton number admit complex extended solutions which take values in the 'algebraic loop group' $\Lambda^{\mathrm{alg}}\mathrm{GL}_n\mathbb{C}$, and for this group the orbits of $\Lambda_+\mathrm{GL}_n\mathbb{C}$ are easy to describe: we have two decompositions

$$\begin{aligned} \Lambda^{\mathrm{alg}}\mathrm{GL}_n\mathbb{C} &= \bigsqcup_k \Lambda_-^{\mathrm{alg}}\mathrm{GL}_n\mathbb{C}\, \gamma_k\, \Lambda_+^{\mathrm{alg}}\mathrm{GL}_n\mathbb{C} \\ &= \bigsqcup_k \Lambda_+^{\mathrm{alg}}\mathrm{GL}_n\mathbb{C}\, \gamma_k\, \Lambda_+^{\mathrm{alg}}\mathrm{GL}_n\mathbb{C} \end{aligned}$$

which are indexed by the same set (see [102]).

The calculations in section 8.6 take place in the intersection of a finite-dimensional orbit $\Lambda_+^{\mathrm{alg}}\mathrm{GL}_n\mathbb{C}\,\gamma_k H_+^{(n)}$ with the (translate by γ_k of the) big cell. Translating by γ_k^{-1}, this set is

$$\gamma_k^{-1}\Lambda_+^{\mathrm{alg}}\mathrm{GL}_n\mathbb{C}\,\gamma_k H_+^{(n)} \cap \Lambda_-^{\mathrm{alg}}\mathrm{GL}_n\mathbb{C}H_+^{(n)}.$$

In Example 8.16, for example, the complex extended solution

$$(\gamma^{-1}L)_- = \begin{pmatrix} 1 & & \\ \frac{1}{\lambda}\alpha_1 & 1 & \\ \frac{1}{\lambda^2}\beta_1 & \frac{1}{\lambda}\alpha_2 & 1 \end{pmatrix}$$

gives $W = (\gamma^{-1}L)_- H_+^{(3)} =$

$$\mathrm{Span}\left\{\begin{pmatrix} 1 \\ \frac{1}{\lambda}\alpha_1 \\ \frac{1}{\lambda^2}\beta_1 \end{pmatrix}, \lambda\begin{pmatrix} 1 \\ \frac{1}{\lambda}\alpha_1 \\ \frac{1}{\lambda^2}\beta_1 \end{pmatrix}, \begin{pmatrix} 0 \\ 1 \\ \frac{1}{\lambda}\alpha_2 \end{pmatrix}, \begin{pmatrix} 0 \\ 0 \\ 1 \end{pmatrix}, \lambda\begin{pmatrix} 0 \\ 0 \\ 1 \end{pmatrix}, \lambda\begin{pmatrix} 0 \\ 1 \\ 0 \end{pmatrix}, \lambda^2 H_+^{(3)}\right\}.$$

Orthogonal projection of W on $H_+^{(3)}$ gives $H_+^{(3)}$.

The big cell of $\mathrm{Gr}^{\mathrm{scalar},(n)}$ (which appears in Theorem 8.12) can be characterized in a similar way as

$$\{W \in \mathrm{Gr}^{\mathrm{scalar},(n)} \mid \pi_+|_W : W \to H_+^{\mathrm{scalar}} \text{ is an isomorphism}\}.$$

If G is compact, the real form ΛG (of $\Lambda G^{\mathbb{C}}$) acts transitively on $\Lambda G^{\mathbb{C}}/\Lambda_+ G^{\mathbb{C}}$. In other words, we have

$$\Lambda G^{\mathbb{C}} = \Lambda G\,\Lambda_+ G^{\mathbb{C}}$$

(with $\Lambda G \cap \Lambda_+ G^{\mathbb{C}} = G$). When G is non-compact, however, the action of ΛG on $\Lambda G^{\mathbb{C}}/\Lambda_+ G^{\mathbb{C}}$ is not in general transitive. The orbit decomposition in this situation may be deduced from the Birkhoff decomposition as follows. Let $C : G^{\mathbb{C}} \to G^{\mathbb{C}}$ be the conjugation map of $G^{\mathbb{C}}$ (with fixed points G). Then $C_\Lambda(\gamma)(\lambda) = C(\gamma(1/\bar{\lambda}))$ is the analogous conjugation map of $\Lambda G^{\mathbb{C}}$ (with fixed points ΛG). To find the Iwasawa factorization $\gamma = \gamma_{\mathbb{R}} w \gamma_+$ of $\gamma \in \Lambda G^{\mathbb{C}}$, with $\gamma_{\mathbb{R}} \in \Lambda G$, $w \in G^{\mathbb{C}}$, $\gamma_+ \in \Lambda_+ G^{\mathbb{C}}$, it suffices to find the Birkhoff factorization of $C_\Lambda(\gamma)^{-1}\gamma$, which is

$$\begin{aligned} C_\Lambda(\gamma)^{-1}\gamma &= C_\Lambda(\gamma_+)^{-1} C_\Lambda(w)^{-1} C_\Lambda(\gamma_{\mathbb{R}})^{-1}\gamma_{\mathbb{R}}\, w\, \gamma_+ \\ &= C_\Lambda(\gamma_+)^{-1} C_\Lambda(w)^{-1}\, w\, \gamma_+, \end{aligned}$$

as $C_\Lambda(\gamma_+)^{-1} \in \Lambda_- G^{\mathbb{C}}$. Based on this, a general theory of Iwasawa decompositions for loop groups has been developed by P. Kellersch in

[80]. This generalizes the well known Iwasawa decomposition for the finite-dimensional group $G^{\mathbb{C}}$ with respect to the real form G.

A geometrical description of the orbits of the action of $U_{p,q}$ on the finite-dimensional Grassmannian $\mathrm{Gr}_k(\mathbb{C}^n)$ has been given in [130]; the orbits of $\Lambda U_{p,q}$ on $\mathrm{Gr}^{(n)}$ can be described in a similar way (see [17]). In particular, the union of the open orbits can be characterized as

$$\left\{W \in \mathrm{Gr}^{(n)} \mid \mathrm{Herm}|_{W\cap(\lambda W)^{\perp}} \text{ is non-degenerate}\right\}$$

where Herm is the Hermitian inner product on $H^{(n)} = \mathrm{Map}(S^1, \mathbb{C}^n)$ defined by integrating the Hermitian inner product of signature (p, q) on $\mathbb{C}^n$ over S^1.

9 Quantum cohomology as an integrable system

We have seen that the quantum cohomology of a manifold can be described as a flat connection or D-module. We have hinted that this puts it on the same footing as a solution of an integrable system. In this chapter we shall describe the WDVV (Witten–Dijkgraaf–Verlinde–Verlinde) equations, which provide suitable partial differential equations.

It is important to keep in mind that the question 'Of which integrable system is quantum cohomology a solution?' does not have a unique answer. A flat connection or D-module may be regarded as a solution of various integrable systems—just as a real number may be regarded as a solution of various algebraic equations—and it would be absurd to consider them all. However, the properties of quantum cohomology and the form of its dependence on the spectral parameter restrict the possibilities.

As we have already noted in section 7.5, the flat connection

$$d + \frac{1}{\hbar}\omega$$

associated to the quantum cohomology of a manifold corresponds to a pluriharmonic map. Therefore, we can say that the quantum cohomology of a manifold is a solution to the pluriharmonic map equation, and this qualifies it as 'integrable' in some sense. However, it is a very special solution, and the real question is 'How are the pluriharmonic maps corresponding to quantum cohomology characterized?'. This is another version of the question 'How are the D-modules corresponding to quantum cohomology characterized?'.

We shall say more about quantum cohomology and pluriharmonic maps in Chapter 10, but in this chapter we describe another candidate, the WDVV equations. Solutions of these equations correspond (roughly speaking) to Frobenius manifolds, and in fact quantum cohomology is a special type of Frobenius manifold. One can then ask how these special Frobenius manifolds are characterized.

While these characterization questions are very difficult and still far from being solved, concepts such as pluriharmonic map and Frobenius manifold are of interest in their own right. They arise naturally in different areas of mathematics, and it is of great interest to investigate whether the links with quantum cohomology are merely superficial or of deeper significance.

9.1 Large quantum cohomology

The pluriharmonic map corresponding to the quantum cohomology of a manifold is somewhat hidden, but the corresponding solution to the WDVV equations has a direct expression in terms of Gromov–Witten invariants. It is simply the generating function or 'potential'

$$\mathcal{F}^M(t) = \sum_{l\geq 3,D} \frac{1}{l!}\langle \underbrace{T|\dots|T}_{l}\rangle_D.$$

As usual we make use of the l-point Gromov–Witten invariants $\langle X_1|X_2|\dots|X_l\rangle_D$ only as motivation, so we regard the series as a formal series. Before introducing the WDVV equations we shall discuss some properties of $\mathcal{F}^M$.

An important observation is that it is natural to take $t \in H^*M$ rather than $t \in H^2M$ in the definition of $\mathcal{F}^M$; that is, we extend the domain to the space of all $(s+1)$-tuples $t = (t_0, \dots, t_s)$, or an open subset thereof. As usual, $T \in H_*M$ denotes the homology class which is Poincaré dual to t.

Since Gromov–Witten invariants are multilinear, it follows immediately that

$$\partial_i \mathcal{F}^M(t) = \sum_{l\geq 2,D} \frac{1}{l!}\langle B_i|\underbrace{T|\dots|T}_{l}\rangle_D$$

and hence

$$\partial_i\partial_j\partial_k \mathcal{F}^M(t) = \sum_{l\geq 0,D} \frac{1}{l!}\langle B_i|B_j|B_k|\underbrace{T|\dots|T}_{l}\rangle_D,$$

an expression which is sometimes denoted by $\langle\langle B_i|B_j|B_k\rangle\rangle$. In the special case where $t \in H^2M$, it can be shown that

$$\langle A|B|C|\underbrace{T|\dots|T}_{l}\rangle_D = \langle A|B|C\rangle_D \langle t, D\rangle^l$$

hence the formula for $\partial_i\partial_j\partial_k\mathcal{F}^M$ reduces to the formula (from Chapter 2) for $(b_i \circ_t b_j, b_k)$:

$$\partial_i\partial_j\partial_k\mathcal{F}^M(t)\,|_{H^2M} = \sum_{l\geq 0,D} \frac{1}{l!}\langle B_i|B_j|B_k\rangle_D e^{\langle t,D\rangle}$$
$$= (b_i \circ_t b_j, b_k).$$

Thus, the quantum product is expressable directly in terms of the generating function $\mathcal{F}^M$ (though not, in general, in terms of $\mathcal{F}^M|_{H^2M}$).

Because of this, it is natural to extend the definition of the quantum product:

Definition 9.1 *For any $t \in H^*M$ we define a product operation $\circ_t$ on H^*M by $(b_i \circ_t b_j, b_k) = \partial_i\partial_j\partial_k\mathcal{F}^M(t)$ for all $i, j, k \in \{0, \dots, s\}$. Equivalently, $(a \circ_t b, c) = a\, b\, c\, \mathcal{F}^M(t)$ where $a, b, c \in H^*M$ are regarded as (constant) vector fields on H^*M.*

This extended quantum product is called the *large quantum product,* the previous[1] one (with $t \in H^2M$) being the *small quantum product.* In terms of counting rational curves, the large quantum product involves Gromov–Witten invariants of arbitrary length, which makes it much more difficult to compute than the small quantum product. (We shall discuss later to what extent the large quantum product contains more information than the small one; the answer is not straightforward.) The definition shows immediately that it is a commutative product operation on H^*M (assuming that all cohomology classes are even-dimensional). It has the same identity element $b_0 = 1 \in H^0M$ as the small quantum product. Associativity of this new product is not clear at all from the definition, though it can be proved from the properties of Gromov–Witten invariants, as in the case of small quantum cohomology.

This suggests an obvious question, which leads to the WDVV equations. For *any* (smooth or analytic) function $\mathcal{F} : H^*M \to \mathbb{C}$, we can define a product operation $*_t$ in the same way:

$$(b_i *_t b_j, b_k) = \partial_i\partial_j\partial_k\mathcal{F}(t).$$

Definition 9.2 *The WDVV equations are the system of third-order non-linear partial differential equations for $\mathcal{F}$ given by the associativity condition*

$$(b_i *_t b_j) *_t b_k = b_i *_t (b_j *_t b_k).$$

[1] Sometimes the terms ‘small’ and ‘very small’ are used here instead of ‘large’ and ‘small’.

Since the quantum product is known to be associative, the generating function $\mathcal{F}^M$ of quantum cohomology satisfies the WDVV equations. As in the case of quantum cohomology, these equations have a zero curvature formulation, so the WDVV equations can be regarded as an integrable p.d.e. (we shall explain this in the next section).

Example 9.3 The large quantum cohomology of $\mathbb{C}P^1$ is extremely simple because the only 'new' part is $H^0\mathbb{C}P^1$, which is governed entirely by the fact that $b_0 = 1$ is the identity element for the large quantum product. However, let us write down all relevant quantities for later reference.

First, the large quantum product is specified by

$$b_0 \circ_t b_0 = b_0, \quad b_0 \circ_t b_1 = b_1, \quad b_1 \circ_t b_1 = q_1 b_0$$

where $t = (t_0, t_1)$, and so the connection form is

$$\omega = \begin{pmatrix} 1 & 0 \\ 0 & 1 \end{pmatrix} dt_0 + \begin{pmatrix} 0 & q_1 \\ 1 & 0 \end{pmatrix} dt_1.$$

The large quantum cohomology algebra is

$$\mathbb{C}[b_0, b_1, q_0, q_1]/(b_1^2 - q_1, b_0 - 1)$$

and the large quantum D-module is given by the naive quantization

$$D^{\mathrm{h}}/((\mathrm{h}\partial_1)^2 - q_1, \mathrm{h}\partial_0 - 1).$$

The generating function $\mathcal{F}^{\mathbb{C}P^1}$ can be computed directly (in addition to the triple products for ordinary cohomology, one just needs the Gromov–Witten invariants $\langle B_1 | \dots | B_1 \rangle_1 = 1$), and the answer is

$$\begin{aligned} \mathcal{F}^{\mathbb{C}P^1}(t) &= \tfrac{1}{2} t_0^2 t_1 + \tfrac{1}{3!} t_1^3 + \tfrac{1}{4!} t_1^4 + \cdots \\ &= \tfrac{1}{2} t_0^2 t_1 + e^{t_1} - \left(1 + t_1 + \tfrac{1}{2} t_1^2\right). \end{aligned}$$

The WDVV equations are vacuous here. □

The role of t_0 is clearly very minor. In fact, for any M, t_0 only appears in the 'classical' part $\frac{1}{3!}\langle T|T|T\rangle_0$ of $\mathcal{F}^M$, which is a polynomial function (in the above example, $\frac{1}{2}t_0^2 t_1$). The WDVV equations are primarily concerned with $t_{r+1}, \dots, t_s$.

Example 9.4 The large quantum cohomology of $\mathbb{C}P^2$ is much more interesting (see [85]). The potential turns out to be

$$\mathcal{F}^{\mathbb{C}P^2}(t_0, t_1, t_2) = \tfrac{1}{2}(t_0 t_1^2 + t_0^2 t_2) + \sum_{d \geq 1} N_d \, e^{d t_1} \frac{t_2^{3d-1}}{(3d-1)!}$$

where

$$N_1 = 1, \quad N_2 = 1, \quad N_3 = 12, \quad N_4 = 620, \quad N_5 = 87304, \ \ldots$$

are determined recursively by $N_1 = 1$ and

$$N_d = \sum_{i+j=d} \left(\binom{3d-4}{3i-2} i^2 j^2 - i^3 j \binom{3d-4}{3i-1} \right) N_i N_j .$$

The only non-zero Gromov–Witten invariants with $d \geq 2$ (see also section 8.3.2 of [26]) are

$$\langle B_1 | \ldots | B_1 | B_2 | \ldots | B_2 \rangle_d = d^k \langle B_2 | \ldots | B_2 \rangle_d = d^k N_d,$$

where B_1 occurs k times and B_2 occurs $3d-1$ times. For $d = 1$ we have $\langle B_1 | B_2 | B_2 \rangle_1 = 1 = N_1$. For $d = 0$ the contribution to $\mathcal{F}^{\mathbb{C}P^2}$ is $\frac{1}{3!}\langle T|T|T\rangle_0$, the above polynomial of degree three.

The positive integer N_d is the number of rational curves[2] of degree d in $\mathbb{C}P^2$ which hit $3d-1$ generic points. However, it is not practical to compute N_d directly; the formula was obtained by working backwards and solving the WDVV equations (we shall give the equations later in Example 9.8). □

This example illustrates the 'Reconstruction Theorem' of Kontsevich and Manin (see [73] and [85]), which says that the large quantum cohomology is *determined by* the small quantum cohomology if H^2M generates H^*M. Equivalently, all i-point Gromov–Witten invariants (with $i \geq 3$) are determined by the 3-point Gromov–Witten invariants. This is proved by an inductive argument in [85], using the axioms satisfied by Gromov–Witten invariants. In this situation, therefore, it could be said that large quantum cohomology contains the same information as small quantum cohomology. However, the relation (given by the axioms for Gromov–Witten invariants or the WDVV equation) is highly non-trivial.

It is possible to extend the domain even further, using the gravitational descendant Gromov–Witten invariants $\langle \tau_{d_1} X_1 | \tau_{d_2} X_2 | \ldots | \tau_{d_i} X_i \rangle_D$ of Chapter 5. Namely, one introduces the sequences of variables

$$t_i = t_i^{(0)}, t_i^{(1)}, t_i^{(2)}, \ldots \quad (0 \leq i \leq s)$$

[2] As N_1 determines all N_d, Euclid knew the large quantum cohomology of $\mathbb{C}P^2$ as well, cf. the footnote to Example 2.5.

and considers the analogous generating function. Writing

$$\mathbf{t} = \sum_{0 \le i \le s,\, d \ge 0} t_i^{(d)} \, \tau_d b_i$$

with respect to the basis of (formal) symbols $\tau_d b_i$, the generating function is

$$\mathcal{F}^M(\mathbf{t}) = \sum_{l \ge 3, D} \frac{1}{l!} \langle \underbrace{\mathbf{T} | \dots | \mathbf{T}}_{l} \rangle_D .$$

The product defined using $\mathcal{F}^M(\mathbf{t})$ does not in general admit an identity element, so we do not consider it, but the function $\mathcal{F}^M(\mathbf{t})$ has an important role to play.

The $\langle\!\langle \ \rangle\!\rangle$ notation extends to this situation as follows:

$$\begin{aligned}\langle\!\langle \tau_{d_i} B_i | \tau_{d_j} B_j | \tau_{d_k} B_k \rangle\!\rangle &= \sum_{l \ge 0, D} \frac{1}{l!} \langle \tau_{d_i} B_i | \tau_{d_j} B_j | \tau_{d_k} B_k | \underbrace{\mathbf{T} | \dots | \mathbf{T}}_{l} \rangle_D \\ &= \frac{\partial^3}{\partial t_i^{(d_i)} \partial t_j^{(d_j)} \partial t_k^{(d_k)}} \mathcal{F}^M(\mathbf{t}).\end{aligned}$$

Even in the case $M = \text{point} = B_0$, the gravitational potential is non-trivial. There are no small quantum cohomology parameters (hence no D), just the parameter t_0 and $\mathbf{t} = (t_0^{(0)}, t_0^{(1)}, t_0^{(2)}, \dots)$. Making the abbreviations $\tau_d = \tau_d B_0$ and

$$\langle \underbrace{\tau_0 | \dots | \tau_0}_{n_0} | \dots | \underbrace{\tau_i | \dots | \tau_i}_{n_i} \rangle = \langle \tau_0^{n_0} \dots \tau_i^{n_i} \rangle,$$

and using multilinearity, the 'pure gravity' generating function is

$$\mathcal{F}^{\text{point}}(\mathbf{t}) = \sum_{n_0, \dots, n_i} \frac{(t_0^{(0)})^{n_0} \dots (t_0^{(i)})^{n_i}}{n_0! \dots n_i!} \langle \tau_0^{n_0} \dots \tau_i^{n_i} \rangle.$$

There is one more generalization of the Gromov–Witten invariants that we have to mention, and that is to the case of 'higher genus'. Namely, instead of considering holomorphic maps $\mathbb{C}P^1 \to M$, one considers holomorphic maps $\Sigma_g \to M$ where Σ_g is a compact Riemann surface of genus g. The moduli space of such Riemann surfaces enters into the integral which defines the genus g Gromov–Witten l-point invariant $\langle X_1 | X_2 | \dots | X_l \rangle_{g,D}$. For each g we obtain a genus g potential

$$\mathcal{F}_g^M(t) = \sum_{l \ge 3, D} \frac{1}{l!} \langle \underbrace{T | \dots | T}_{l} \rangle_{g,D} .$$

The gravitational invariants and potential can also be defined for each g.

9.2 Frobenius manifolds

In view of the previous section, it is natural to study families of product operations on vector spaces given by arbitrary 'generating functions' $\mathcal{F}$. Such functions are referred to as *potentials* or *pre-potentials* rather than generating functions (no special meaning is attached to the coefficients of their power series expansions). Thus, we begin to move away from Gromov–Witten invariants and their enumerative meaning, focusing instead on the differential equations involved.

The concept of Frobenius manifold is based on this idea, and in this section we shall give the definition, due to B. Dubrovin (see [34]). The theory of primitive forms in singularity theory, due to K. Saito (see [104]), was the forerunner of this concept, and the first examples of Frobenius manifolds appeared as unfoldings of singularities. The large quantum cohomology of a manifold is, of course, an example of a Frobenius manifold. There are by now enough examples to justify an abstract theory of 'Saito structures' or 'Frobenius structures', whose depth can be appreciated from the books and surveys [36], [67], [70], and [95].

The classical concept of *Frobenius algebra* is the starting point:

Definition 9.5 *Let V be a finite-dimensional complex vector space. (It is possible to use real vector spaces, but we shall consider only the complex case.)*

(i) Let $*$ be a commutative associative product operation with identity element on V, which is bilinear over $\mathbb{C}$.
(ii) Let $(\ ,\ )$ be a non-degenerate symmetric bilinear form on V. (This form is not necessarily positive definite, but it will be referred to as the 'inner product'.)

Assume that the 'Frobenius condition' $(a * b, c) = (b, a * c)$ holds, for all $a, b, c \in V$. Then, the vector space V, with its algebra structure and inner product, is called a Frobenius algebra.

Evidently the ordinary cohomology algebra H^*M is a Frobenius algebra. For each t, the (large or small) quantum cohomology algebra of M is a Frobenius algebra. As in these examples, we have for any Frobenius algebra a linear transformation $\omega(a)$ of V for any $a \in V$, defined by

$$\omega(a)(b) = a * b.$$

In common with other structures in differential geometry, Frobenius structures on manifolds are defined in two steps: first, it is assumed that each tangent space has the structure of a Frobenius algebra, then it is assumed that these structures satisfy an integrability condition.

Definition 9.6 *A complex manifold C is said to be a Frobenius manifold if each tangent space T_xC is a Frobenius algebra, with smoothly (or analytically) varying product operations $*_x$, identity elements 1_x, and inner products $(\ ,\ )_x$, such that the following conditions are satisfied:*

(i) The Levi-Cività connection ∇^g associated to the (pseudo-Riemannian) metric $g = \{(\ ,\ )_x \mid x \in C\}$ has zero curvature. (This implies[3] that the the manifold is locally isometric to a vector space with (indefinite) inner product. With respect to such 'flat' coordinates, we have $\nabla^g = d$.)
(ii) The connection form $\omega = \{\omega_x \mid x \in C\}$ (with respect to flat local coordinates) satisfies $d\omega = 0$. (Since the commutativity and associativity conditions imply $\omega \wedge \omega = 0$, we have $d\omega + \omega \wedge \omega = 0$.)
(iii) The identity vector field $1 = \{1_x \mid x \in C\}$ is constant.
(iv) The product and metric are assumed to be homogeneous, in a sense which will be described in section 9.3.

In fact, these conditions impose extremely strong restrictions on the global topology of the manifold C, and it suffices essentially to consider the case $C = \mathbb{C}^n$. Since this is the case of interest to us, we shall assume $C = \mathbb{C}^n$ (or an open subset) from now on. Because of the difficulties in establishing convergence of various power series expansions, some authors work in a formal category and consider 'germs of formal Frobenius manifolds'. All this seems to make the term 'manifold' redundant, but it is retained in order to emphasize that some parts, at least, of the structure depend very non-trivially on x.

It is possible to rephrase the definition of Frobenius manifold in various ways. For example, let us consider pairs (g, ω) where g is a pseudo-Riemannian metric with Levi–Cività connection ∇^g and ω is an $(n \times n)$ matrix-valued 1-form, such that:

(A) the product operation defined by $a *_x b = \omega_x(a)(b)$ is associative,
(C) $\nabla^g \omega = 0$,
(Z) ∇^g has zero curvature,

and finally

[3] See Example 4.25.

(S) the trilinear form $(a, b, c) \mapsto g(a, b *_x c)$ is symmetric in a, b, c. In particular, $*_x$ is commutative.

Postponing (iii) and (iv) until later, one obtains in this way a reformulation of parts (i) and (ii) of the definition of a Frobenius manifold structure on $\mathbb{C}^n$, in which the (Frobenius) symmetry condition (S) relates the metric and the product, and there are three fundamental integrability conditions: (A) (associativity), (C) (closedness), and (Z) (flatness of the metric).

The reason for pointing out this reformulation is that it accommodates various kinds of integrable p.d.e. in a uniform framework: in a given situation, some parts of the definition are trivially satisfied, while the remaining parts are satisfied if and only if a certain function satisfies a (usually non-linear and non-trivial) differential equation. This is the way in which the definition of Frobenius manifold is usually applied. Rather than facing the integrability conditions all at once, an overwhelming task, one 'squeezes the toothpaste tube'

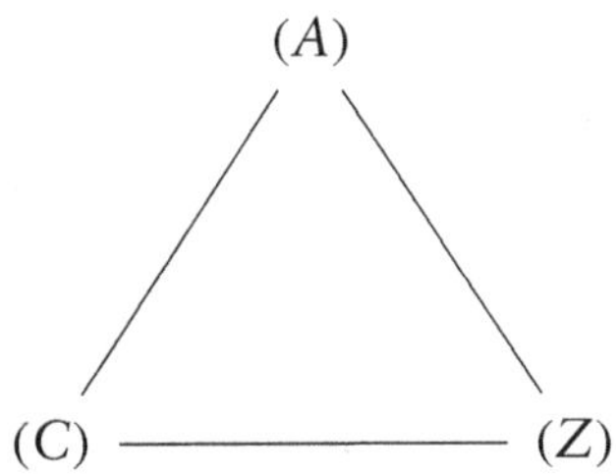

in order to focus on one corner or side. That is, one assumes some of the conditions and studies the system of p.d.e. which expresses the remaining ones. In the literature one can find a bewildering variety of permutations of the conditions, but these are supported by highly non-trivial examples.

For example, in (abstract) quantum cohomology, the flat metric point of view prevails; condition (Z) is built-in and one looks for 1-forms ω satisfying the non-trivial conditions (A) and (C). Let us introduce the 'Dubrovin connection'

$$d + \tfrac{1}{\mathrm{h}}\omega.$$

Then conditions (A) and (C) together are equivalent to the flatness of this connection (for all h), and condition (S) is a self-adjointness condition (as in section 5.3). Thus, the Dubrovin connection contains all relevant information.

From now on we shall investigate further this flat point of view, leading up to a 'potential function' which generalizes the Gromov–Witten potential. We fix a constant metric $g = (\ ,\ )$ on the vector space $\mathbb{C}^n$, together with a family of products $*_x$ with a constant identity element 1. We choose bases $w_1, \dots, w_n$, and $v_1, \dots, v_n$ such that $(v_i, w_j) = \delta_{ij}$ and such that all (w_i, w_j) are constant (in the case of quantum cohomology, these will be the usual bases $b_0, \dots, b_s$ and $a_0, \dots, a_s$).

The structure constants can be expressed in two ways:

$$w_i *_x w_j = \sum_k \alpha_{ijk} w_k, \ \alpha_{ijk} = (w_i *_x w_j, v_k)$$

$$w_i *_x w_j = \sum_k \beta_{ijk} v_k, \ \beta_{ijk} = (w_i *_x w_j, w_k)$$

(thus $w_i = \sum_j (w_i, w_j) v_j$ and $\beta_{ijk} = \sum_l (w_k, w_l) \alpha_{ijl}$). The first way involves the connection form $\omega = \sum_i \omega_i dx_i$, that is,

$$\alpha_{ijk} = (\omega_i)_{kj}.$$

The second way involves a potential $\mathcal{F}$, namely

$$\beta_{ijk} = \partial_i \partial_j \partial_k \mathcal{F}$$

where $\partial_i = \partial / \partial x_i$. To prove this, we shall apply the Poincaré Lemma three times (on a simply connected domain). The following diagram illustrates the procedure; the arrows denote derivatives.

$$\begin{array}{ccccc} \omega & \longleftarrow & A & \longleftarrow & G \\ & & g\Big| & & g\Big| \\ & & B & \longleftarrow & H & \longleftarrow & \mathcal{F} \end{array}$$

First, we have $d\omega = 0$, that is, $\partial_i (\omega_j)_{rs} = \partial_j (\omega_i)_{rs}$. This implies that there exists a matrix-valued map $A = (A_{rs})$ such that $\omega = dA$, that is, $(\omega_i)_{rs} = \partial_i A_{rs}$. Thus we have

$$\partial_i A_{rs} = \alpha_{isr} = (w_i *_x w_s, v_r).$$

Let us introduce $B_{rs} = \sum_p (w_r, w_p) A_{ps}$; as (w_r, w_p) is constant we have

$$\partial_i B_{rs} = \beta_{isr} = (w_i *_x w_s, w_r).$$

By commutativity both formulae are symmetric in i, s. This implies that there exist matrix-valued maps $G = (G_r), H = (H_r)$ such that $A_{rs} = \partial_s G_r, B_{rs} = \partial_s H_r$. We shall now focus on B alone. By the Frobenius condition we can choose B_{rs} so that $B_{rs} = B_{sr}$ (apply ∂_i to both sides and use the definition of B_{rs}), that is, $\partial_s H_r = \partial_r H_s$. This implies that there exists a map

$$\mathcal{F} : \mathbb{C}^n \to \mathbb{C}$$

such that $H_r = \partial_r \mathcal{F}$. It satisfies $\beta_{isr} = \partial_i B_{rs} = \partial_i \partial_s H_r = \partial_i \partial_s \partial_r \mathcal{F}$, as required.

All this can be expressed in coordinate-free fashion by saying that $\omega = dA$ and the symmetric bilinear form $B(a,b) = (A(a),b)$ is the Hessian of $\mathcal{F}$.

Conversely, given $\mathcal{F} : \mathbb{C}^n \to \mathbb{C}$ and a non-degenerate symmetric bilinear form $(\ ,\ )$ on $\mathbb{C}^n$, one obtains maps G, H, A, B, ω and a family of product operations $*_x$.

We have expressed the Dubrovin connection entirely in terms of $\mathcal{F}$. This puts the WDVV equations (and the equations obtained by imposing additional conditions such as (iii) and (iv) in the definition of Frobenius manifold) on a similar footing to other integrable p.d.e. such as the KdV, mKdV, or sine-Gordon equations.

Example 9.7 The case $n = 2$ is easy as the WDVV equations (associativity) impose no condition, but let us investigate Frobenius manifold structures on $\mathbb{C}^2$ in order to see how Example 9.3 (the large quantum cohomology of $\mathbb{C}P^1$) is accommodated by the general theory. To be precise, we shall assume conditions (i), (ii), and (iii), leaving (iv) to the next section.

All non-degenerate symmetric bilinear forms are equivalent over the complex numbers, so let us choose

$$(e_i, e_j) = \begin{cases} 1 & \text{if } i + j = 1 \\ 0 & \text{otherwise} \end{cases}$$

with respect to a basis e_0, e_1 of $\mathbb{C}^2$. We write $x = x_0 e_0 + x_1 e_1$. We assume without loss of generality that e_0 is the identity element for the product operation $*_x$, and we write $e_0 = 1, e_1 = b$. Then $*_x$ is determined entirely by

$$b *_x b = f(x) + g(x)b$$

for some functions $f, g : \mathbb{C}^2 \to \mathbb{C}$. Replacing b by $b - \frac{1}{2}g$ if necessary, we may in fact assume that $g = 0$. We have

$$\omega = \begin{pmatrix} 1 & 0 \\ 0 & 1 \end{pmatrix} dx_0 + \begin{pmatrix} 0 & f(x_0, x_1) \\ 1 & 0 \end{pmatrix} dx_1,$$

and the condition $d\omega = 0$ is satisfied if and only if the derivative of f with respect to x_0 is zero, that is, $f = f(x_1)$. It is easy to verify that, for any such f, all axioms of a Frobenius manifold are satisfied (except possibly for the homogeneity condition, which we shall discuss later). For the quantum cohomology of $\mathbb{C}P^1$, we have $f(x_1) = e^{x_1}$ (up to second-order terms). For the ordinary cohomology of $\mathbb{C}P^1$ we have $f(x_1) = 0$.

The Hessian of the potential $\mathcal{F}$ is equal to

$$B = \begin{pmatrix} x_0 & \int f(x_1)dx_1 \\ x_1 & x_0 \end{pmatrix}$$

hence

$$\mathcal{F}(x_0, x_1) = \tfrac{1}{2}x_0^2 x_1 + \textstyle\iiint f(x_1)dx_1dx_1dx_1.$$

In terms of $\mathcal{F}$, the Dubrovin connection is

$$\frac{1}{\mathrm{h}}\begin{pmatrix} 1 & 0 \\ 0 & 1 \end{pmatrix} dx_0 + \frac{1}{\mathrm{h}}\begin{pmatrix} 0 & \mathcal{F}''' \\ 1 & 0 \end{pmatrix} dx_1.$$

□

Example 9.8 The case $n = 3$ includes the large quantum cohomology of $\mathbb{C}P^2$ (Example 9.4). With this example in mind, let us choose the metric $(\ ,\ )$ and basis e_0, e_1, e_2 so that

$$(e_i, e_j) = \begin{cases} 1 & \text{if } i + j = 2 \\ 0 & \text{otherwise.} \end{cases}$$

We write $x = x_0e_0 + x_1e_1 + x_2e_2$ and assume that $e_0 = 1$ is the identity element for the product operation $*_x$. By the general theory above there exists a potential function $\mathcal{F}$ such that

$$\partial_i\partial_j\partial_k\mathcal{F} = (e_i *_x e_j, e_k) = \beta_{ijk}.$$

Since $e_0 = 1$, we have

$$\beta_{0jk} = \partial_0\partial_j\partial_k\mathcal{F} = (e_0 *_x e_j, e_k) = (e_j, e_k).$$

The dual basis to e_0, e_1, e_2 is $d_0 = e_2, d_1 = e_1, d_2 = e_0$, so $\beta_{ijk} = \alpha_{ij2-k}$, and we have

$$\omega_i = \begin{pmatrix} \alpha_{i00} & \alpha_{i10} & \alpha_{i20} \\ \alpha_{i01} & \alpha_{i11} & \alpha_{i21} \\ \alpha_{i02} & \alpha_{i12} & \alpha_{i22} \end{pmatrix} = \begin{pmatrix} \beta_{i02} & \beta_{i12} & \beta_{i22} \\ \beta_{i01} & \beta_{i11} & \beta_{i21} \\ \beta_{i00} & \beta_{i10} & \beta_{i20} \end{pmatrix}.$$

Thus,

$$\omega = \begin{pmatrix} 1 & 0 & 0 \\ 0 & 1 & 0 \\ 0 & 0 & 1 \end{pmatrix} dx_0 + \begin{pmatrix} 0 & \beta_{112} & \beta_{122} \\ 1 & \beta_{111} & \beta_{121} \\ 0 & 1 & 0 \end{pmatrix} dx_1 + \begin{pmatrix} 0 & \beta_{212} & \beta_{222} \\ 0 & \beta_{211} & \beta_{221} \\ 1 & 0 & 0 \end{pmatrix} dx_2.$$

There is just one non-trivial associativity equation, $(e_1 *_x e_1) *_x e_2 = e_1 *_x (e_1 *_x e_2)$. Using

$$e_1 *_x e_1 = \sum_{i=0}^{2} \beta_{11i} d_i = \sum_{i=0}^{2} \beta_{11i} e_{2-i}$$

$$e_1 *_x e_2 = \sum_{j=0}^{2} \beta_{12j} d_j = \sum_{j=0}^{2} \beta_{12j} e_{2-j}$$

we obtain $\beta_{222} + \beta_{111}\beta_{122} = \beta_{112}^2$. Thus, $\mathcal{F}$ satisfies the WDVV equation

$$\partial_2^3 \mathcal{F} + (\partial_1^3 \mathcal{F})(\partial_1 \partial_2^2 \mathcal{F}) = (\partial_1^2 \partial_2 \mathcal{F})^2.$$

Conversely, let $\mathcal{F}$ be a function which satisfies the WDVV equation and also the p.d.e. $\partial_0 \partial_j \partial_k \mathcal{F} = (e_j, e_k)$. Then we obtain a Frobenius manifold structure for the above choice of metric (omitting the homogeneity condition as usual).

From the 'integrable p.d.e. point of view' (cf. section 7.6), if we choose any function $\mathcal{F}$ and write down the above connection form $\frac{1}{\mathrm{h}}\omega$ (taking β_{ijk} to mean $\partial_i \partial_j \partial_k \mathcal{F}$), then the only condition which is not obvious is the condition $\omega \wedge \omega = 0$, and this is equivalent to the WDVV equation.

To obtain the generating function for Gromov–Witten invariants of $\mathbb{C}P^2$, we write

$$\mathcal{F}(x_0, x_1, x_2) = \tfrac{1}{2}(x_0^2 x_1 + x_0 x_1^2) + a(x_1, x_2),$$

and impose the 'initial condition'

$$\omega|_{x_0 = x_2 = 0} = \begin{pmatrix} 0 & 0 & e^{x_1} \\ 1 & 0 & 0 \\ 0 & 1 & 0 \end{pmatrix} dx_1.$$

It turns out (see section 8.3.2 of [26]) that there is a solution of the form

$$\mathcal{F}(x_0, x_1, x_2) = \tfrac{1}{2}(x_0 x_1^2 + x_0 x_2^2) + \sum_{d \geq 1} N_d \, e^{d x_1} \frac{x_2^{3d-1}}{(3d-1)!},$$

unique up to third-order terms, and this determines the numbers N_d. □

9.3 Homogeneity

Let $p : \mathbb{C}^n \to \mathbb{C}$ be a weighted homogeneous polynomial in $x_1, \dots, x_n$ of total degree a, where the variables $x_1, \dots, x_n$ are assigned degrees

$$|x_1| = a_1, \dots, |x_n| = a_n.$$

Then

$$p(\epsilon^{a_1} x_1, \dots, \epsilon^{a_n} x_n) = \epsilon^a p(x_1, \dots, x_n)$$

for any scalar ϵ. Differentiating this equation with respect to ϵ and putting $\epsilon = 1$, we obtain the Euler relation

$$\sum_{i=1}^{n} a_i x_i \partial_i p = ap.$$

More generally, a function f defined on an open subset of $\mathbb{C}^n$ is said to be homogeneous of total degree a if it satisfies $Ef = 0$, where E is the 'Euler vector field'

$$E = \sum_{i=1}^{n} a_i x_i \partial_i - a.$$

Assume now that f is a homogeneous solution of a p.d.e. defined by a zero curvature condition, in which the connection matrices are also homogeneous (e.g. because they arise from a D-module with homogeneous relations—see below). Then the p.d.e. admits a 'scaling symmetry', and f is a 'scaling invariant' or 'self-similar' solution of it. We shall give some examples of this phenomenon.

Example 9.9 For the KdV equation, we have seen (section 7.1) that if $u(x,t)$ is a solution then so is $\epsilon^2 u(\epsilon x, \epsilon^3 t)$, that is, the KdV equation admits the scaling symmetry $u \mapsto \epsilon^2 u(\epsilon x, \epsilon^3 t)$. Therefore, it makes sense to consider solutions which are scaling invariant in the sense that $u(x,t) = \epsilon^2 u(\epsilon x, \epsilon^3 t)$. This means that we consider homogeneous solutions with respect to the Euler vector field

$$E = \pm(x\partial_x + 3t\partial_t + 2).$$

For the version which involves the spectral parameter λ, the scaling invariant condition is $u(x,t,\lambda) = \epsilon^2 u(\epsilon x, \epsilon^3 t, \epsilon^{-2}\lambda)$, and the Euler vector field is $E = \pm(x\partial_x + 3t\partial_t - 2\lambda\partial_\lambda + 2)$.

The origin of such solutions is a homogeneity property of the underlying D-module (or flat connection). The simplest way to express this is to say that the ideal of relations consists of homogeneous differential operators. Let us assign the degrees as follows:

$$|x| = -1, \; |t| = -3, \; |\lambda| = 2, \text{ and } |u| = 2.$$

Thus, $Eu = 0$, where $E = -x\partial_x - 3t\partial_t - 2$.

Let us examine the flat connection. First, we note that $\partial_x, \partial_t, \partial_\lambda$ have degrees $1, 3, -2$ respectively, so the usual relations $L - \lambda, \partial_t - P$ are homogeneous of degrees $2, 3$. Let us choose the usual basis $[1], [\partial_x]$. Then the flat connection

$$\Omega^\lambda = \begin{pmatrix} 0 & \lambda - u \\ 1 & 0 \end{pmatrix} dx + \begin{pmatrix} -u_x/4 & \lambda^2 - u\lambda/2 - u^2/2 - u_{xx}/4 \\ \lambda + u/2 & u_x/4 \end{pmatrix} dt$$

(section 4.3) satisfies the scaling invariant condition

$$\Omega^{\epsilon^{-2}\lambda}(\epsilon x, \epsilon^3 t) = \begin{pmatrix} 1 & 0 \\ 0 & \epsilon \end{pmatrix} \Omega^\lambda(x,t) \begin{pmatrix} 1 & 0 \\ 0 & \epsilon \end{pmatrix}^{-1},$$

the form of the diagonal matrix coming from the degrees $0, 1$ of the basis elements. Note that dx, dt have degrees $-1, -3$, so the condition on Ω^λ can be expressed by saying that the degrees of its entries are as shown in the boxes below:

$$\begin{pmatrix} \boxed{0} & \boxed{1} \\ \boxed{-1} & \boxed{0} \end{pmatrix}.$$

This condition is equivalent to the existence of the scaling symmetry $u \mapsto \epsilon^2 u(\epsilon x, \epsilon^3 t)$.

More generally, let us consider the D-module $D_{x,t}/(L, \partial_t - P)$ as in Chapter 7. If $P = f + g\partial_x$ with $|f| = j = |g| + 1$, we find that the components of the matrices $\Omega_1^\lambda, \Omega_2^\lambda$ (with respect to the basis given by 1, ∂_x, which are of degrees $0, 1$) are homogeneous with degrees shown in the boxes below:

$$\Omega_1^\lambda : \begin{pmatrix} \boxed{1} & \boxed{2} \\ \boxed{0} & \boxed{1} \end{pmatrix}, \qquad \Omega_2^\lambda : \begin{pmatrix} \boxed{j} & \boxed{j+1} \\ \boxed{j-1} & \boxed{j} \end{pmatrix}.$$

This leads to a p.d.e. with the scaling symmetry $u \mapsto \epsilon^2 u(\epsilon x, \epsilon^j t)$ (cf. Example 7.4). □

Example 9.10 We have already discussed the homogeneity properties of the (small) quantum cohomology D-module in Chapters 5 and 6. Here we assign the degrees of the variables as follows:

$$|q_i| = 2\langle c_1(TM), A_i\rangle \ (1 \le i \le r), \quad |\mathrm{h}| = 2.$$

The relations of the quantum D-module (and quantum cohomology algebra) are then homogeneous. Note that each $\partial_i = \partial/\partial t_i = q_i\partial/\partial q_i$ has degree zero.

With respect to a monomial basis, the connection Ω^{h} satisfies the scaling invariant condition

$$\Omega^{\epsilon^{-2}\mathrm{h}}(\epsilon^{-|q_1|}q_1, \dots, \epsilon^{-|q_r|}q_r)$$
$$= \mathrm{diag}(I, \epsilon^2 I, \dots, \epsilon^{2v} I)\ \Omega^{\mathrm{h}}(q_1, \dots, q_r)\ \mathrm{diag}(I, \epsilon^2 I, \dots, \epsilon^{2v} I)^{-1},$$

where we use the block matrix notation of Chapter 6. At this point there is no integrable p.d.e. under consideration. □

Example 9.11 For large quantum cohomology we assign degrees as follows:

$$\begin{aligned} &|t_0| = 2 \\ &|q_i| = 2\langle c_1(TM), A_i\rangle \ (1 \le i \le r) \ (\text{and } |t_i| = 0) \\ &|t_i| = 2 - |b_i| \ (r+1 \le i \le s) \end{aligned}$$

with $|\mathrm{h}| = 2$ as usual. This looks awkward, but it arises from the definition of Gromov–Witten invariants as follows. First, the potential function $\mathcal{F}$ breaks up into terms of the form

$$\sum_j \frac{1}{j!} \langle t_{k_1} B_{k_1} | \dots | t_{k_j} B_{k_j} | t_{k_{j+1}} B_{k_{j+1}} | \dots | t_{k_i} B_{k_i} \rangle_D$$

where $b_{k_1}, \dots, b_{k_j}$ have degree 2 and $b_{k_{j+1}}, \dots, b_{k_i}$ have degrees other than 2. If the Gromov–Witten invariant in the sum is non-zero then

$$|b_{k_1}| + \cdots + |b_{k_i}| = 2 \dim_{\mathbb{C}} M + 2\langle c_1(TM), D\rangle + 2(i-3)$$

(see section 2.1 and appendix 5.4). Now, as a function of $t_{k_1}, \dots, t_{k_i}$, we obtain a series with terms of the form

$$q_{k_1}^{d_1} \cdots q_{k_j}^{d_j} t_{k_{j+1}} \cdots t_{k_i}.$$

It is easily verified that each such term has the same degree $-2(\dim_{\mathbb{C}} M - 3)$ if we define the degrees of q_i, t_j as above. Thus, the potential function $\mathcal{F}^M$

is homogeneous of degree $-2(\dim_{\mathbb{C}} M - 3)$. The Euler vector field for the potential function is, therefore,

$$E = 2t_0\partial_0 + \sum_{i=1}^{r} 2\langle c_1(TM), A_i\rangle\partial_i + \sum_{i=r+1}^{s} (2 - |b_i|)t_i\partial_i + 2(\dim_{\mathbb{C}} M - 3).$$

Here we ignore any third-order (or lower) terms of $\mathcal{F}^M$ which are annihilated by the WDVV equation; thus $|\mathcal{F}^{\mathbb{C}P^1}| = 4$ and $|\mathcal{F}^{\mathbb{C}P^2}| = 2$, in agreement with the formulae of section 9.1. □

Now we can explain the homogeneity condition (iv) in Definition 9.6. The product and metric are tensors (sections of $T^*C \otimes T^*C \otimes TC$ and $T^*C \otimes T^*C$, respectively), and, with respect to a fixed choice of Euler vector field, the homogeneity condition is obtained by replacing the action of a vector field on functions by the action of the Lie derivative on tensors.

One application of homogeneity is that it allows us to consider λ as a variable having equal status with $x_1, \dots, x_n$; we can extend the D-module as in section 4.4, by adding the (λ-version of the) Euler vector field as a relation. Thus, we have an $(x_1, \dots, x_n)$-family of λ-o.d.e. rather than a λ-family of $(x_1, \dots, x_n)$-p.d.e. While this may seem a case of the tail wagging the dog, it is in fact a very important point of view. Namely, if the point $\lambda = 0$ is a singular point, then the absence of monodromy in Ω^λ is equivalent to the constancy of monodromy in the fundamental solution matrix H: this follows from

$$\tilde{\Omega}^\lambda = d\tilde{H}\tilde{H}^{-1} = dHH^{-1} + HdMM^{-1}H^{-1}$$

where the monodromy transformation is given by $H \mapsto \tilde{H} = HM$, $\Omega^\lambda \mapsto \tilde{\Omega}^\lambda$ (see the end of section 4.1).

Many integrable systems (and Frobenius manifolds) arise from this 'isomonodromy' condition (see [13], [43], and [74]).

9.4 Semisimple Frobenius manifolds

We have seen in section 6.7 that a D-module (modelled on quantum cohomology) can sometimes be 'normalized' by making a suitable change of variables. D-modules modelled on large quantum cohomology respond even more favorably to this treatment. This enabled B. Dubrovin to classify semisimple Frobenius manifolds (see [34] and [36]); semisimple means that the quantum products by cohomology classes may be simultaneously diagonalized. We shall just give a very simple example, the large quantum cohomology of $\mathbb{C}P^1$. As we have seen, this hardly involves large quantum cohomology at all, but the example illustrates the procedure, which is very similar to that in section 6.7.

Recall (Example 9.3) that the large quantum product of $\mathbb{C}P^1$ is specified by

$$b_0 \circ_t b_0 = b_0, \quad b_0 \circ_t b_1 = b_1, \quad b_1 \circ_t b_1 = q_1 b_0.$$

The non-trivial connection matrix

$$\begin{pmatrix} 0 & q_1 \\ 1 & 0 \end{pmatrix}$$

can be diagonalized by making the following change of basis:

$$\hat{b}_0 = b_0 - \frac{1}{\sqrt{q_1}} b_1, \quad \hat{b}_1 = b_0 + \frac{1}{\sqrt{q_1}} b_1$$

(the eigenvalues are $-\sqrt{q_1}$ with eigenvector $f(q_1)(\sqrt{q_1}, -1)$ and $\sqrt{q_1}$ with eigenvector $g(q_1)(\sqrt{q_1}, 1)$; we have chosen particular functions f, g for reasons that will become clear later). We use $\sqrt{q_1}$ as a synonym for $\mathrm{e}^{t_1/2}$. The quantum products of the new basis vectors are even simpler:

$$\hat{b}_0 \circ_t \hat{b}_0 = 2\hat{b}_0, \quad \hat{b}_0 \circ_t \hat{b}_1 = 0, \quad \hat{b}_1 \circ_t \hat{b}_1 = 2\hat{b}_1.$$

Note that the coefficients are all constant.

It is possible to make a corresponding transformation of the quantum D-module, by making the change of variables

$$\hat{q}_0 = \sqrt{q_0}/e^{\sqrt{q_1}}, \quad \hat{q}_1 = \sqrt{q_0}\, e^{\sqrt{q_1}},$$

which is equivalent to

$$q_0 = \hat{q}_0 \hat{q}_1, \quad q_1 = \frac{1}{4}(\log \hat{q}_1/\hat{q}_0)^2$$

(on some domain). In terms of t_0, t_1 we have

$$\hat{t}_0 = \frac{1}{2}t_0 - \mathrm{e}^{t_1/2} \qquad t_0 = \hat{t}_0 + \hat{t}_1$$

$$\hat{t}_1 = \frac{1}{2}t_0 + \mathrm{e}^{t_1/2} \qquad t_1 = 2\log\frac{1}{2}(\hat{t}_1 - \hat{t}_0).$$

For convenience we list all further relevant transformation formulae before proceeding any further. Differential operators transform as

$$\hat{\partial}_0 = \partial_0 - \frac{1}{\sqrt{q_1}}\partial_1 \qquad \partial_0 = \frac{1}{2}(\hat{\partial}_0 + \hat{\partial}_1)$$

$$\hat{\partial}_1 = \partial_0 + \frac{1}{\sqrt{q_1}}\partial_1 \qquad \partial_1 = \frac{1}{2}(-\sqrt{q_1}\hat{\partial}_0 + \sqrt{q_1}\hat{\partial}_1)$$

while differential forms transform as

$$d\hat{t}_0 = \frac{1}{2}(dt_0 - \sqrt{q_1}dt_1) \qquad dt_0 = d\hat{t}_0 + d\hat{t}_1$$

$$d\hat{t}_1 = \frac{1}{2}(dt_0 + \sqrt{q_1}dt_1) \qquad dt_1 = \frac{1}{\sqrt{q_1}}(-d\hat{t}_0 + d\hat{t}_1).$$

The connection form

$$\Omega^{\mathrm{h}} = \frac{1}{\mathrm{h}}\begin{pmatrix}1 & 0\\ 0 & 1\end{pmatrix} dt_0 + \frac{1}{\mathrm{h}}\begin{pmatrix}0 & q_1\\ 1 & 0\end{pmatrix} dt_1$$

transforms to

$$\hat{\Omega}^{\mathrm{h}} = \frac{1}{\mathrm{h}}\begin{pmatrix}2 & 0\\ 0 & 0\end{pmatrix} d\hat{t}_0 + \frac{1}{\mathrm{h}}\begin{pmatrix}0 & 0\\ 0 & 2\end{pmatrix} d\hat{t}_1 + \frac{1}{4\sqrt{q_1}}\begin{pmatrix}1 & -1\\ -1 & 1\end{pmatrix} d\hat{t}_0 + \frac{1}{4\sqrt{q_1}}\begin{pmatrix}-1 & 1\\ 1 & -1\end{pmatrix} d\hat{t}_1.$$

The relations $(\mathrm{h}\partial_1)^2 - q_1, \mathrm{h}\partial_0 - 1$ of the large quantum D-module transform to

$$(\mathrm{h}\hat{\partial}_0)^2 + (\mathrm{h}\hat{\partial}_1)^2 - 2(\mathrm{h}\hat{\partial}_0)(\mathrm{h}\hat{\partial}_1) + \frac{\mathrm{h}}{\sqrt{q_1}}\mathrm{h}\hat{\partial}_1 - \frac{\mathrm{h}}{\sqrt{q_1}}\mathrm{h}\hat{\partial}_0 - 4, \quad \mathrm{h}\hat{\partial}_0 + \mathrm{h}\hat{\partial}_1 - 2.$$

These relations de-quantize to the commutative ring

$$\mathbb{C}[\hat{b}_0, \hat{b}_1, \hat{q}_0, \hat{q}_1]/((\hat{b}_0 - \hat{b}_1)^2 - 4, \hat{b}_0 + \hat{b}_1 - 2)$$

or just

$$\mathbb{C}[\hat{b}_0, \hat{b}_1, \hat{q}_0, \hat{q}_1]/(\hat{b}_0\hat{b}_1, \hat{b}_0 + \hat{b}_1 - 2)$$

which is simply another way of writing the large quantum cohomology algebra

$$\mathbb{C}[b_0, b_1, q_0, q_1]/(b_1^2 - q_1, b_0 - 1)$$

(except that the change of variables is not defined over the polynomial ring).

Note that the appropriate quantizations of the relations $\hat{b}_0\hat{b}_1, \hat{b}_0+\hat{b}_1-2$ are $\mathrm{h}\hat{\partial}_0\mathrm{h}\hat{\partial}_1 + \mathrm{h}(\frac{1}{4\sqrt{q_1}}\mathrm{h}\hat{\partial}_0 - \frac{1}{4\sqrt{q_1}}\mathrm{h}\hat{\partial}_1), \mathrm{h}\hat{\partial}_0 - \mathrm{h}\hat{\partial}_1 - 2$, that is, not the naive quantizations.

The similarity of $\hat{\Omega}^{\mathrm{h}}$ with the connection matrix for the KdV equation has been used by S. Barannikov in [11] to construct a hierarchy of equations. Somewhat different hierarchies, based on higher genus Gromov–Witten invariants, have been constructed by A. Givental (see [53]) and by B. Dubrovin and Y. Zhang (see [38]).

10 Integrable systems and quantum cohomology

In this final chapter we look at the map $W : \mathbb{C}^r \to \mathrm{Gr}^{(s+1)}$ in the case of quantum cohomology and, more generally, abstract quantum cohomology. That is, we apply the fundamental construction of Chapter 8 to the situation of Chapters 5 and 6.

It should not be surprising that this yields benefits, as the map W is a geometrical manifestation of the quantum cohomology D-module, and we have seen already that W is of great importance in other integrable systems such as the KdV equation and the harmonic map equation. In fact these expectations will be exceeded handsomely: in the case of the quantum cohomology of a Calabi–Yau manifold, W can be interpreted as a variation of Hodge structure (VHS) on a 'mirror dual' Calabi–Yau manifold. In other words, the map W reveals the phenomenon of mirror symmetry!

In this case, the image of the map W lies in a finite-dimensional subspace of $\mathrm{Gr}^{(s+1)}$, in fact a twistor space of the type considered in section 8.6, so $\mathrm{Gr}^{(s+1)}$ is not strictly necessary. However, as we shall see, it is most natural to discuss quantum cohomology within this infinite-dimensional framework.

We give a brief summary of the essential features of mirror symmetry for Calabi–Yau manifolds in sections 10.1 and 10.2, working entirely in the conventional finite-dimensional framework. Then, in sections 10.3 and 10.4, we embed the finite-dimensional discussion so far into $\mathrm{Gr}^{(s+1)}$. There are two reasons for doing this. One is for the purpose of generalization beyond Calabi–Yau manifolds (and, indeed, beyond quantum cohomology). The other is purely aesthetic. Just as the Grassmannian framework clarifies the description of harmonic maps of finite uniton number, the same applies to quantum cohomology or VHS for Calabi–Yau manifolds (which is to be expected, since the latter give particular examples of pluriharmonic maps of finite uniton number). For example, the symplectic/orthogonal dichotomy is eliminated in the loop group language, and the Riemann bilinear relations can be described more conceptually. The mirror transformation arises naturally by the procedure outlined in Chapter 6.

For manifolds which are not Calabi–Yau—indeed, for Fano manifolds, the very simplest type of manifolds whose quantum cohomology we encountered early in the book—the image of W does not in general lie in a finite-dimensional subspace of $\mathrm{Gr}^{(s+1)}$. Mirror symmetry and VHS are less well developed in this case, but the same relation appears to exist. Namely, mirror duals of certain Fano manifolds were proposed by A. Givental in [50], and generalized VHS for such varieties were proposed by S. Barannikov in [9]–[11]. These VHS are quite different from those appearing in classical algebraic geometry; they are 'semi-infinite' (in the same sense as elements of $\mathrm{Gr}^{(s+1)}$) and they have to be interpreted in terms of non-commutative geometry. Although the theories of VHS and quantum cohomology developed quite separately, they both lead to the same very natural (and very difficult) question: when do abstract VHS or abstract quantum cohomology arise 'from geometry'? Mirror symmetry does not address this problem directly, but it links the two theories in a very promising way, specifically by means of the map W.

In section 10.5 we sketch the ingredients of 'abstract mirror symmetry', or, at least, a differential geometric version of it (i.e. ignoring the arithmetic aspects). It is natural to use the term 'abstract VHS' for the counterpart of abstract quantum cohomology here, and we shall do so, although the term 'abstract VHS' is already used in the literature specifically for the finite-dimensional situation. The abstract VHS in section 10.5 can refer to finite-dimensional or infinite-dimensional flag manifolds.

Our construction method (from scalar equations) for abstract quantum cohomology in Chapter 6 can equally well be regarded as a construction method for abstract mirror symmetry. We shall give some simple examples beyond genuine mirror symmetry. This leads to a more general outlook—there is no reason to focus solely on quantum cohomology or VHS.

In section 10.6 we discuss briefly some related situations and further developments. It is clear that the definition of abstract mirror symmetry in section 10.5 is related to the axioms of a Frobenius manifold; indeed, this was the context of Barannikov's work on semi-infinite VHS. A systematic treatment of Frobenius-like structures arising from VHS has been given by C. Hertling (see [67] and [68]), in which specific geometric structures are given by specific lists of axioms.

10.1 Motivation: variations of Hodge structure (VHS)

The relation between quantum cohomology of Calabi–Yau manifolds and VHS on 'mirror dual' Calabi–Yau manifolds is well known. The case of the quintic 3-fold—a non-singular hypersurface in $\mathbb{C}P^4$ given by a homogeneous polynomial equation of degree five—brought this phenomenon into mathematics in the 1990's. Although it remained conjectural until the foundations of quantum cohomology were sufficiently developed, and until the relevant 'Mirror Theorem' was proved, mirror symmetry predicted correctly *the number of rational curves in one Calabi–Yau manifold* (thus, quantum cohomology data, or in physics terminology, the A-model) in terms of *the coefficients of power series solutions of a differential equation associated with another Calabi–Yau manifold* (VHS data, or the B-model).

Only a few of these numbers were amenable to calculation by algebraic geometric methods, yet the mirror conjecture produced all of them at once. This was a spectacular advance, and an indication of deeper phenomena waiting to be discovered. Of course, the relation between Gromov–Witten invariants and solutions of quantum differential equations is exactly of this type, but the quantum differential equations in Chapters 3 and 5 were introduced in a tautological manner. Mirror symmetry predicts an independent geometrical meaning for these differential equations. A comprehensive review of the subject, with many references, can be found in [26].

While the underlying mirror symmetry between manifolds is still far from understood, the relation between quantum cohomology data and VHS data can be formulated precisely. In this section we sketch the concept of a VHS, then discuss an example of mirror symmetry in detail in the next section.

Let V_z be a family of compact Kähler manifolds of complex dimension n, parametrized holomorphically by $z = (z_1, \dots, z_r)$ in $\mathbb{C}^r$ or an open subset thereof. P. A. Griffiths studied the associated 'period map'

$$F : \mathbb{C}^r \to U \subseteq G^{\mathbb{C}}/P$$

where U is an open orbit of the action of a non-compact real form of $G^{\mathbb{C}}$ on the compact generalized flag manifold $G^{\mathbb{C}}/P$ (see [58], and [127] for a recent survey).

This map arises because the Hodge decomposition of $H^i(V_z; \mathbb{C})$ depends non-trivially on the parameter z. Although the topological invariant $H^i(V_z; \mathbb{C})$ itself may be identified with a fixed $E = H^i(V_{z_0}; \mathbb{C})$, as z varies, the subspace

$$H^{p,q}(V_z; \mathbb{C}) \oplus H^{p+1,q-1}(V_z; \mathbb{C}) \oplus \cdots$$

varies holomorphically, defining a holomorphic subbundle of a trivial bundle.

Let us consider the middle-dimensional cohomology group $H^n(V_z;\mathbb{C})$. We define

$$E^{n-p,p}(z) = H^{n-p,p}(V_z;\mathbb{C}), \quad h^p = \dim E^{n-p,p}(z)$$

and

$$F_{h_p} = E^{n,0} \oplus E^{n-1,1} \oplus \cdots \oplus E^{n-p,p}, \quad h_p = h^0 + h^1 + \cdots + h^p.$$

Our F_{h_p} corresponds to F^p in [58] (and F^{n-p} in more recent references) but we prefer to indicate the dimension explicitly here.

For simplicity we assume that $h^p \neq 0$ for all $0 \leq p \leq n$. We obtain a holomorphic map

$$F : z \mapsto (F_{h_0}(z), F_{h_1}(z), \ldots, F_{h_{n-1}}(z))$$

into the flag manifold

$$F_{h_0,h_1,\ldots,h_{n-1}}(E) \cong \frac{\mathrm{U}_{h_n}}{\mathrm{U}_{h^0} \times \cdots \times \mathrm{U}_{h^n}},$$

and this is called the period map for the VHS (of weight n) associated to the family V_z.

In [58] the 'infinitesimal bilinear relations'

$$\partial_{z_i} F_{h_p} \subseteq F_{h_{p+1}}, \quad 1 \leq i \leq r$$

were established, a property which nowadays is called Griffiths transversality. In the language of section 8.6, this property says that F is a superhorizontal holomorphic map into the total space of the twistor fibration

$$\pi : \frac{\mathrm{U}_{h_n}}{\mathrm{U}_{h^0} \times \cdots \times \mathrm{U}_{h^n}} \longrightarrow \frac{\mathrm{U}_{h_n}}{\mathrm{U}_{h^0+h^2+\cdots} \times \mathrm{U}_{h^1+h^3+\cdots}}$$

where π is given by

$$(F_{h_0}, \ldots, F_{h_{n-1}}) \longmapsto E^{n,0} \oplus E^{n-2,2} \oplus \cdots$$

from which it follows that $\pi \circ F$ is a pluriharmonic map into a Grassmannian.

In addition to Griffiths transversality, there are pointwise conditions on the map F, the 'first and second Riemann bilinear relations', which imply that its image lies in a smaller homogeneous space. This makes F a period map for a variation of 'polarized' Hodge structures. The conditions arise from the following two additional structures.

(i) Real structure: We have a natural real structure, namely, the conjugation map $B : x \mapsto \bar{x}$ with respect to the real form $H^n(V_{z_0};\mathbb{R})$ of $H^n(V_{z_0};\mathbb{C})$. This satisfies $\overline{E^{p,q}} = E^{q,p}$.

(ii) Non-degenerate bilinear form: The Poincaré intersection form

$$Q(x,y) = \int_{V_0} x \wedge y$$

on $H^n(V_{z_0};\mathbb{R})$ (and its $\mathbb{C}$-linear extension) is non-degenerate. When n is even, Q is symmetric and we can write $Q(x,y) = (x,y)$ as usual. When n is odd, Q is skew-symmetric; it is a symplectic bilinear form.

For all n one has a non-degenerate Hermitian form

$$H(x,y) = (-i)^n Q(x,\bar{y}).$$

The map F together with the Hermitian form H constitutes a variation of polarized Hodge structures of weight n.

The key property relating (i) and (ii) is $E^{p,q} \perp E^{i,j}$ if $(i,j) \neq (n-p,n-q)$. Hence

$$E^{n-p,p} \perp E^{n-q,q} \text{ if } q \neq n-p,$$

where we write $x \perp y$ to mean $Q(x,y) = 0$.

This gives the first Riemann bilinear relations:

$$F_{h_p} \perp F_{h_{n-p-1}}.$$

For dimensional reasons, it follows that these two subspaces are orthogonal complements of each other. Conversely, the orthogonality of these two subspaces implies the key property above, since $E^{n-p,p} = F_{h_p} \cap \overline{F}^{\perp}_{h_{p-1}} = F_{h_p} \cap \overline{F}_{h_p}$. The following diagram may be helpful in visualizing such relations:

F_{h_p}	$\boxed{E^{n,0}}$	$\cdots$	$\boxed{E^{n-p,p}}$	$E^{n-p-1,p+1}$	$\cdots$	$\cdots$	$E^{0,n}$
$\overline{F}^{\perp}_{h_p}$	$E^{n,0}$	$\cdots$	$E^{n-p,p}$	$\boxed{E^{n-p-1,p+1}}$	$\cdots$	$\cdots$	$\boxed{E^{0,n}}$
$F^{\perp}_{h_p} = F_{h_{n-p-1}}$	$E^{0,n}$	$\cdots$	$E^{p,n-p}$	$\boxed{E^{p+1,n-p-1}}$	$\cdots$	$\cdots$	$\boxed{E^{n,0}}$

In each row, the boxes in the right column correspond to the subspace in the left column. Note that the ordering is reversed in the last row.

To summarize the resulting conditions on F it is necessary to consider the cases n even and n odd separately.

The case $n = 2m$.

If $n = 2m$ then F_{h_p} satisfies

$$Q(F_{h_p}, F_{h_p}) = 0$$

for $p \le m - 1$, that is, F_{h_p} (and each $E^{n-p,p}$) is isotropic with respect to the symmetric form Q. The middle space $E^{m,m}$ satisfies $E^{m,m} = \overline{E^{m,m}}$, that is, it is real. For $p \ge m + 1$, we have $F_{h_p} = F^{\perp}_{h_{n-p-1}}$. Thus, F is determined by the m mutually orthogonal isotropic subspaces $E^{2m,0}, \ldots, E^{m+1,m-1}$ and the real subspace $E^{m,m}$. The signature of Q is (a, b) where

$$a = h^0 + h^2 + \cdots + h^{2m}, \quad b = h^1 + h^3 + \cdots + h^{2m-1}$$

but the conditions just stated relate only to the $\mathbb{C}$-linear extension of Q and therefore the signature is irrelevant at this stage. Thus, F can be identified with a superhorizontal holomorphic map into the (compact) total space of the twistor fibration

$$G/H = \frac{\mathrm{SO}_{a+b}}{\mathrm{U}_{h^0} \times \cdots \times \mathrm{U}_{h^{m-1}} \times \mathrm{SO}_{h^m}} \longrightarrow G/K = \frac{\mathrm{SO}_{a+b}}{\mathrm{SO}_a \times \mathrm{SO}_b}.$$

The case $n = 2m + 1$.

If $n = 2m + 1$ then F_{h_p} is isotropic for $p \le m$, and again $F_{h_p} = F^{\perp}_{h_{n-p-1}}$ for $p \ge m + 1$. This time Q is skew-symmetric, and under a change of basis it is equivalent over $\mathbb{C}$ to the standard skew-symmetric form on $\mathbb{C}^{2c}$ where $h_n = 2c = 2(h^0 + h^1 + \cdots + h^m)$. Thus, F can be identified with a superhorizontal holomorphic map into the (compact) total space of the twistor fibration

$$G/H = \frac{\mathrm{Sp}_c}{\mathrm{U}_{h^0} \times \cdots \times \mathrm{U}_{h^m}} \longrightarrow G/K = \frac{\mathrm{Sp}_c}{\mathrm{U}_c}.$$

This concludes our discussion of the first Riemann bilinear relations.

The second Riemann bilinear relations say that

$$(-1)^p H \text{ is positive definite on } E^{n-p,p}.$$

In particular, H is a Hermitian form on E of signature (c, c). It follows that F takes values in a specific orbit of a real form[1] $G' = \mathrm{SO}_{a,b}$ or $\mathrm{Sp}_{2c}\mathbb{R}$, where G' acts on the flag manifold $G^{\mathbb{C}}/P$ in the natural way. The orbit is a homogeneous space of the form G'/H' which we shall describe in detail later. It is important to note that, whereas the compact real form G acts transitively on the flag manifold $G^{\mathbb{C}}/P$, the non-compact real form G' does not, and the second Riemann bilinear relations pick out a particular orbit. This orbit is open, but not dense.

There are associated twistor fibrations

$$\pi' : G'/H' = \frac{\mathrm{SO}_{a,b}}{\mathrm{U}_{h^0} \times \cdots \times \mathrm{U}_{h^{m-1}} \times \mathrm{SO}_{h^m}} \longrightarrow G'/K' = \frac{\mathrm{SO}_{a,b}}{\mathrm{SO}_a \times \mathrm{SO}_b}$$

or

$$\pi' : G'/H' = \frac{\mathrm{Sp}_{2c}\mathbb{R}}{\mathrm{U}_{h^0} \times \cdots \times \mathrm{U}_{h^m}} \longrightarrow G'/K' = \frac{\mathrm{Sp}_{2c}\mathbb{R}}{\mathrm{U}_c},$$

and the twistor construction produces a pluriharmonic map $\pi' \circ F$ into the non-compact symmetric space G'/K' (see section 8.6).

(We refer to [130] for the general theory of orbits of non-compact real groups on flag manifolds. In general there are finitely many orbits, precisely one of which is closed, and the number of open orbits can be predicted Lie-algebraically. In rare cases, listed in [131], the non-compact group acts transitively, and in other rare cases there is precisely one open orbit. Neither of these holds in our situation above; the simplest case is the action of $\mathrm{Sp}_2\mathbb{R} = \mathrm{SL}_2\mathbb{R} \cong \mathrm{SU}_{1,1}$ on the two-sphere, where there are two open orbits—the upper and lower hemispheres—and one closed orbit—the equator.)

Classically the connection (or D-module) associated to F is called the Gauss–Manin connection. It has a purely cohomological description (as a ‘local system’). The map F is called the period map because it can be represented explicitly by the matrix whose (i, j)-th entry is $\int_{\gamma_i} \omega_j$, where the γ_i’s are a basis of n-cycles and the ω_j’s are a basis of n-dimensional cohomology classes compatible with the flag F. That is,

$$F = [\psi]$$

[1] A list of definitions of matrix groups used here can be found at the end of the section.

where the j-th column of the matrix ψ is (the transpose of)

$$(\int_{\gamma_1} \omega_j, \int_{\gamma_2} \omega_j, \ldots).$$

The components of this vector are simply the coefficients in the expansion of ω_j with respect to the dual basis $\gamma_1^*, \gamma_2^*, \ldots$ (defined by $\gamma_i^*(\gamma_j) = \delta_{ij}$). In this sense, the j-th column of ψ is the vector ω_j, and we may write

$$\psi = \begin{pmatrix} | & | & \\ \omega_1 & \omega_2 & \cdots \\ | & | & \end{pmatrix}$$

to indicate this. The first h_p column vectors span F_{h_p}.

This illustrates the principle mentioned in section 8.6, that a natural representative ψ of the superhorizontal map F can be expected to arise from the underlying geometry—in this case the period integrals. As in the case of harmonic maps, the Birkhoff factorization produces a canonical modification of such a ψ. We shall give a fundamental but very simple example here, in order to set the scene for the next section.

Example 10.1 Let $V = M_g$ be a compact connected Riemann surface of genus g. Thus $n = 1$ and $h^0 = h^1 = g$. The map F assigns to each complex structure z the g-space $H^{1,0}(V_z; \mathbb{C})$ in a (fixed) $2g$-space $E = H^1(V_{z_0}; \mathbb{C})$. If we choose a basis $\omega_1(z), \ldots, \omega_g(z)$ of holomorphic 1-forms, then $F(z)$ is spanned by $\omega_1(z), \ldots, \omega_g(z)$.

The Riemann bilinear relations here are usually written as

$$Q(\omega_i, \omega_j) = 0 \text{ for all } i, j$$
$$\sqrt{-1}\, Q(\omega, \bar{\omega}) > 0 \text{ for all non-zero } \omega \text{ in } H^{1,0}(V_z; \mathbb{C}).$$

The second condition says that the Hermitian form $\sqrt{-1}\, Q(x, \bar{y})$ is positive definite on $H^{1,0}(V_z; \mathbb{C})$ (although it is $H(x, y) = -\sqrt{-1}\, Q(x, \bar{y})$ which is positive definite in the general theory sketched earlier; this is just a matter of convention).

It is well known that there is a basis $\alpha_1, \ldots, \alpha_g, \beta_1, \ldots, \beta_g$ of the space of (real) 1-cycles of M_g such that

$$\alpha_i \cap \beta_j = -\beta_j \cap \alpha_i = \delta_{ij}.$$

With respect to the dual basis, the matrix of the symplectic form Q is

$$\begin{pmatrix} & I_g \\ -I_g & \end{pmatrix}.$$

Since Q is non-degenerate, it is possible to choose $\omega_1, \ldots, \omega_g$ such that $\int_{\alpha_i} \omega_j = \delta_{ij}$. The $g \times 2g$ classical period matrix

$$\begin{pmatrix} \int_{\gamma_1} \omega_1 & \int_{\gamma_2} \omega_1 & \cdots & \int_{\gamma_{2g}} \omega_1 \\ \vdots & \vdots & & \vdots \\ \int_{\gamma_1} \omega_g & \int_{\gamma_2} \omega_g & \cdots & \int_{\gamma_{2g}} \omega_g \end{pmatrix} \quad \text{or simply} \quad \begin{pmatrix} - & \omega_1 & - \\ & \vdots & \\ - & \omega_g & - \end{pmatrix}$$

can then be written $(I_g \ Z)$ where the (i, j)-th entry of Z is $\int_{\beta_j} \omega_i$. The matrix Z is symmetric ($Z^t = Z$) and its imaginary part represents a positive definite quadratic form ('$\operatorname{Im} Z > 0$').

The map F here is a holomorphic map to the homogeneous space

$$\mathrm{Sp}_g/\mathrm{U}_g \cong \mathrm{Sp}_{2g}\mathbb{C}/P \cong \mathrm{Sp}_{2g}\mathbb{C} \cdot \mathbb{C}^g$$

and there are various choices for a map ψ representing it. For values of z where the columns are linearly independent one could try

$$\psi = \begin{pmatrix} | & & | & | & & | \\ \omega_1 & \cdots & \omega_g & \omega_1' & \cdots & \omega_g' \\ | & & | & | & & | \end{pmatrix}$$

where prime denotes derivative with respect to z. The left hand $2g \times g$ block of this matrix is the transpose of the classical period matrix. Using the basis $\alpha_1^*, \ldots, \alpha_g^*, \beta_1^*, \ldots, \beta_g^*$ we have

$$\psi = \begin{pmatrix} I_g & 0 \\ Z & Z' \end{pmatrix}.$$

Evidently $F = \psi \cdot \mathbb{C}^g$. However, ψ does not necessarily take values in $\mathrm{Sp}_{2g}\mathbb{C}$ here. A matrix $\begin{pmatrix} P & Q \\ R & S \end{pmatrix}$ belongs to $\mathrm{Sp}_{2g}\mathbb{C}$ if and only if

$$S^t P - Q^t R = I, \quad S^t Q = Q^t S, \quad R^t P = P^t R,$$

so

$$\psi = \begin{pmatrix} I_g & 0 \\ Z & I_g \end{pmatrix} \text{ or } \begin{pmatrix} I_g & -2Z^{-1} \\ Z & -I_g \end{pmatrix}$$

are $\mathrm{Sp}_{2g}\mathbb{C}$-valued maps which represent F. The canonical representative

$$\psi = \begin{pmatrix} I_g & 0 \\ Z & I_g \end{pmatrix}$$

can be obtained (e.g. from any of the representatives above) by performing a lower/upper triangular factorization $\psi = \psi_- \psi_+$. By the uniqueness of the factorization, this holds even when ψ itself does not take values in $\mathrm{Sp}_{2g}\mathbb{C}$ (i.e. if one knows that an $\mathrm{Sp}_{2g}\mathbb{C}$-valued representative exists, then ψ_- is always such a representative, independent of the nature of ψ itself).

A matrix $\begin{pmatrix} P & Q \\ R & S \end{pmatrix}$ from $\mathrm{Sp}_{2g}\mathbb{C}$ belongs to $\mathrm{Sp}_{2g}\mathbb{R}$ if and only if P, Q, R, S are all real. The condition $\mathrm{Im}\, Z > 0$ is precisely the condition that F takes values in the $\mathrm{Sp}_{2g}\mathbb{R}$-orbit of

$$\begin{aligned}\mathrm{Span}\{e_1 + ie_{g+1}, \ldots, e_g + ie_{2g}\} &= c \cdot \mathrm{Span}\{e_1, \ldots, e_g\} \\ &= c \cdot \mathbb{C}^g \in \mathrm{Gr}_g(\mathbb{C}^{2g}),\end{aligned}$$

and this says that ψ takes values in the product

$$\mathrm{Sp}_{2g}\mathbb{R}\; c\; \Delta$$

where Δ is the group of upper-triangular matrices in $\mathrm{Sp}_{2g}\mathbb{C}$. In other words, we can write

$$\psi = \psi_{\mathbb{R}}\; c\; \psi_{\Delta}$$

where $\psi_{\mathbb{R}}, \psi_{\Delta}$ take values in $\mathrm{Sp}_{2g}\mathbb{R}, \Delta$. The fact that ψ admits such a factorization is a strong condition, quite different from the existence of the factorization $\psi = \psi_-\psi_+$.

Let us summarize this discussion (whose only purpose was to express the period map and bilinear relations more explicitly). The map

$$F : z \mapsto H^{1,0}(V_z; \mathbb{C})$$

is a holomorphic map into the Grassmannian $\mathrm{Gr}_g(\mathbb{C}^{2g})$; the first Riemann bilinear relation says that the image lies in the smaller 'symplectic Grassmannian' $\mathrm{Sp}_g/\mathrm{U}_g \cong \mathrm{Sp}_{2g}\mathbb{C}/P$, and the second Riemann bilinear relation says that the image lies in a certain open subspace, a particular orbit of $\mathrm{Sp}_{2g}\mathbb{R}$, which can be identified with the Siegel upper half-space $\mathrm{Sp}_{2g}\mathbb{R}/\mathrm{U}_g$.

Note that ψ_- represents the holomorphic data efficiently, while $\psi_{\mathbb{R}}$ takes into account the polarization given by the Hermitian form. These procedures are special cases of the Bruhat and Iwasawa factorizations respectively, and, as such, extend to the case of loop groups, to which we shall soon return.

The twistor fibration is the identity map and there is no Griffiths transversality/superhorizontality condition here—this is a non-compact analogue of Example 8.15. □

Example 10.2 Let us write out the case $g = 1$ explicitly, as we shall refer to it later. Let α, β be the basis described above, and let ω be a holomorphic

1-form such that $\int_\alpha \omega = 1$. We have

$$\psi = \begin{pmatrix} \int_\alpha \omega & \int_\alpha \omega' \\ \int_\beta \omega & \int_\beta \omega' \end{pmatrix} = \begin{pmatrix} 1 & 0 \\ \int_\beta \omega & \int_\beta \omega' \end{pmatrix} = \begin{pmatrix} 1 & 0 \\ \int_\beta \omega & 1 \end{pmatrix} \begin{pmatrix} 1 & 0 \\ 0 & \int_\beta \omega' \end{pmatrix},$$

and therefore

$$\psi_- = \begin{pmatrix} 1 & 0 \\ \int_\beta \omega & 1 \end{pmatrix}.$$

The Riemann bilinear relations (with the conventions of the previous example) are

$$Q(\omega, \omega) = 0$$
$$2 \operatorname{Im} \textstyle\int_\beta \omega = \sqrt{-1}\, Q(\omega, \bar{\omega}) > 0.$$

The matrix of Q is $\begin{pmatrix} & 1 \\ -1 & \end{pmatrix}$.

The first bilinear relation $Q(\omega, \omega) = 0$ is vacuous as Q is skew-symmetric (we have $\mathrm{Sp}_2\mathbb{R} = \mathrm{SL}_2\mathbb{R}$), but the second says that

$$z \mapsto \psi_-(z) \begin{bmatrix} 1 \\ 0 \end{bmatrix}$$

takes values in the orbit

$$\mathrm{Sp}_2\mathbb{R} \cdot \begin{bmatrix} 1 \\ i \end{bmatrix} = \left\{ \begin{bmatrix} x \\ y \end{bmatrix} \in \mathbb{C}P^1 \;\middle|\; \operatorname{Im} \frac{y}{x} > 0 \right\} \cong \frac{\mathrm{Sp}_2\mathbb{R}}{\mathrm{U}_1}.$$

The function $\int_\beta \omega$ is well known: with appropriate choices it is equal to the (multi-valued) inverse of the modular elliptic function $J : \mathrm{Sp}_2\mathbb{R}/\mathrm{U}_1 \to \mathbb{C}$ (see Example 10.11). □

For reference purposes, we conclude this section by stating the definitions of the classical groups used in this chapter, together with some elementary linear algebra notation which will be useful.

1. Orthogonal groups.

$$\mathrm{SO}_{a,b} = \left\{ A \in \mathrm{SL}_{a+b}\mathbb{R} \;\middle|\; \begin{pmatrix} I_a & \\ & -I_b \end{pmatrix}^{-1} (A^t)^{-1} \begin{pmatrix} I_a & \\ & -I_b \end{pmatrix} = A \right\}$$

$$\mathrm{SO}_{a+b} = \mathrm{SO}_{a+b,0} = \left\{ A \in \mathrm{SL}_{a+b}\mathbb{R} \;\middle|\; (A^t)^{-1} = A \right\}$$

The complexifications $\mathrm{SO}_{a,b}\mathbb{C}$, $\mathrm{SO}_{a+b}\mathbb{C}$ are defined in the same way, replacing $\mathrm{SL}_{a+b}\mathbb{R}$ by $\mathrm{SL}_{a+b}\mathbb{C}$. In both cases the (Lie group) conjugation map is $C(A) = \bar{A}$; the fixed points of C give the original real group. The complexifications are isomorphic as complex Lie groups (but not, of course, as 'complex Lie groups with conjugation maps').

2. *Symplectic groups.*

$$\mathrm{Sp}_{2c}\mathbb{R} = \left\{ A \in \mathrm{GL}_{2c}\mathbb{R} \,\middle|\, \begin{pmatrix} & I_c \\ -I_c & \end{pmatrix}^{-1} (A^t)^{-1} \begin{pmatrix} & I_c \\ -I_c & \end{pmatrix} = A \right\}$$

$$\mathrm{Sp}_c = \left\{ A \in \mathrm{GL}_c\mathbb{H} \,\middle|\, (\bar{A}^t)^{-1} = A \right\}$$

The complexification of $\mathrm{Sp}_{2c}\mathbb{R}$ is defined similarly as

$$\mathrm{Sp}_{2c}\mathbb{C} = \left\{ A \in \mathrm{GL}_{2c}\mathbb{C} \,\middle|\, \begin{pmatrix} & I_c \\ -I_c & \end{pmatrix}^{-1} (A^t)^{-1} \begin{pmatrix} & I_c \\ -I_c & \end{pmatrix} = A \right\}$$

This is also the complexification of Sp_c, since the latter can be identified with $\mathrm{Sp}_{2c}\mathbb{C} \cap \mathrm{U}_{2c}$. The conjugation map $C(A) = \bar{A}$ has fixed point set $\mathrm{Sp}_{2c}\mathbb{R}$, while the conjugation map $C(A) = (\bar{A}^t)^{-1}$ has fixed point set Sp_c.

3. *Unitary groups.*

$$\mathrm{SU}_{a,b} = \left\{ A \in \mathrm{SL}_{a+b}\mathbb{C} \,\middle|\, \begin{pmatrix} I_a & \\ & -I_b \end{pmatrix}^{-1} (\bar{A}^t)^{-1} \begin{pmatrix} I_a & \\ & -I_b \end{pmatrix} = A \right\}$$

$$\mathrm{SU}_{a+b} = \mathrm{SU}_{a+b,0} = \left\{ A \in \mathrm{SL}_{a+b}\mathbb{C} \,\middle|\, (\bar{A}^t)^{-1} = A \right\}$$

The complexification of each of these groups is $\mathrm{SL}_{a+b}\mathbb{C}$. In each case, the equation involving A should be read as $C(A) = A$.

More generally, we shall be concerned with orthogonal and symplectic groups of the following type:

$$\mathrm{SO}^M_{a+b}\mathbb{C} = \{A \in \mathrm{SL}_{a+b}\mathbb{C} \mid M^{-1}(A^t)^{-1}M = A\} \quad \text{where } M^t = M$$

$$\mathrm{Sp}^M_{2c}\mathbb{C} = \{A \in \mathrm{GL}_{2c}\mathbb{C} \mid M^{-1}(A^t)^{-1}M = A\} \quad \text{where } M^t = -M.$$

That is, we consider linear transformations A which preserve a bilinear form $Q(x, y) = x^t M y$.

We shall also be concerned with real forms of these complex groups which are defined in a similar way:

$$\mathrm{SO}^{M,N}_{a,b}\mathbb{R} = \{A \in \mathrm{SO}^{M}_{a+b}\mathbb{C} \mid N\bar{A}N^{-1} = A\}$$
$$\mathrm{Sp}^{M,N}_{2c}\mathbb{R} = \{A \in \mathrm{Sp}^{M}_{2c}\mathbb{C} \mid N\bar{A}N^{-1} = A\},$$

where N is a matrix which satisfies $N\bar{N} = I$. This means that the linear transformation preserves the real structure

$$B(x) = N\bar{x}$$

of the underlying complex vector space. These real forms are the fixed point sets for the conjugation maps $C(A) = N\bar{A}N^{-1}$. We assume that $Q(B(x), B(y)) = \overline{Q(x, y)}$, that is, $\bar{M} = N^t MN$, so that the maps are well defined.

The forms defined by

$$H(x, y) = \begin{cases} \pm Q(x, B(y)) & \text{if } Q \text{ is symmetric} \\ \pm iQ(x, B(y)) & \text{if } Q \text{ is skew-symmetric} \end{cases}$$

are then Hermitian, that is, $H(y, x) = \overline{H(x, y)}$.

Different choices of M, N arise naturally when we consider variations of Hodge structures, and we need all this notation in order to distinguish them.

We remind the reader that, under a change of basis from $v_1, v_2, \ldots$ to $\hat{v}_1, \hat{v}_2, \ldots$, the matrices transform as follows

$$\hat{A} = P^{-1}AP, \quad \hat{M} = P^t MP, \quad \hat{N} = P^{-1}N\bar{P}$$

where $P = (p_{ij})$ is defined by $\hat{v}_i = \sum_j p_{ji}v_j$, which we abbreviate to $\hat{v}_i = P \cdot v_i$. Thus, we have

$$\mathrm{SO}^{\hat{M}}_{a+b}\mathbb{C} = P^{-1}\, \mathrm{SO}^{M}_{a+b}\mathbb{C}\, P, \quad \mathrm{Sp}^{\hat{M}}_{2c}\mathbb{C} = P^{-1}\, \mathrm{Sp}^{M}_{2c}\mathbb{C}\, P,$$

and

$$\mathrm{SO}^{\hat{M},\hat{N}}_{a,b}\mathbb{R} = P^{-1}\, \mathrm{SO}^{M,N}_{a,b}\mathbb{R}\, P, \quad \mathrm{Sp}^{\hat{M},\hat{N}}_{2c}\mathbb{R} = P^{-1}\, \mathrm{Sp}^{M,N}_{2c}\mathbb{R}\, P.$$

10.2 Mirror symmetry: an example

Calabi–Yau manifolds can be defined as compact Kähler manifolds V of dimension n such that $\wedge^n TV$ is a trivial line bundle, that is, $c_1(TV) = 0$

(a detailed explanation with more refined definitions can be found in [77]). They have a particularly well behaved deformation theory. If V_z is a family of Calabi–Yau manifolds, mirror symmetry predicts the existence of another family V_z° of Calabi–Yau manifolds such that the two families are related in a specific way. One aspect of this relationship is the 'mirror symmetry' of the Hodge numbers:

$$\dim H^{p,q}(V_z;\mathbb{C}) = \dim H^{n-p,q}(V_z^\circ;\mathbb{C}).$$

A deeper aspect, and the one we shall focus on, is the equivalence of the VHS for V_z (the 'B-model VHS', represented by the map F) with the quantum cohomology of V_z° (the 'A-model VHS', personified by our map W). In other words, the VHS D-module for V_z (or Gauss–Manin connection) is equivalent to the quantum D-module for V_z° (or Dubrovin/Givental connection). This equivalence involves a non-trivial coordinate transformation, known as the mirror map, which can be interpreted as a map from a space of deformations of complex structure of V to a space of deformations of Kähler structure of V°. (An even deeper aspect is the relation between the manifolds V_z and V_z° themselves, but there appears to be no general description of this at present.)

As an example, we shall explain this for the case of a non-singular hypersurface of degree k in $\mathbb{C}P^{k-1}$, where everything can be computed in detail. In this section we give the standard version, which is a series of verifiable but somewhat unconnected calculations. In the following two sections we indicate how the loop group and Grassmannian model reveal a simpler picture.

Let M_N^k denote a smooth hypersurface in $\mathbb{C}P^{N-1}$ of degree k. We have[2] $\dim H^2 M_N^k = 1$ and $\dim H^\sharp M_N^k = N-1$. When $1 \le k \le N-1$, M_N^k is Fano, and $QH^\sharp M_N^k$ may be defined naively as in Chapter 2. When $k = N$, M_N^k is Calabi–Yau. In this case (and in the case $k > N$), the naive interpretation of Gromov–Witten invariants fails, as rational curves in M_N^k do not have the expected properties. Nevertheless, Gromov–Witten invariants can be defined, and they 'count' rational curves in a sense which can be justified.

An alternative way to obtain these Gromov–Witten invariants when $k = N$ is to use mirror symmetry, as we shall now explain. Starting with a VHS for a mirror manifold

$$V = \tilde{M}_k^k$$

[2] For the case $(N,k) = (4,2)$, where $M_N^k \cong \mathbb{C}P^1 \times \mathbb{C}P^1$, we have $\dim H^2 M_N^k = 2$.

of M_k^k, we shall construct an abstract quantum D-module, which turns out to coincide with the quantum D-module of

$$V^\circ = M_k^k.$$

However, since we start with the subspace $H^n \tilde{M}_k^k$ of $H^* \tilde{M}_k^k$, we obtain the subspace $QH^\sharp M_k^k$ of $QH^* M_k^k$.

With the notation of section 10.1 we have $n = k-2$, and all spaces $E^{n-p,p}$ are one-dimensional. With respect to a basis of the vector space $H^n(\tilde{M}_k^k;\mathbb{C})$, the essentially unique holomorphic $(k-2)$-form ω on V (see [77]) produces a natural representative

$$\psi = \begin{pmatrix} | & & | \\ \omega & \cdots & \omega^{(k-2)} \\ | & & | \end{pmatrix}$$

of the period map F, with

$$F_{h_p} = F_{p+1} = \operatorname{Span}\{\omega, \omega', \ldots, \omega^{(p)}\}.$$

The case $k = 3$, a hypersurface in $\mathbb{C}P^2$ of degree 3, that is, elliptic curve, has already appeared in Example 10.2 in the guise of M_g with $g = 1$.

This is a non-compact analogue of the situation for harmonic maps to $\mathbb{C}P^2$ in Example 8.16. However, in the current situation there are additional restrictions from the Riemann bilinear relations, which lead to a VHS in the form of a holomorphic superhorizontal map

$$z \mapsto F(z) \in \begin{cases} \mathrm{Sp}_{k-1}\mathbb{R}/\underbrace{\mathrm{U}_1 \times \cdots \times \mathrm{U}_1}_{\frac{k-1}{2}} & (k \text{ odd}) \\ \mathrm{SO}_{\frac{k}{2},\frac{k}{2}-1}/\underbrace{\mathrm{U}_1 \times \cdots \times \mathrm{U}_1}_{\frac{k}{2}-1} & (k \text{ even}). \end{cases}$$

By composing with the canonical twistor fibration, we obtain a harmonic map to the symmetric space

$$\mathrm{Sp}_{k-1}\mathbb{R}/\,U_{\frac{k-1}{2}} \quad \text{or} \quad \mathrm{SO}_{\frac{k}{2},\frac{k}{2}-1}/\,\mathrm{SO}_{\frac{k}{2}} \times \mathrm{SO}_{\frac{k}{2}-1}.$$

We shall discuss the case $k = 5$ and its relation with quantum cohomology in detail, then summarize the situation for general k at the end.

Following the general scheme of section 10.1, let us choose a basis $\hat{x}_0, \hat{x}_1, \hat{x}_2, \hat{x}_3$ of $H^3(\tilde{M}_5^5;\mathbb{C})$ with $\hat{x}_p \in H^{3-p,p}(\tilde{M}_5^5;\mathbb{C})$ such that

$$(-1)^p H(\hat{x}_p, \hat{x}_p) = 1 \text{ and } B(\hat{x}_i) = \hat{x}_{3-i}.$$

With respect to this basis, the matrices $\hat{M}, \hat{N}$ of the symplectic form Q and the real structure B (see the end of section 10.1) are

$$\hat{M} = \begin{pmatrix} & & & -i \\ & & i & \\ & -i & & \\ i & & & \end{pmatrix}, \qquad \hat{N} = \begin{pmatrix} & & & 1 \\ & & 1 & \\ & 1 & & \\ 1 & & & \end{pmatrix}.$$

We may regard this basis as $\hat{x}_0 = \frac{1}{\sqrt{2}}(e_0 + ie_3)$, $\hat{x}_1 = \frac{1}{\sqrt{2}}(e_1 + ie_2)$, $\hat{x}_2 = \frac{1}{\sqrt{2}}(e_1 - ie_2)$, $\hat{x}_3 = \frac{1}{\sqrt{2}}(e_0 - ie_3)$, where e_0, e_1, e_2, e_3 is a basis of the real vector space $H^3(\tilde{M}_5^5; \mathbb{R})$. The matrices of Q, B with respect to e_0, e_1, e_2, e_3 are

$$M = \begin{pmatrix} & & & 1 \\ & & -1 & \\ & 1 & & \\ -1 & & & \end{pmatrix} = J \text{ (say)}, \qquad N = \begin{pmatrix} 1 & & & \\ & 1 & & \\ & & 1 & \\ & & & 1 \end{pmatrix} = I.$$

The groups $\mathrm{Sp}_4^J\mathbb{C}, \mathrm{Sp}_4^{J,I}\mathbb{R}$ are therefore almost the same as the standard groups $\mathrm{Sp}_4\mathbb{C}, \mathrm{Sp}_4\mathbb{R}$ defined in section 10.1; the only difference is that the matrix of the symplectic form has alternating 1's and -1's instead of two blocks $I, -I$.

For the calculations it will be convenient to use the 'more symmetrical' basis $\tilde{x}_0 = \frac{1}{\sqrt{2}}(e_0 + ie_3)$, $\tilde{x}_1 = \frac{1}{\sqrt{2}}(e_1 + ie_2)$, $\tilde{x}_2 = \frac{1}{\sqrt{2}}(ie_1 + e_2)$, $\tilde{x}_3 = \frac{1}{\sqrt{2}}(ie_0 + e_3)$ which is given by applying the matrix

$$c = \frac{1}{\sqrt{2}} \begin{pmatrix} 1 & 0 & 0 & i \\ 0 & 1 & i & 0 \\ 0 & i & 1 & 0 \\ i & 0 & 0 & 1 \end{pmatrix}$$

to e_0, e_1, e_2, e_3. We collect here some simple properties of this matrix:

Lemma 10.3 *Let*

$$S = \begin{pmatrix} & & & 1 \\ & & 1 & \\ & 1 & & \\ 1 & & & \end{pmatrix}.$$

Then we have

(i) $c^2 = iS$, $\bar{c} = c^{-1} = -icS$, $cS = Sc$,
(ii) $c^{-1}J = Jc$, $JS = -SJ$.

It follows that the matrices $\tilde{M}, \tilde{N}$ of Q, B with respect to $\tilde{x}_0, \tilde{x}_1, \tilde{x}_2, \tilde{x}_3$ are

$$\tilde{M} = c^t M c = c^t J c = J, \qquad \tilde{N} = c^{-1} N \bar{c} = c^{-1} \bar{c} = -iS.$$

From the remarks at the end of section 10.1 we obtain

$$\mathrm{Sp}_4^{\tilde{M}} \mathbb{C} = \mathrm{Sp}_4^J \mathbb{C}$$

(the lemma implies that $c \in \mathrm{Sp}_4^J \mathbb{C}$), and

$$\mathrm{Sp}_4^{\tilde{M},\tilde{N}} \mathbb{R} = c^{-1}\ \mathrm{Sp}_4^{J,I} \mathbb{R}\ c.$$

We may write $F = [\psi]$. The first bilinear relation says that ψ takes values in the complex symplectic group $\mathrm{Sp}_4^{\tilde{M}} \mathbb{C}$. The second bilinear relation says that

$$c^{-1}\psi \in \mathrm{Sp}_4^{\tilde{M},\tilde{N}} \mathbb{R}\ \Delta \quad \text{that is,} \quad \psi \in \mathrm{Sp}_4^{I,J} \mathbb{R}\ c\ \Delta$$

where Δ denotes the group of upper-triangular matrices. In other words, the map $F = [\psi]$ takes values in the $\mathrm{Sp}_4^{J,I} \mathbb{R}$-orbit of the 'standard flag' $[c]$.

An explicit formula for ψ_- arises from its interpretation in terms of period integrals. The Gauss–Manin D-module is cyclic with generator ω, and can be represented as $D/(\mathrm{PF})$ where PF denotes the Picard–Fuchs differential operator. This is the differential operator of order 4 which annihilates the component functions of ω (the period integrals). The standard[3] choice of V (see Chapter 5 of [26]) produces the Picard–Fuchs operator

$$\partial_\theta^4 - 5z(5\partial_\theta + 4)(5\partial_\theta + 3)(5\partial_\theta + 2)(5\partial_\theta + 1)$$

where $z = e^\theta$ and $\partial_\theta = z\frac{\partial}{\partial z}$.

From the Frobenius method (for solving an o.d.e. in a neighbourhood of a regular singular point, applied here at the point $z = 0$, where the indicial equation is $s^4 = 0$; cf. appendix 5.4), it follows that there exists a basis of solutions u_0, u_1, u_2, u_3 of the form

$$\begin{aligned}
u_0 &= f_0 \\
u_1 &= f_0 \log z + f_1 \\
u_2 &= f_0 \tfrac{1}{2!}(\log z)^2 + f_1 \log z + f_2 \\
u_3 &= f_0 \tfrac{1}{3!}(\log z)^3 + f_1 \tfrac{1}{2!}(\log z)^2 + f_2 \log z + f_3
\end{aligned}$$

[3] This choice turns out to be slightly inappropriate, as we shall explain in section 10.5.

where f_0, f_1, f_2, f_3 are holomorphic at $z = 0$. We may choose these functions so that $f_0(0) = 1$ and $f_1(0) = f_2(0) = f_3(0) = 0$. This defines them uniquely, and their power series expansions may be calculated explicitly.

Let us put $u = (u_0, u_1, u_2, u_3)$,

$$\psi = \begin{pmatrix} | & | & | & | \\ u & u' & u'' & u''' \\ | & | & | & | \end{pmatrix},$$

and $F = [\psi]$. (The dash denotes derivative with respect to θ here.) This is a fundamental solution matrix for the linear system with (transposed) coefficient matrix

$$\psi^{-1}\psi' = \begin{pmatrix} 0 & & & 120/(1-5^5z) \\ 1 & 0 & & 1250/(1-5^5z) \\ & 1 & 0 & 4375/(1-5^5z) \\ & & 1 & 6250/(1-5^5z) \end{pmatrix}.$$

We shall need the normalized form of ψ, obtained by performing the lower/upper triangular factorization $\psi = \psi_-\psi_+$. This is

$$\psi_- = \begin{pmatrix} 1 & 0 & 0 & 0 \\ A_1 & 1 & 0 & 0 \\ A_2 & B_2 & 1 & 0 \\ A_3 & B_3 & C_3 & 1 \end{pmatrix}$$

where $A_i = u_i/u_0$, $B_i = A_i'/A_1$, and $C_i = B_i'/B_2'$. We obtain

$$\psi_-^{-1}\psi_-' = \begin{pmatrix} 0 & & & \\ v_1 & 0 & & \\ & v_2 & 0 & \\ & & v_3 & 0 \end{pmatrix},$$

where

$$v_1 = A_1', \quad v_2 = B_2', \quad v_3 = C_3',$$

as explained in appendix 4.6. Then the transformation $\psi \mapsto \psi_-$ corresponds to a gauge transformation of the Picard–Fuchs equation, and the transformed operator is

$$(1 - 5^5 z)\, v_0 v_1 v_2 v_3\, \partial_\theta \frac{1}{v_3} \partial_\theta \frac{1}{v_2} \partial_\theta \frac{1}{v_1} \partial_\theta \frac{1}{v_0}.$$

For the same reason as in Example 10.1, ψ_- must take values in $\mathrm{Sp}_4^{\tilde{M}}\mathbb{C}$, hence $\psi_-^{-1}\psi_-'$ is in $\mathfrak{sp}_4^{\tilde{M}}\mathbb{C}$. The latter consists of matrices X which satisfy $\tilde{M}^{-1}X^t\tilde{M} = -X$, from which it follows that $v_1 = v_3$.

Mirror symmetry predicts that the reflection of the 'Hodge diamond' for $V = \tilde{M}_5^5$ in the NE-SW diagonal is the Hodge diamond for $V^\circ = M_5^5$. Writing $h^{p,q} = \dim_{\mathbb{C}} H^{p,q}(M_5^5;\mathbb{C})$, it is known that

$$
\begin{array}{ccccccc}
 & & & h^{3,3} & & & \\
 & & h^{3,2} & & h^{2,3} & & \\
 & h^{3,1} & & h^{2,2} & & h^{1,3} & \\
h^{3,0} & & h^{2,1} & & h^{1,2} & & h^{0,3} \\
 & h^{2,0} & & h^{1,1} & & h^{0,2} & \\
 & & h^{1,0} & & h^{0,1} & & \\
 & & & h^{0,0} & & &
\end{array}
\quad = \quad
\begin{array}{ccccccc}
 & & & 1 & & & \\
 & & 0 & & 0 & & \\
 & 0 & & 1 & & 0 & \\
1 & & 101 & & 101 & & 1 \\
 & 0 & & 1 & & 0 & \\
 & & 0 & & 0 & & \\
 & & & 1 & & &
\end{array}
$$

Under the mirror symmetry reflection, the 'vertical' four-dimensional vector space $H^\sharp(M_5^5;\mathbb{C})$ in the above Hodge diamond for M_5^5 corresponds to the 'horizontal' four-dimensional vector space $E = H^3(\tilde{M}_5^5;\mathbb{C})$ in the Hodge diamond for $\tilde{M}_5^5$. This is indicated in the diagram below: the boxes in the left hand diagram represent $E = H^3(\tilde{M}_5^5;\mathbb{C})$, and those in the right hand diagram represent $H^\sharp(M_5^5;\mathbb{C})$.

$$
\begin{array}{ccccccc}
 & & & 1 & & & \\
 & & 0 & & 0 & & \\
 & 0 & & 101 & & 0 & \\
\boxed{1} & & \boxed{1} & & \boxed{1} & & \boxed{1} \\
 & 0 & & 101 & & 0 & \\
 & & 0 & & 0 & & \\
 & & & 1 & & &
\end{array}
\qquad\qquad
\begin{array}{ccccccc}
 & & & \boxed{1} & & & \\
 & & 0 & & 0 & & \\
 & 0 & & \boxed{1} & & 0 & \\
1 & & 101 & & 101 & & 1 \\
 & 0 & & \boxed{1} & & 0 & \\
 & & 0 & & 0 & & \\
 & & & \boxed{1} & & &
\end{array}
$$

Miraculously, *we obtain the quantum cohomology* $QH^\sharp M_5^5$ *if we identify the basis* $\tilde{x}_0, \tilde{x}_1, \tilde{x}_2, \tilde{x}_3$ *of* $H^3(\tilde{M}_5^5;\mathbb{C})$ *with the basis* $1, b, b^2, b^3$ *of* $H^\sharp(M_5^5;\mathbb{C})$*, and regard the Picard–Fuchs equation as the quantum differential equation*—at least, after making a suitable coordinate change from $z = e^\theta$ to $q = e^t$. It is 'miraculous' because there is no a priori relation between $H^3(\tilde{M}_5^5;\mathbb{C})$ and $H^\sharp(M_5^5;\mathbb{C})$.

Before verifying this, we recall some properties of the quantum product. First, we have $|q| = 0$ here, as $c_1(TM_5^5) = 0$. Denoting the generator of

$H^2M_5^5$ by b as usual we must have

$$\begin{aligned} b \circ 1 &= b \\ b \circ b &= \langle B|B|\tfrac{1}{5}B\rangle_0\, b^2 + \langle B|B|\tfrac{1}{5}B\rangle_1\, qb^2 + \langle B|B|\tfrac{1}{5}B\rangle_2\, q^2b^2 + \cdots \\ b \circ b^2 &= b^3 \\ b \circ b^3 &= 0, \end{aligned}$$

the $\frac{1}{5}$ arising because it is $1, b, b^2, b^3$ and $\frac{1}{5}b^3, \frac{1}{5}b^2, \frac{1}{5}b, \frac{1}{5}$ that are dual bases with respect to the intersection form. To produce this, the quantum differential operator would have to be of the form $\partial_t^2 * \partial_t^2$ where $b \circ b = (1/*)\, b^2$.

Such a form can be achieved by the factorized Picard–Fuchs equation if one makes the coordinate change (mirror transformation)

$$t = \frac{u_1}{u_0},$$

because this implies

$$\frac{1}{v_1}\partial_\theta = \partial_t,$$

and the Picard–Fuchs operator becomes

$$(1 - 5^5 z)\, v_0 v_1^2 v_2 v_3 \partial_t \frac{v_1}{v_3}\, \partial_t \frac{v_1}{v_2}\, \partial_t\, \partial_t \frac{1}{v_0}.$$

As remarked earlier, we have $v_1 = v_3$. Thus, taking

$$\partial_t^2 \frac{v_1}{v_2}\, \partial_t^2$$

(by an act of faith) as the quantum differential operator, we obtain the prediction

$$\begin{aligned} b \circ 1 &= b \\ b \circ b &= (v_2/v_1)\, b^2 \\ b \circ b^2 &= b^3 \\ b \circ b^3 &= 0. \end{aligned}$$

From an explicit computation of the solutions u_0, u_1, u_2, u_3 (for which we shall give a short cut in the next section), we have

$$\begin{aligned} \frac{v_2}{v_1} &= 1 + 575z + 1418125z^2 + \cdots \\ &= 1 + 575q + 975375q^2 + \cdots \end{aligned}$$

that is

$$\langle B|B|\tfrac{1}{5}B\rangle_0 = 1, \quad \langle B|B|\tfrac{1}{5}B\rangle_1 = 575, \quad \langle B|B|\tfrac{1}{5}B\rangle_2 = 975375, \ldots.$$

This turns out to be the correct answer, although the enumerative interpretation of the Gromov–Witten invariants $\langle B|B|B\rangle_d$ is non-trivial. In fact (see [26])

$$\sum_{d\geq 0} \langle B|B|B\rangle_d q^d = 5 + \sum_{d\geq 1} n_d\, d^3 \frac{q^d}{1-q^d}$$

where n_d is the number of degree d rational curves in M_5^5. Assuming this we obtain $n_1 = 2875, n_2 = 609250, \ldots$. It is remarkable that these numbers of rational curves arise from the solution of a differential equation.

We conclude with a brief indication of the changes that are needed to treat M_k^k for any k.

The case $k = 2l + 1$.

This is very similar to the case $l = 2$. We take a basis $\hat{x}_0, \ldots, \hat{x}_{2l-1}$ of $H^{k-2}(\tilde{M}_k^k; \mathbb{C})$ such that $(-1)^p H(\hat{x}_p, \hat{x}_p) = 1$ and $B(\hat{x}_i) = \hat{x}_{2l-i-1}$. The Hermitian form is $H(x) = (-i)^{2l-1} Q(x, B(x))$. Introducing $\tilde{x}_0, \ldots, \tilde{x}_{2l-1}$ and

$$c = \frac{1}{\sqrt{2}} \begin{pmatrix} 1 & & & i \\ & \ddots & \iddots & \\ & \iddots & \ddots & \\ i & & & 1 \end{pmatrix}$$

in the same way as before, we find that the matrices of Q and B are

$$\tilde{M} = (-1)^l J, \quad \tilde{N} = -iS$$

where

$$S = \begin{pmatrix} & & & 1 \\ & & 1 & \\ & \iddots & & \\ 1 & & & \\ 1 & & & \end{pmatrix}, \quad J = \begin{pmatrix} & & & 1 \\ & & -1 & \\ & \iddots & & \\ 1 & & & \\ -1 & & & \end{pmatrix}.$$

(With respect to the basis $e_0, \ldots, e_{2l-1}$ the matrices are $M = (-1)^l J, N = I$, and with respect to $\hat{x}_0, \ldots, \hat{x}_{2l-1}$ they are $\hat{M} = -i(-1)^l J, \hat{N} = S$.) The matrix of the Hermitian form H with respect to $\tilde{x}_0, \dots, \tilde{x}_{2l-1}$ is

$$(-i)^{2l-1}\tilde{M}\tilde{N} = D = \mathrm{diag}(1, -1, \dots, 1, -1).$$

The formulae of Lemma 10.3 continue to hold, and we have

$$\mathrm{Sp}^{\tilde{M}}_{2l}\mathbb{C} = \mathrm{Sp}^{J}_{2l}\mathbb{C}, \quad \mathrm{Sp}^{\tilde{M},\tilde{N}}_{2l}\mathbb{R} = c^{-1}\,\mathrm{Sp}^{J,I}_{2l}\mathbb{R}\; c.$$

Exactly as in the case $l = 2$, the first bilinear relation says that ψ takes values in $\mathrm{Sp}^{\tilde{M}}_{2l}\mathbb{C}$, and the second bilinear relation says that $c^{-1}\psi$ takes values in the product $\mathrm{Sp}^{\tilde{M},\tilde{N}}_{2l}\mathbb{R}\;\Delta$, or (equivalently) ψ takes values in $\mathrm{Sp}^{J,I}_{2l}\mathbb{R}\; c\;\Delta$.

The case $k = 2l$.

The same principles apply here, but with somewhat different details. As a basis of $H^{k-2}(\tilde{M}^k_k; \mathbb{C})$ we take

$$\hat{x}_0 = \tfrac{1}{\sqrt{2}}(e_0 + ie_{2l-2}), \dots, \hat{x}_{l-2} = \tfrac{1}{\sqrt{2}}(e_{l-2} + ie_l),$$
$$\hat{x}_{l-1} = e_{l-1},$$
$$\hat{x}_l = \tfrac{1}{\sqrt{2}}(e_{l-2} - ie_l), \dots, \hat{x}_{2l-2} = \tfrac{1}{\sqrt{2}}(e_0 - ie_{2l-2}).$$

We also need the following $(2l-1) \times (2l-1)$ matrices:

$$S = \begin{pmatrix} & & & & 1 \\ & & & 1 & \\ & & \cdot^{\cdot^{\cdot}} & & \\ & 1 & & & \\ 1 & & & & \end{pmatrix}, \quad K = \begin{pmatrix} & & & & 1 \\ & & & -1 & \\ & & \cdot^{\cdot^{\cdot}} & & \\ & -1 & & & \\ 1 & & & & \end{pmatrix}.$$

The matrices of Q and B are

$$\hat{M} = (-1)^{l+1}K, \quad \hat{N} = S,$$

and the matrix of the Hermitian form H is

$$(-i)^{2l-2}\hat{M}\hat{N} = D = \mathrm{diag}(1, -1, \dots, -1, 1).$$

Thus, S, K, D are the natural odd-dimensional versions of the previous S, J, D (for $k = 2l+1$), but we use new notation for K because it represents a symmetric form, rather than a skew-symmetric one.

Omitting the calculations, we obtain

$$\mathrm{SO}^{\hat{M}}_{l,l-1}\mathbb{C} = p^{-1}\mathrm{SO}^{D}_{l,l-1}\mathbb{C}\, p, \quad \mathrm{SO}^{\hat{M},\hat{N}}_{l,l-1}\mathbb{R} = p^{-1}\mathrm{SO}^{D,I}_{l,l-1}\mathbb{R}\, p,$$

where p is the matrix which expresses $\hat{x}_0, \ldots, \hat{x}_{2l-2}$ in terms of $e_0, \ldots, e_{2l-2}$ (the analogue of c). The first bilinear relation says that ψ takes values in $\mathrm{SO}^{\hat{M}}_{l,l-1}\mathbb{C}$, and the second bilinear relation says that $p^{-1}\psi$ takes values in the product $\mathrm{SO}^{\hat{M},\hat{N}}_{l,l-1}\mathbb{R}\ \Delta$, or (equivalently) ψ takes values in $\mathrm{SO}^{D,I}_{l,l-1}\mathbb{R}\ p\ \Delta$.

10.3 *h*-version

Despite the finite-dimensional nature of mirror symmetry for the Calabi–Yau case, it fits naturally into the loop group (and infinite-dimensional Grassmannian) framework. Embedding into the loop group has advantages, as the difference between the odd and even cases disappears, and the calculations of the previous section can be done more systematically by the method of Chapter 6. More significantly, however, it provides a framework for the non-Calabi–Yau case, where infinite-dimensional phenomena appear and the VHS aspects are less clear.

In this section we prepare for the infinite-dimensional version by introducing the parameter h. With generalizations to 'abstract VHS' in mind, the discussion in this section is based entirely on the differential equation, ignoring the period integral interpretation.

We begin with the 'h-version' of the Picard–Fuchs[4] o.d.e., multiplying through by h^{k-1} in order to obtain an operator in the ring D^{h} of Chapter 6:

$$\mathrm{PF} = (\mathrm{h}\partial_\theta)^{k-1} - kz\mathrm{h}^{k-1}(k\partial_\theta + k - 1)(k\partial_\theta + k - 2)\ldots(k\partial_\theta + 1).$$

Let us write this in the form $a_{k-1}(\mathrm{h}\partial_\theta)^{k-1}+a_{k-2}(\mathrm{h}\partial_\theta)^{k-2}+\cdots+a_0$. Then, we apply the procedure of Chapter 6 to the D-module $D^{\mathrm{h}}/(\mathrm{PF})$. With respect to the monomial basis $[1], [\mathrm{h}\partial_\theta], \ldots, [(\mathrm{h}\partial_\theta)^{k-2}]$ we obtain

$$\Omega^{\mathrm{h}} = \frac{1}{\mathrm{h}}\begin{pmatrix} 0 & & & & p_0 \\ 1 & 0 & & & p_1 \\ & 1 & 0 & & p_2 \\ & & \ddots & \ddots & \vdots \\ & & & 1 & p_{k-2} \end{pmatrix},$$

where $p_i = -a_i/a_{k-1}$.

The introduction of h converts the fundamental solution ψ to

$$L = \gamma^{-1}\psi\gamma$$

[4] This o.d.e. has been mentioned already in Examples 5.8 and 6.29.

where

$$\psi = \begin{pmatrix} | & & | \\ u & \cdots & u^{(k-2)} \\ | & & | \end{pmatrix}, \quad \gamma = \begin{pmatrix} 1 & & & \\ & \mathrm{h} & & \\ & & \ddots & \\ & & & \mathrm{h}^{k-2} \end{pmatrix}.$$

As in the previous section, the dash denotes derivative with respect to θ.

The differential operator PF is homogeneous of degree $2k-2$, where h has weight 2 and z has weight zero. This leads (see section 9.3) to the homogeneity condition

$$\Omega^{\epsilon^{-2}\mathrm{h}}(z) = \mathrm{diag}(1, \epsilon, \dots, \epsilon^{2k-2})\ \Omega^{\mathrm{h}}(z)\ \mathrm{diag}(1, \epsilon, \dots, \epsilon^{2k-2})^{-1},$$

which forces L_+ (see section 6.7) to have the form $L_+ = Q_0(I + \mathrm{h}Q_1 + \cdots + \mathrm{h}^{k-2}Q_{k-2})$, where the only non-zero entries of Q_i lie on its i-th diagonal. In particular, Q_0 has the form

$$Q_0 = \mathrm{diag}(\alpha_0, \alpha_1, \dots, \alpha_{k-2}).$$

We obtain

$$\hat{\Omega}^{\mathrm{h}} = \frac{1}{\mathrm{h}} Q_0 \begin{pmatrix} 0 & & & & \\ 1 & 0 & & & \\ & 1 & 0 & & \\ & & \ddots & \ddots & \\ & & & 1 & 0 \end{pmatrix} Q_0^{-1} = \frac{1}{\mathrm{h}} \begin{pmatrix} 0 & & & & \\ \frac{\alpha_1}{\alpha_0} & 0 & & & \\ & \frac{\alpha_2}{\alpha_1} & 0 & & \\ & & \ddots & \ddots & \\ & & & \frac{\alpha_{k-2}}{\alpha_{k-3}} & 0 \end{pmatrix}.$$

As explained in Chapter 6, the transformation from Ω^{h} to $\hat{\Omega}^{\mathrm{h}}$ is achieved by the gauge transformation L_+^{-1}; equivalently, by the base change from $P_i = (\mathrm{h}\partial_\theta)^i$ to $\hat{P}_i = L_+^{-1} \cdot P_i$. Calculation would give L_+ and $\hat{P}_i$ as in Chapter 6 (we shall do this for M_3^3 in Example 10.11). However, it is simpler to observe directly that the transformed linear system is

$$\mathrm{h} \begin{pmatrix} \hat{P}_0\phi \\ \hat{P}_1\phi \\ \vdots \\ \hat{P}_{k-2}\phi \end{pmatrix}' = \begin{pmatrix} 0 & & & & \\ \frac{\alpha_1}{\alpha_0} & 0 & & & \\ & \frac{\alpha_2}{\alpha_1} & 0 & & \\ & & \ddots & \ddots & \\ & & & \frac{\alpha_{k-2}}{\alpha_{k-3}} & 0 \end{pmatrix}^t \begin{pmatrix} \hat{P}_0\phi \\ \hat{P}_1\phi \\ \vdots \\ \hat{P}_{k-2}\phi \end{pmatrix}$$

from which it follows that

$$\hat{P}_i = \frac{\alpha_{i-1}}{\alpha_i}\, \mathrm{h}\partial_\theta \dots \mathrm{h}\partial_\theta\, \frac{\alpha_0}{\alpha_1}\, \partial_\theta\, \frac{1}{\alpha_0}$$

(and L_+ can be read off from this).

The transformed Picard–Fuchs equation is

$$(1 - k^k z)\, \alpha_{k-2}\, \partial_\theta \frac{\alpha_{k-3}}{\alpha_{k-2}}\, \partial_\theta \quad \ldots \quad \partial_\theta \frac{\alpha_0}{\alpha_1}\, \partial_\theta \frac{1}{\alpha_0}\, \phi = 0.$$

The coordinate transformation determined by

$$\frac{\alpha_0}{\alpha_1} \partial_\theta = \partial_t$$

transforms $\hat{\Omega}^{\mathrm{h}}$ to

$$\frac{1}{\mathrm{h}} \begin{pmatrix} 0 & & & & \\ 1 & 0 & & & \\ & \frac{\alpha_0 \alpha_2}{\alpha_1^2} & 0 & & \\ & & \ddots & \ddots & \\ & & & \frac{\alpha_0 \alpha_{k-2}}{\alpha_1 \alpha_{k-3}} & 0 \end{pmatrix}$$

and the Picard–Fuchs equation to

$$(\mathrm{h}\partial_t) \frac{\alpha_1 \alpha_{k-3}}{\alpha_0 \alpha_{k-2}} (\mathrm{h}\partial_t) \quad \ldots \quad (\mathrm{h}\partial_t) \frac{\alpha_1^2}{\alpha_0 \alpha_2} (\mathrm{h}\partial_t)^2 \frac{1}{\alpha_0} \phi = 0.$$

We have not yet shown how to compute $\alpha_0, \ldots, \alpha_{k-2}$ explicitly. This will follow from an analysis of L_- in the Birkhoff factorization $L = L_- L_+$. For Fano manifolds, as we have seen in Chapter 6, the Birkhoff factorization $L = L_- L_+$ has the special feature that L_+ is polynomial in h and may be calculated easily 'by quadrature', while L_- is an infinite series in $1/\mathrm{h}$. For Calabi–Yau manifolds both L_+ and L_- are polynomial, and their computation requires roughly the same effort, and this effort turns out to be equivalent to solving the original Picard–Fuchs equation. Let us focus on L_-. By the homogeneity property, the coefficient of h^{-i} can have non-zero entries only on the $(-i)$-th diagonal. This is easy to see directly, as $\hat{\Omega}$ is nilpotent, so the equation $L_-^{-1} L_-' = \hat{\Omega}$ can be solved by quadrature, exactly as in Example 8.16. To carry this out, let us introduce new functions $v_0, \ldots, v_{k-2}$ by

$$\alpha_i = v_0 \ldots v_i$$

so that

$$\hat{\Omega} = \tfrac{1}{\mathrm{h}} \begin{pmatrix} 0 & & & & \\ v_1 & 0 & & & \\ & v_2 & 0 & & \\ & & \ddots & \ddots & \\ & & & v_{k-2} & 0 \end{pmatrix}.$$

Then we obtain

$$L_- = \gamma^{-1} \begin{pmatrix} 0 & & & & \\ \int v_1 & 0 & & & \\ \ddots & \int v_2 & 0 & & \\ \ddots & \ddots & \ddots & \ddots & \\ \ddots & \ddots & \ddots & \int v_{k-2} & 0 \end{pmatrix} \gamma,$$

where the $(-i)$-th diagonal of the lower-triangular matrix here has entries

$$\int v_1 \int v_2 \int \dots \int v_i, \quad \int v_2 \int v_3 \int \dots \int v_{i+1}, \quad \dots \quad \int v_{k-i-1} \int v_{k-i} \int \dots \int v_{k-2}.$$

Here $\int a \int b$ means $\int (a \int b \, d\theta) \, d\theta$, and so on.

The functions $v_0, \dots, v_{k-2}$ are the same as the ones appearing in section 10.2, and they play exactly the same role as in appendix 4.6 (Frobenius symbolic factors). They are related to a basis $u_0, \dots, u_{k-2}$ of solutions to the Picard–Fuchs equation by

$$u_0 = v_0, \quad u_1 = v_0 \int v_1, \quad \dots \quad u_{k-2} = v_0 \int v_1 \int v_2 \int \dots \int v_{k-2}.$$

However, in appendix 4.6 (where $z = z_0$ was a regular point) this basis was determined by the initial condition $u_i^{(j)}(z_0) = \delta_{ij}$. Here (where $z = 0$ is a regular singular point of maximal unipotent monodromy) we use the 'Frobenius basis' defined uniquely by the property

$$\begin{pmatrix} u_0 \\ u_1 \\ \vdots \\ u_{k-2} \end{pmatrix} = z^N \begin{pmatrix} f_0 \\ f_1 \\ \vdots \\ f_{k-2} \end{pmatrix}, \quad N = \begin{pmatrix} 0 & & & \\ 1 & \ddots & & \\ & \ddots & \ddots & \\ & & 1 & 0 \end{pmatrix},$$

where each f_i is holomorphic at $z = 0$ and $f_0(0) = 1$, $f_1(0) = \dots = f_{k-2}(0) = 0$. It follows that $v_0, \dots, v_{k-2}$ are determined uniquely by $v_0(0) = \dots = v_{k-2}(0) = 1$. We conclude that $\alpha_0, \dots, \alpha_{k-2}$ are determined uniquely by the basepoint conditions $\alpha_0(0) = \dots = \alpha_{k-2}(0) = 1$ as well.

The resulting formula for L_- is

$$L_- = \gamma^{-1} \begin{pmatrix} 1 & \dots \\ u_1/u_0 & \dots \\ \vdots & \\ u_{k-2}/u_0 & \dots \end{pmatrix} \gamma,$$

where the omitted columns are obtained by successively differentiating, then triangularizing. Making use of the relations, $u = z^N f, u' = z^N(Nf + f'), \ldots$ we can rewrite L_- as

$$L_- = \gamma^{-1} z^N \gamma \ \gamma^{-1} \begin{pmatrix} 1 & \cdots \\ f_1/f_0 & \cdots \\ \vdots & \\ f_{k-2}/f_0 & \cdots \end{pmatrix} \gamma = z^{N/h}\, S$$

where S is holomorphic at $z = 0$ and satisfies $S(0) = I$.

We have not yet used any properties of the original Picard–Fuchs equation, beyond the fact that $z = 0$ is a regular singular point of maximal unipotent monodromy. Any such o.d.e. gives rise to a fundamental solution matrix L_- of the above type. In the language of Chapter 8 this is a complex extended solution for a harmonic map of finite uniton number (cf. Examples 8.16, 8.17).

Example 10.4 Let us calculate everything explicitly in the case $k = 5$. We need only the functions f_1, f_0 (see the previous section). By substituting power series for $u_0 = f_0$, $u_1 = f_0 \log z + f_1$ in the o.d.e., it is easy to verify that

$$f_0(z) = 1 + 120z + 113400z^2 + \cdots$$

$$f_1(z) = 770z + 810225z^2 + \cdots$$

hence the coordinate change $q = ze^{f_1(z)/f_0(z)}$ is given by

$$q = z + 770z^2 + 1014275z^3 + \cdots$$

$$z = q - 770q^2 + 171525q^3 + \cdots .$$

Now we make use of the symmetry property

$$(1 - 5^5 z)v_0 v_1 v_2 v_3 = \frac{1}{v_0}, \quad \frac{1}{v_3} = \frac{1}{v_1}$$

which follows from the fact that the Picard–Fuchs operator satisfies $\overline{\text{PF}}^* = \text{PF}$, that is, it is self-adjoint in the sense of section 6.3. For regular points this phenomenon was explained in appendix 4.6; for a regular singular point of maximal unipotent monodromy the same method applies, because of the uniqueness of the Frobenius basis of solutions.

We obtain

$$
\begin{aligned}
v_0 &= f_0 \\
v_1 &= 1 + (f_1/f_0)' = 1 + 770z + 1435650z^2 + \cdots \\
v_2 &= \left(\left(1 - 5^5 z\right) v_0^2 v_1^2\right)^{-1} = 1 + 1345z + 3296525z^2 + \cdots \\
v_3 &= v_1.
\end{aligned}
$$

This gives

$$\frac{v_2}{v_1} = 1 + 575z + 1418125z^2 + \cdots$$

and if we substitute $z = q - 770q^2 + 171525q^3 + \cdots$ we have

$$\frac{v_2}{v_1} = 1 + 575q + 975375q^2 + \cdots$$

in agreement with the assertions in the previous section. □

10.4 Loop group version

In this section we give a loop group framework for mirror symmetry. This is based on the specific groups which occur for the VHS of $V = \tilde{M}_k^k$ and the quantum cohomology of $V^\circ = M_k^k$, but analogous statements could easily be made for other Calabi–Yau manifolds. The results are summarized in the theorem at the end of the section.

On the group $\mathrm{SL}_{k-1}\mathbb{C}$ let us define an involution σ and a conjugation map C by

$$
\begin{aligned}
\sigma(A) &= S(A^t)^{-1}S \\
C(A) &= D\,(\bar{A}^t)^{-1}\,D,
\end{aligned}
$$

where D and S are as in section 10.2. The real form defined by C (i.e. its fixed point set) is the group

$$\mathrm{SU}_{k-1}^D = \{A \in \mathrm{SL}_{k-1}\mathbb{C} \mid D^{-1}(\bar{A}^t)^{-1}D = A\}.$$

For consistency with our conventions for SO and Sp, we should write $\mathrm{SU}_{l,l}^D$ when $k = 2l+1$ and $\mathrm{SU}_{l,l-1}^D$ when $k = 2l$, but SU_{k-1}^D is a convenient abbreviation covering both cases.

On loops $\alpha \in \Lambda \mathrm{SL}_{k-1}\mathbb{C}$ we obtain an involution σ_Λ and a conjugation map C_Λ as follows:

$$\sigma_\Lambda(\alpha) = \sigma(\alpha(-\mathrm{h})) = S(\alpha(-\mathrm{h})^t)^{-1}S$$

$$C_\Lambda(\alpha) = C(\alpha(1/\bar{\mathrm{h}})) = D\,(\overline{\alpha(1/\bar{\mathrm{h}})}^{\,t})^{-1}\,D.$$

The fixed points of σ_Λ constitute the twisted loop group

$$(\Lambda \mathrm{SL}_{k-1}\mathbb{C})_\sigma = \{\alpha \in \Lambda \mathrm{SL}_{k-1}\mathbb{C} \mid \sigma(\alpha) = \alpha(-\mathrm{h})\},$$

and the fixed points of C_Λ constitute the real form $(\Lambda \mathrm{SU}^D_{k-1})_\sigma$ of $(\Lambda \mathrm{SL}_{k-1}\mathbb{C})_\sigma$.

We shall make use of the embedding

$$\mathrm{SL}_{k-1}\mathbb{C} \to \Lambda \mathrm{SL}_{k-1}\mathbb{C}, \quad A \mapsto \gamma^{-1}A\gamma$$

where γ is as in section 10.3. An easy calculation gives:

Lemma 10.5 *The maps induced by $\sigma_\Lambda, C_\Lambda$ on loops of the form $\gamma^{-1}A\gamma$ are given (respectively) by*

$$A \mapsto (SD)^{-1}(A^t)^{-1}(SD)$$

$$A \mapsto D(\bar{A}^t)^{-1}D.$$

It follows that $\gamma^{-1}A\gamma$ is a twisted loop if and only if A preserves the bilinear form whose matrix is

$$SD = \begin{pmatrix} & & & & \cdot^{\cdot^{\cdot}} \\ & & & -1 & \\ & & 1 & & \\ & -1 & & & \\ 1 & & & & \end{pmatrix}.$$

This is the key observation, as the subgroup preserving this bilinear form is isomorphic to $\mathrm{Sp}_{2l}\mathbb{C}$ if k is odd or $\mathrm{SO}_{l,l-1}\mathbb{C}$ if k is even. More precisely we have:

Proposition 10.6 *We have $\gamma^{-1}A\gamma \in (\Lambda SL_{k-1}\mathbb{C})_\sigma$ if and only if $A \in Sp^{\tilde{M}}_{2l}\mathbb{C} = Sp^{J}_{2l}\mathbb{C}$ (when $k = 2l+1$) or $A \in SO^{\hat{M}}_{l,l-1}\mathbb{C} = SO^{K}_{l,l-1}\mathbb{C}$ (when $k = 2l$).*

In section 5.3 we saw that the condition for $L = \gamma^{-1}\psi\gamma$ to take values in the twisted loop group $(\Lambda \mathrm{SL}_{k-1}\mathbb{C})_\sigma$ is just the Frobenius property of quantum cohomology. Thus, *the first bilinear relation for $\tilde{M}^k_k$ corresponds,* *via this loop group embedding, to the Frobenius property of the quantum cohomology of* M_k^k.

Let us write $G^{\mathbb{C}} = \{A \mid \sigma_\Lambda(\gamma^{-1}A\gamma) = \gamma^{-1}A\gamma\}$. (Thus, $G^{\mathbb{C}}$ is $\mathrm{Sp}_{2l}^{\tilde{M}}\mathbb{C}$ or $\mathrm{SO}_{l,l-1}^{\hat{M}}\mathbb{C}$.)

Lemma 10.7 *The maps induced by* σ, C *on*

$$G^{\mathbb{C}} = \{A \in SL_{k-1}\mathbb{C} \mid \gamma^{-1}A\gamma \in (\Lambda SL_{k-1}\mathbb{C})_\sigma\}$$

are given (respectively) by

$$A \mapsto DAD$$

$$A \mapsto S\bar{A}S.$$

We call these maps σ_J, C_J *when k is odd, and* σ_K, C_K *when k is even.*

Proof This follows directly from the previous lemma, since we have $(A^t)^{-1} = (SD)A(SD)^{-1}$ when $A \in G^{\mathbb{C}}$. □

The conjugation map $A \mapsto S\bar{A}S$ defines a real form of $G^{\mathbb{C}}$, namely

$$G = \{A \in G^{\mathbb{C}} \mid S\bar{A}S = A\}.$$

Since $\tilde{N}$ and $\hat{N}$ are multiples of S, we have

$$G = \mathrm{Sp}_{2l}^{\tilde{M},\tilde{N}}\mathbb{R} \text{ or } \mathrm{SO}_{l,l-1}^{\hat{M},\hat{N}}\mathbb{R}.$$

On G we have the involution $A \mapsto DAD$, and this defines a symmetric space, which we shall consider later.

We also have $G = G^{\mathbb{C}} \cap \mathrm{SU}_{k-1}^D$ and $\gamma^{-1}G\gamma = \gamma^{-1}G^{\mathbb{C}}\gamma \cap (\Lambda\mathrm{SU}_{k-1}^D)_\sigma$. The latter gives:

Proposition 10.8 *Let* $A \in G^{\mathbb{C}}$. *Then we have* $\gamma^{-1}A\gamma \in (\Lambda SU_{k-1}^D)_\sigma$ *if and only if* $A \in G$, *that is,* $A \in Sp_{2l}^{\tilde{M},\tilde{N}}\mathbb{R}$ *(when* $k = 2l+1$*) or* $A \in SO_{l,l-1}^{\hat{M},\hat{N}}\mathbb{R}$ *(when* $k = 2l$*).*

In addition to $\gamma^{-1}G\gamma = \gamma^{-1}G^{\mathbb{C}}\gamma \cap (\Lambda\mathrm{SU}_{k-1}^D)_\sigma$, it is obvious that $\gamma^{-1}\Delta\gamma = \gamma^{-1}G^{\mathbb{C}}\gamma \cap (\Lambda_+\mathrm{SL}_{k-1}\mathbb{C})_\sigma$, where Δ denotes the upper-triangular subgroup of $G^{\mathbb{C}}$. These two statements imply that the Iwasawa decomposition for $G^{\mathbb{C}}$ is compatible with the Iwasawa decomposition for $(\Lambda\mathrm{SL}_{k-1}\mathbb{C})_\sigma$.

In section 10.2 we saw that ψ satisfies the second bilinear relation if and only if $c^{-1}\psi$ takes values in $\mathrm{Sp}_{2l}^{\tilde{M},\tilde{N}}\mathbb{R}\,\Delta$, that is, it admits an

Iwasawa factorization $c^{-1}\psi = (c^{-1}\psi)_{\mathbb{R}}(c^{-1}\psi)_{\Delta} \in \mathrm{Sp}_{2l}^{\tilde{M},\tilde{N}}\mathbb{R}\,\Delta$. Therefore, *$\psi$ satisfies the second bilinear relation if and only if $(\gamma^{-1}c\gamma)^{-1}L$ takes values in $(\Lambda SU_{l,l}^{D})_{\sigma}(\Lambda_{+}SL_{2l}\mathbb{C})_{\sigma}$*, that is, it admits an Iwasawa factorization $(\gamma^{-1}c\gamma)^{-1}L = ((\gamma^{-1}c\gamma)^{-1}L)_{\mathbb{R}}((\gamma^{-1}c\gamma)^{-1}L)_{+}$. A similar statement applies for $p^{-1}\psi$ and the group $\mathrm{SO}_{l,l-1}^{\hat{M},\hat{N}}\mathbb{R}$.

The Griffiths transversality condition on ψ is, in the language of Chapter 8, the condition that ψ defines a superhorizontal holomorphic map into a flag manifold. As we have seen, this is the condition that $L = \gamma^{-1}\psi\gamma$ is a complex extended solution:

Proposition 10.9 *ψ satisfies the Griffiths transversality condition if and only if $L = \gamma^{-1}\psi\gamma$ is a complex extended solution, that is, $L^{-1}L'$ is of the form*

$$L^{-1}L' = \tfrac{1}{\mathrm{h}}\omega + \theta^{(0)} + \mathrm{h}\theta^{(1)} + \cdots .$$

In particular, if $L = L_{-}L_{+}$ is the Birkhoff factorization, $L_{-}^{-1}L_{-}'$ is of the form

$$L_{-}^{-1}L_{-}' = \tfrac{1}{\mathrm{h}}\hat{\omega}.$$

By the theory of Chapter 8, any complex extended solution with values in the twisted loop group $(\mathrm{SL}_{k-1}\mathbb{C})_{\sigma}$ defines a harmonic map to the symmetric space $\mathrm{SU}_{k-1}^{D}/\mathrm{Fix}(\sigma)$, where $\mathrm{Fix}(\sigma) = \{A \in \mathrm{SU}_{k-1}^{D} \mid \sigma(A) = A\} \cong \mathrm{SO}_{k-1}$.

On the other hand, Griffiths transversality and the Riemann bilinear relations say that the harmonic map given by the VHS arises from a certain twistor fibration. Our discussion so far embeds this twistor fibration into the loop group framework in the following way:

$$\begin{array}{ccc}
\dfrac{\mathrm{SO}_{l,l-1}^{\hat{M},\hat{N}}\mathbb{R}}{\mathrm{U}_1\times\cdots\times\mathrm{U}_1} \text{ or } \dfrac{\mathrm{Sp}_{2l}^{\tilde{M},\tilde{N}}\mathbb{R}}{\mathrm{U}_1\times\cdots\times\mathrm{U}_1} & \longrightarrow & (\Lambda\mathrm{SU}_{k-1}^{D})_{\sigma} \\
\downarrow & & \downarrow \\
\dfrac{\mathrm{SO}_{l,l-1}^{\hat{M},\hat{N}}\mathbb{R}}{\mathrm{SO}_l\times\mathrm{SO}_{l-1}} \text{ or } \dfrac{\mathrm{Sp}_{2l}^{\tilde{M},\tilde{N}}\mathbb{R}}{\mathrm{U}_l} & \longrightarrow & \dfrac{\mathrm{SU}_{k-1}^{D}}{\mathrm{SO}_{k-1}}.
\end{array}$$

The left hand vertical maps are the twistor projections (as in section 10.1), and the right hand one is given by evaluation at $\mathrm{h} = 1$. The 'discrepancy' between the groups on the left and those on the right is caused by the fact that the loop γ does not belong to $\Lambda\mathrm{SO}_{l,l-1}^{\hat{M},\hat{N}}\mathbb{R}$ or $\Lambda\mathrm{Sp}_{2l}^{\tilde{M},\tilde{N}}\mathbb{R}$. In our examples of harmonic maps in section 8.6 we considered only the unitary group, and this kind of discrepancy did not arise.

Let us summarize the loop group versions of the VHS properties that we have seen in this section:

Theorem 10.10 *The $\Lambda SL_{k-1}\mathbb{C}$-valued holomorphic map $L = \gamma^{-1}\psi\gamma$, which was obtained in sections 10.2 and 10.3, satisfies the following properties.*

(1) Griffiths transversality: L is a complex extended solution.
(2) First Riemann bilinear relation: L takes values in the twisted loop group $(\Lambda SL_{k-1}\mathbb{C})_\sigma$, and admits a Birkhoff factorization with respect to the product $(\Lambda_- SL_{k-1}\mathbb{C})_\sigma\ (\Lambda_+ SL_{k-1}\mathbb{C})_\sigma$.
(3) Second Riemann bilinear relation: For some fixed loop $\gamma_0 = \gamma^{-1}x\gamma$, $\gamma_0^{-1}L$ admits an Iwasawa factorization with respect to the product $(\Lambda SU^D_{k-1})_\sigma\ (\Lambda_+ SL_{k-1}\mathbb{C})_\sigma$.

It is possible to express these conditions in terms of the Grassmannian model for $(\Lambda \mathrm{SL}_{k-1})_\sigma/(\Lambda_+ \mathrm{SL}_{k-1}\mathbb{C})_\sigma$. S. Barannikov gave versions of (1) and (2) in [9].

Quantum cohomology and variations of polarized Hodge structure of Calabi–Yau manifolds give rise to harmonic maps of finite uniton number into symmetric spaces, and we have described such maps in terms of their extended solutions, which are holomorphic maps W into (finite-dimensional flag submanifolds of) infinite-dimensional Grassmannians. Mirror symmetry can be observed by looking at such harmonic maps from two different viewpoints: the D-module aspect (represented by a holomorphic vector-valued function J which generates W) reveals quantum cohomology, while the map W itself represents the VHS.

We conclude with a simple example:

Example 10.11 As mentioned earlier in this chapter, M_3^3 is an elliptic curve, that is, a compact Riemann surface of genus 1. It is the simplest non-trivial example of type M_k^k, as discussed in this section, and the simplest non-trivial example of type M_g, as discussed in section 10.1. Mirror symmetry appears to be uninteresting here as it is known that $\tilde{M}_3^3$ and M_3^3 are both tori. However, some non-trivial remnants of the general picture survive—amounting to the relation between the 'cubic curve in $\mathbb{C}P^2$' and 'quotient of the plane by a lattice' points of view.

The mirror diagram has the following simple form:

$$\begin{array}{ccc} & h^{1,1} & \\ h^{1,0} & & h^{0,1} \\ & h^{0,0} & \end{array} = \begin{array}{ccc} & 1 & \\ 1 & & 1 \\ & 1 & \end{array}$$

The boxes in the left hand diagram below represent $H^1(\tilde{M}_3^3, \mathbb{C})$, and we take $\tilde{x}_0 = (e_0 + ie_1)/\sqrt{2}, \tilde{x}_1 = (ie_0 + e_1)/\sqrt{2}$ as basis vectors. Those in the right hand diagram represent $H^\sharp(M_3^3, \mathbb{C})$, and we take $b_0 = 1, b_1 = b$ as basis vectors.

$$\begin{array}{ccc} & 1 & \\ \boxed{1} & & \boxed{1} \\ & 1 & \end{array} \qquad \begin{array}{ccc} & \boxed{1} & \\ 1 & & 1 \\ & \boxed{1} & \end{array}$$

We have

$$c = \frac{1}{\sqrt{2}} \begin{pmatrix} 1 & i \\ i & 1 \end{pmatrix}, \quad \tilde{M} = \begin{pmatrix} & -1 \\ 1 & \end{pmatrix}, \quad \tilde{N} = -i \begin{pmatrix} & 1 \\ 1 & \end{pmatrix}, \quad S = \begin{pmatrix} & 1 \\ 1 & \end{pmatrix}.$$

As we are dealing with 2×2 matrices, there are further simplifications. First, we have $\mathrm{Sp}_2^{\tilde{M}}\mathbb{C} = \mathrm{Sp}_2\mathbb{C} = \mathrm{SL}_2\mathbb{C}$ (and $\mathrm{Sp}_2\mathbb{R} = \mathrm{SL}_2\mathbb{R}$). Next, for any $A \in \mathrm{GL}_2\mathbb{C}$, we have

$$(A^t)^{-1} = \begin{pmatrix} & 1 \\ -1 & \end{pmatrix} A \begin{pmatrix} & 1 \\ -1 & \end{pmatrix},$$

from which it follows that $\mathrm{Sp}_2^{\tilde{M},\tilde{N}}\mathbb{R} = \mathrm{SU}_{1,1}$. Thus, the relevant embedding is

$$\mathrm{SU}_{1,1} \to (\Lambda \mathrm{SU}_{1,1})_\sigma, \quad A \mapsto \gamma^{-1} A \gamma = \begin{pmatrix} 1 & \\ & \mathrm{h} \end{pmatrix}^{-1} A \begin{pmatrix} 1 & \\ & \mathrm{h} \end{pmatrix}.$$

The Picard–Fuchs operator is

$$(\mathrm{h}\partial_\theta)^2 - 3z\mathrm{h}^2(3\partial_\theta + 2)(3\partial_\theta + 1) = \mathrm{h}^2(1 - 27z)(\partial_\theta^2 - u\partial_\theta - v)$$

where $u = 27z/(1 - 27z)$ and $v = 6z/(1 - 27z)$. With respect to the basis $[1], [\mathrm{h}\partial_\theta]$ of the D-module, we have

$$\Omega^{\mathrm{h}} = \frac{1}{\mathrm{h}} \begin{pmatrix} 0 & 0 \\ 1 & 0 \end{pmatrix} + \begin{pmatrix} 0 & 0 \\ 0 & v \end{pmatrix} + \mathrm{h} \begin{pmatrix} 0 & u \\ 0 & 0 \end{pmatrix}.$$

The calculation of L_- was explained in this section: we have

$$\begin{aligned} L_- &= I + \frac{1}{\mathrm{h}} \begin{pmatrix} 1 & 0 \\ \int v_1 & 1 \end{pmatrix} \\ &= \begin{pmatrix} 1 & \\ & \mathrm{h} \end{pmatrix}^{-1} \begin{pmatrix} 1 & 0 \\ \int v_1 & 1 \end{pmatrix} \begin{pmatrix} 1 & \\ & \mathrm{h} \end{pmatrix} \end{aligned}$$

where $v_0 = u_0, v_1 = (u_1/u_0)'$, and u_0, u_1 is the Frobenius basis. The original Picard–Fuchs operator factors as $(1 - 27z)v_0 v_1 \mathrm{h}\partial_\theta \frac{1}{v_1} \mathrm{h}\partial_\theta \frac{1}{v_0}$.

The calculation of L_+ is a special case of Example 6.5. We have

$$L_+ = \begin{pmatrix} \alpha_0 & 0 \\ 0 & \alpha_1 \end{pmatrix} \left(I + \mathrm{h} \begin{pmatrix} 0 & \alpha_0'/\alpha_0 \\ 0 & 0 \end{pmatrix} \right)$$

where $\alpha_0 = v_0, \alpha_1 = v_0 v_1$.

The self-adjointness of the operator gives $\alpha_0\alpha_1 = 1/(1-27z)$. Thus, all quantities are easily computed from α_0 alone, and this has the explicit power series representation

$$\alpha_0(z) = \sum_{n\geq 0} \frac{(3n)!}{(n!)^3} z^n = 1 + 6z + 90z^2 + \cdots .$$

In particular, the mirror coordinate transformation is

$$t = \int v_1(z) d\theta = \frac{u_1(z)}{u_0(z)} = \log z + \frac{f_1(z)}{f_0(z)} = \log z + 15z + \frac{333}{2}z^2 + \cdots ,$$

which gives

$$q = z e^{f_1(z)/f_0(z)} = z + 15z^2 + 279z^3 + \cdots .$$

It is a remarkable fact, and a general feature of mirror symmetry, that the series for q (and its inverse $z = q - 15q^2 + 171q^3 - \cdots$) has integer coefficients. In this example, however, there is a classical explanation. Namely, it is known that the inverse of the 'Schwarz map' u_1/u_0 is related to the modular elliptic function $J : \mathrm{Sp}_2\mathbb{R}/\mathrm{U}_1 \to \mathbb{C}$. The precise relation (which depends on the representation of the elliptic curve as a hypersurface, see [83]) is as follows:

$$\begin{aligned} J &= \frac{1}{1728}\left(q^{-1} + 744 + 196884q + \cdots\right) \\ &= \frac{1}{1728} \frac{(1+216z)^3}{z(1-27z)^3}. \end{aligned}$$

□

10.5 Integrable systems of mirror symmetry type

It is a very difficult problem to characterize quantum cohomology (or variations of Hodge structure) purely in differential equation-theoretic terms, but less difficult and perhaps of broader interest to identify a class of integral systems to which quantum cohomology belongs. In this section we shall just make a start in this direction by formulating the 'differential geometric ingredients of mirror symmetry', which should apply to any such integrable system. We shall indicate how our construction procedure of Chapter 6 produces candidates for such objects, and show that there are non-trivial examples beyond quantum cohomology. Finally, at the end of the section, we make some comments on the additional arithmetic properties enjoyed by quantum cohomology.

Definition 10.12 *Let G be a real semisimple Lie group. Let σ be an involution on G; we denote also by σ its extension to the complexified group $G^{\mathbb{C}}$. Let U be some open subset of $\mathbb{C}^r$.*

We say that a flat connection $d + \Omega^{\mathrm{h}}$ (or the associated D-module) is of differential geometric mirror symmetry type if

(i) there exists a complex extended solution $L : U \to (\Lambda G^{\mathbb{C}})_\sigma$ such that $\Omega^{\mathrm{h}} = L^{-1}dL$,

(ii) L admits a Birkhoff factorization $L = L_- L_+$ with respect to the big cell $(\Lambda_- G^{\mathbb{C}})_\sigma\ (\Lambda_+ G^{\mathbb{C}})_\sigma \subseteq (\Lambda G^{\mathbb{C}})_\sigma$, and

(iii) for some element $\gamma_0 \in (\Lambda G^{\mathbb{C}})_\sigma$, the map $\gamma_0^{-1}L$ admits an Iwasawa factorization $\gamma_0^{-1}L = (\gamma_0^{-1}L)_{\mathbb{R}}(\gamma_0^{-1}L)_+$ with respect to the open cell $(\Lambda G)_\sigma\ (\Lambda_+ G^{\mathbb{C}})_\sigma \subseteq (\Lambda G^{\mathbb{C}})_\sigma$.

The complex extended solution condition means that Ω^{h} is a holomorphic map to the twisted loop algebra $(\Lambda \mathfrak{g}^{\mathbb{C}})_\sigma$ and has the form $\frac{1}{\mathrm{h}}\omega + \theta^{(0)} + \mathrm{h}\theta^{(1)} + \cdots$; it follows that $\hat{\Omega}^{\mathrm{h}} = L_-^{-1}dL_-$ has the form $\frac{1}{\mathrm{h}}\hat{\omega}$.

This is a very weak definition as no conditions are imposed on the domain U, yet any map may be translated into the intersection of the two open (Birkhoff and Iwasawa) cells on a sufficiently small domain. In practice, the 'global' properties of L are important, in particular the singularities (of the factorized map) which occur if L leaves either of these cells. However, the definition is a convenient starting point for discussion, with properties (i), (ii), and (iii) playing the roles of Griffiths transversality and the first and second bilinear relations. Supplementary conditions (distinguished cyclic element, homogeneity, monodromy conditions, etc.) may be imposed according to taste.

The homogeneous space $(\Lambda G^{\mathbb{C}})_\sigma/(\Lambda_+ G^{\mathbb{C}})_\sigma$ is an infinite-dimensional (Kac–Moody) generalized complex flag manifold; we have discussed this explicitly in Chapter 8 in the case $G = \mathrm{SU}_n$, and the same general principles apply to any G (see [102]). The real form $(\Lambda G)_\sigma$ acts on this homogeneous space, transitively when G is compact, but generally with an infinite number of orbits when G is non-compact. A finite number of these orbits are open, and one of these is the orbit of the identity coset. We refer to [80] for an initial investigation of these orbits; it is likely that the theory and applications of such orbits will eventually parallel those in the finite-dimensional case.

In this description, L_- represents the (abstract) quantum cohomology, and $(\gamma_0^{-1}L)_{\mathbb{R}}$ represents the (abstract) VHS. The Birkhoff and Iwasawa factorizations allow us to pass from one to the other:

$$L_- \xrightarrow{(\gamma_0^{-1}L)_{\mathbb{R}}=(\gamma_0^{-1}L_-L_+)_{\mathbb{R}}=(\gamma_0^{-1}L_-)_{\mathbb{R}}} (\gamma_0^{-1}L)_{\mathbb{R}}$$

$$L_- \xleftarrow{L_-=(\gamma_0(\gamma_0^{-1}L)_{\mathbb{R}}(\gamma_0^{-1}L)_+)_-=(\gamma_0(\gamma_0^{-1}L)_{\mathbb{R}})_-} (\gamma_0^{-1}L)_{\mathbb{R}}.$$

When G is compact, we may assume $\gamma_0 = 1$, and the simpler correspondence

$$L_- \xrightarrow{L_{\mathbb{R}}=(L_-)_{\mathbb{R}}} L_{\mathbb{R}}$$

$$L_- \xleftarrow{L_-=(L_{\mathbb{R}})_-} L_{\mathbb{R}}$$

holds. This is a useful and well known point of view in the theory of harmonic maps (emphasized by J. Dorfmeister; see [28]), where L_- represents the 'Weierstrass data' and $(\gamma_0^{-1}L)_{\mathbb{R}}$ represents the extended harmonic map. From this point of view the above definition (without any 'supplementary conditions') is simply the definition of a (pluri)harmonic map from U into the symmetric space $G/\mathrm{Fix}(\sigma)$.

Example 10.13 This example is a generalization of mirror symmetry for M_k^k, but different from the generalization given in section 10.4. For simplicity we consider only the case $k = 5$. In section 10.4 we used the embedding

$$\gamma^{-1}\mathrm{Sp}_4^{\tilde{M},\tilde{N}}\mathbb{R}\,\gamma \to \left(\Lambda \mathrm{SU}_{2,2}^D\right)_\sigma,$$

but we could instead use the embedding

$$\gamma^{-1}\mathrm{Sp}_4^{\tilde{M},\tilde{N}}\mathbb{R}\,\gamma \to (\Lambda c^{-1}\mathrm{SL}_4\mathbb{R}c)_\sigma.$$

That is, we use the same involution σ, but we choose a different real form of $\mathrm{SL}_4\mathbb{C}$ and $(\Lambda\mathrm{SL}_4\mathbb{C})_\sigma$, given by

$$C(A) = S\bar{A}S$$

on $\mathrm{SL}_4\mathbb{C}$. We obtain $\mathrm{Sp}_4^{\tilde{M},\tilde{N}}\mathbb{R} = \mathrm{Sp}_4^{\tilde{M}}\mathbb{C} \cap c^{-1}\mathrm{SL}_4\mathbb{R}\,c$ and

$$\gamma^{-1}\mathrm{Sp}_4^{\tilde{M},\tilde{N}}\mathbb{R}\,\gamma = \gamma^{-1}\mathrm{Sp}_4^{\tilde{M}}\mathbb{C}\,\gamma \cap (\Lambda c^{-1}\mathrm{SL}_4\mathbb{R}\,c)_\sigma.$$

The finite-dimensional situation (for $L = \gamma^{-1}\psi\gamma$) is exactly the same as before, giving harmonic maps into $\mathrm{Sp}_4^{\tilde{M},\tilde{N}}\mathbb{R}/\mathrm{U}_2$, but for more general $(\Lambda\mathrm{SL}_4\mathbb{C})_\sigma$-valued complex extended solutions L we obtain harmonic maps into $c^{-1}\mathrm{SL}_4\mathbb{R}c/\mathrm{Fix}(\sigma)$. Now, $c^{-1}Xc$ is a fixed point of σ if and only if $X^t = X^{-1}$, so we obtain harmonic maps into the symmetric space

$$\frac{c^{-1}\mathrm{SL}_4\mathbb{R}c}{c^{-1}\mathrm{SO}_4 c} \cong \frac{\mathrm{SL}_4\mathbb{R}}{\mathrm{SO}_4}.$$

Harmonic (or rather, pluriharmonic) maps of this type arise from the Frobenius manifolds of 'topological-antitopological fusion' in [35]. □

Example 10.14 The Gauss hypergeometric differential operator

$$\partial_\theta(\partial_\theta + c - 1) - z(\partial_\theta + a)(\partial_\theta + b)$$

provides the simplest example of 'differential geometric mirror symmetry'. Here we take $\partial_\theta = z\partial/\partial z$ as usual, and a, b, c are real numbers, so the operator can be expanded as

$$z(1-z)\frac{\partial^2}{\partial z^2} + (c - (a+b+1)z)\frac{\partial}{\partial z} - ab.$$

When $a = \frac{1}{3}$, $b = \frac{2}{3}$, $c = \frac{3}{3}$, and z is replaced by $3^3 z$, we obtain the Picard–Fuchs operator of the elliptic curve M_3^3 (see Example 10.11). It is, therefore, of interest to see what happens for other a, b, c.

To obtain an operator in D^{h} we multiply through by h^2 and define

$$\mathrm{G} = \mathrm{h}^2\partial_\theta(\partial_\theta + c - 1) - z\mathrm{h}^2(\partial_\theta + a)(\partial_\theta + b).$$

It is easy to verify that

$$\bar{\mathrm{G}}^* = \mathrm{G} \iff a + b = 1 \text{ and } c = 1.$$

More generally, for arbitrary a, b, c we have $(\overline{f\mathrm{G}})^* = f\mathrm{G}$ where $f(z) = z^{c-1}(z-1)^{a+b-c}$.

The hypergeometric differential equation has regular singular points at $z = 0, 1, \infty$. The eigenvalues of the monodromy matrices at these points are, respectively,

$$1, e^{2\pi i(1-c)}; \quad 1, e^{2\pi i(c-a-b)}; \quad e^{2\pi i a}, e^{2\pi i b}.$$

It is a well known property of this equation that the quotient u_1/u_0 of any two linearly independent solutions is a holomorphic function on the upper half-plane, whose image is a 'Schwarz curvilinear triangle' with interior angles

$$\pi|1-c|, \quad \pi|c-a-b|, \quad \pi|a-b|.$$

Analytic continuation gives rise to a (possibly multi-valued) function on $\mathbb{C}-\{0,1,\infty\}$. In the special case where

$$|1-c| = \frac{1}{N_0}, \quad |c-a-b| = \frac{1}{N_1}, \quad |a-b| = \frac{1}{N_\infty}$$

with N_0, N_1, N_∞ positive integers (or infinity), the triangles fit together (without overlapping) at each vertex. The image of all possible analytic continuations is then either

(a) the whole of $\mathbb{C}\cup\{\infty\}$, when $1/N_0 + 1/N_1 + 1/N_\infty > 1$,

(b) the plane $\mathbb{C}$, when $1/N_0 + 1/N_1 + 1/N_\infty = 1$, or

(c) a proper open 'Schwarz disk' in $\mathbb{C}$, when $1/N_0 + 1/N_1 + 1/N_\infty < 1$.

On these domains the inverse of u_1/u_0 is a single-valued function which is invariant under the action of the projective monodromy group.

Let us consider first the case where $a+b=1=c$. Then the operator is

$$\begin{aligned} G &= (\mathrm{h}\partial_\theta)^2 - z\mathrm{h}^2(\partial_\theta + a)(\partial_\theta + b) \\ &= (1-z)(\mathrm{h}\partial_\theta)^2 - \mathrm{h}z(a+b)\mathrm{h}\partial_\theta - \mathrm{h}^2 zab \\ &= (1-z)v_0v_1\mathrm{h}\partial_\theta\frac{1}{v_1}\mathrm{h}\partial_\theta\frac{1}{v_0} \end{aligned}$$

where v_0, v_1 are as in Example 6.39. That is, $v_0 = u_0$ and $v_1 = (u_1/u_0)' = \partial(u_1/u_0)/\partial\theta$. Since $c=1$ we may (near $z=0$) choose solutions $u_0 = f_0$, $u_1 = f_0 \log z + f_1$ where the holomorphic functions f_0, f_1 are normalized by $f_0(0)=1, f_1(0)=0$.

As in section 10.3 it is natural to take

$$L = \gamma^{-1}\psi\gamma, \quad \text{where } \psi = \begin{pmatrix} u_0 & u_0' \\ u_1 & u_1' \end{pmatrix}, \; \gamma = \begin{pmatrix} 1 & \\ & \mathrm{h} \end{pmatrix}.$$

Since $u_0(0) \neq 0$, we have a Birkhoff factorization $L = L_- L_+$ (on the entire domain of definition of u_0, u_1), with

$$L_- = I + \frac{1}{\mathrm{h}}\begin{pmatrix} 0 & 0 \\ u_1/u_0 & 0 \end{pmatrix}$$

and

$$L_+ = \begin{pmatrix} v_0 & 0 \\ 0 & v_0 v_1 \end{pmatrix} \left(I + \mathrm{h} \begin{pmatrix} 0 & v_0'/v_0 \\ 0 & 0 \end{pmatrix} \right)$$

(see Example 10.11).

Let us now restrict attention to the case where the image of u_1/u_0 is a Schwarz disk. In addition, let us restrict to the case where

$$(N_0, N_1, N_\infty) = (\infty, \infty, N)$$

for some $N = 2, 3, \ldots$. Thus we have $a = \frac{N-1}{2N}$, $b = \frac{N+1}{2N}$. There are $2N$ 'large' hyperbolic triangles which meet at the vertex $u_1/u_0\,(\infty)$, together with infinitely many 'smaller' triangles (obtained by reflecting these $2N$ triangles in their sides) which fill out the disk.

In fact, with our particular choice of u_0 and u_1, the disk is precisely the half-plane (in $\mathbb{C}$) consisting of all complex numbers with negative real part. We can now observe (a fortiori) that

$$\psi \begin{bmatrix} 1 \\ 0 \end{bmatrix} \in \mathrm{SL}_2\mathbb{R} \cdot \begin{bmatrix} 1 \\ -1 \end{bmatrix} = \mathrm{SL}_2\mathbb{R}\, d \cdot \begin{bmatrix} 1 \\ 0 \end{bmatrix}$$

where

$$d = \frac{1}{\sqrt{2i}} \begin{pmatrix} 1 & i \\ -1 & i \end{pmatrix}.$$

Then L_- satisfies all the conditions of Definition 10.12, with $\gamma_0 = \gamma^{-1} d\gamma$, on its entire domain $\mathbb{C} - \{0, 1, \infty\}$.

The group $\mathrm{SU}^D_{1,1}$ in section 10.4 has the property that

$$\mathrm{SU}^D_{1,1} \cdot \begin{bmatrix} 1 \\ 0 \end{bmatrix} = \text{unit disk},$$

and d is needed to convert this to the appropriate half-plane. Equivalently, we are converting $\mathrm{SU}^D_{1,1}$ to the group of matrices which preserve the Hermitian form whose matrix is

$$(d^{-1})^t D \bar{d}^{-1} = \begin{pmatrix} & -1 \\ -1 & \end{pmatrix}. \qquad \square$$

For mirror symmetry of Calabi–Yau manifolds, we need only the special case where L is of the form $L = \gamma^{-1}\psi\gamma$, and indeed only the case where ψ takes values in the orthogonal or symplectic group. Mirror symmetry for M^k_k fits into the scheme of Definition 10.12 providing we take a suitable constant loop $\gamma_0 = \gamma^{-1}d\gamma$. This adjustment, from the loop $\gamma_0 = \gamma^{-1}c\gamma$ of section 10.4, is necessary in order to match up the particular Picard–Fuchs equation with the general VHS framework.

From the point of view taken in this book, a more natural approach would be to regard the Picard–Fuchs equation or the quantum D-module—or more generally any abstract quantum D-module—as the fundamental object, and construct the data of Definition 10.12 directly.

Let us sketch how Chapter 6 gives a method for constructing objects of this type, starting from a self-adjoint D-module $\mathcal{M} = D^{\mathrm{h}}/(T_1, \dots, T_u)$ of rank n. First, we must assume that the D-module is of the type described in section 6.1, so that we obtain a flat connection satisfying the Griffiths transversality condition. We do not immediately need to assume the conditions which ensure the existence of an abstract quantum product, but let us assume them anyway, because we need to assume that the self-adjoint condition arises as in the case of quantum cohomology from a distinguished cyclic element of $\bar{\mathcal{M}}^*$ (cf. the comments at the end of appendix 4.6). We also need to assume that the matrix $\Omega^{\mathrm{h}}|_{z=0}$ is maximally nilpotent, in order to be able to normalize L_- as in section 5.3. Finally, we assume that there is an analogue of the Poincaré intersection form, and we denote by S its matrix with respect to the basis obtained by de-quantizing the distinguished basis $[\hat{P}_1], \dots, [\hat{P}_n]$ of $\mathcal{M}$. As explained in section 8.2, this basis gives an identification

$$D^{\mathrm{h}}/(T_1, \dots, T_u) \cong \mathcal{H}_z^n \otimes \mathcal{H}_{\mathrm{h}}^+ \cong \Gamma W \quad (\subseteq \mathcal{H}_z^n \otimes \mathcal{H}_{\mathrm{h}})$$

$$[P] = [\sum_1^n f_i \hat{P}_i] \longleftrightarrow P \cdot \begin{pmatrix} 1 \\ 0 \\ \vdots \\ 0 \end{pmatrix} = \begin{pmatrix} f_1 \\ f_2 \\ \vdots \\ f_n \end{pmatrix} \longleftrightarrow PJ$$

under which the pairing $\langle\!\langle\ ,\ \rangle\!\rangle$ corresponds to the pairing $(\!(\ ,\ )\!)$ (defined using S). The complex loop group here is $(\Lambda \mathrm{SL}_n\mathbb{C})_\sigma$ where σ is the involution $A \mapsto S(A^t)^{-1}S$. As its real form we take $(\Lambda \mathrm{SU}_n^S)_\sigma$. The first bilinear relation is satisfied because of the self-adjointness condition. The second bilinear relation is satisfied, at least in a neighbourhood of $z = 0$. In particular, this applies to the quantum differential equations, where we have the usual cyclic element and the Poincaré intersection form.

This sketch is a precise description in the case

$$\mathcal{M} = D^{\mathrm{h}}/(T)$$

which we have considered in detail in Chapter 6. Here T is an ordinary differential operator of order n (of the type described in section 6.1), the distinguished cyclic element of $\bar{\mathcal{M}}^*$ is δ_{n-1} (projection on $(\mathrm{h}\partial)^{n-1}$), S is the matrix with (i,j)-th element $\delta_{i+j,n}$, and we assume $\bar{T}^* = T$ (or more generally $(\overline{fT})^* = fT$).

Example 10.15 The simplest example is the D-module of rank 2 given by the hypergeometric differential operator, so let us reconsider Example 10.14 from this new point of view. The main point of interest is whether the second bilinear relation holds over the largest possible domain $U = \mathbb{C} - \{0, 1, \infty\}$.

The basis $\hat{P}_0 = \frac{1}{v_0}$, $\hat{P}_1 = \frac{1}{v_1}\mathrm{h}\partial_\theta \frac{1}{v_0}$ satisfies $\langle\!\langle [P_i], [P_j] \rangle\!\rangle^{1-z} = \delta_{i+j,1}$ and gives the above identification

$$\mathcal{M} \cong \Gamma W.$$

The map W is given explicitly by

$$W = L_- H_+^{(2)} = \gamma^{-1}\psi_-\gamma H_+^{(2)}, \quad \text{where } \psi_- = \begin{pmatrix} 1 & 0 \\ \delta & 1 \end{pmatrix}, \gamma = \begin{pmatrix} 1 & \\ & \mathrm{h} \end{pmatrix}$$

and $\delta = u_1/u_0$ (as in Example 10.14). This W takes values in the infinite-dimensional twisted Grassmannian $\mathrm{Gr}_\sigma^{(2)} = (\Lambda \mathrm{SL}_2)_\sigma H_+^{(2)}$; the second bilinear relation is satisfied if and only if the image of W is contained in the open orbit $(\Lambda \mathrm{SU}_{1,1}^S)_\sigma H_+^{(2)}$.

To investigate this condition, we use the isomorphism $\mathrm{Gr}_\sigma^{(2)} \cong \mathrm{Fl}_{1,1}^{(2)}$ which is induced by the map $T : \Lambda \mathrm{SL}_2 \to (\Lambda \mathrm{SL}_2)_\sigma, \alpha \mapsto \gamma^{-1}\alpha(\mathrm{h}^2)\gamma$ (see Example 8.18 and Chapter 19 of [60]). (As we have pointed out in Example 10.11,

$$(A^t)^{-1} = \begin{pmatrix} & 1 \\ -1 & \end{pmatrix} A \begin{pmatrix} & 1 \\ -1 & \end{pmatrix},$$

so the involution σ coincides with that used in Example 8.18 and [60].) Since $T^{-1}(L_-) = \psi_-$ we obtain the $\mathrm{Fl}_{1,1}^{(2)}$-valued map $W_2 \subseteq W_1 \subseteq W_0$, where

$$W_2 = \mathrm{h}H_+^{(2)} \quad W_1 = \begin{bmatrix} 1 \\ \delta \end{bmatrix} \oplus \mathrm{h}H_+^{(2)}, \quad W_0 = H_+^{(2)}.$$

This is, of course, essentially the $\mathbb{C}P^1$-valued map from Example 10.14. However, we are considering the orbits of the group $(\Lambda \mathrm{SU}_{1,1}^S)_\sigma$ on $\mathrm{Gr}_\sigma^{(2)}$, or equivalently the orbits of $\Lambda \mathrm{SU}_{1,1}^S$ on $\mathrm{Fl}_{1,1}^{(2)}$, so we need some information on these.

The open orbits of the action of $\Lambda \mathrm{SU}_{p,q}$ on $\mathrm{Gr}^{(n)}$ were discussed briefly in appendix 8.8. It turns out (see [80]) that the action of $\Lambda \mathrm{SU}_{1,1}$ on $\mathrm{Gr}^{(2)}$ has just one open orbit, but the action of $\Lambda \mathrm{SU}_{1,1}$ on $\mathrm{Fl}_{1,1}^{(2)}$ has two. Moreover (see [17]), these orbits can be described geometrically as follows. Consider the Hermitian inner product on $H^{(2)} = \mathrm{Map}(S^1, \mathbb{C}^2)$ defined (as in [102]) by

$$(f, g) \mapsto \int_{S^1} f^t \, S \, B(g) \, \tfrac{d\mathrm{h}}{\mathrm{h}},$$

where $B(g)$ temporarily denotes the usual complex conjugate of the map g (to distinguish it from $\bar{g} = g(-\mathrm{h})$). Then, the point $W_2 \subseteq W_1 \subseteq W_0$ belongs to one of the two open orbits if and only if this Hermitian inner product is non-degenerate with signature $(1,1)$ on $W_0 \cap (\mathrm{h} W_0)^\perp$ and either positive definite, or negative definite, on the line corresponding to W_1.

In our (very simple) example we have

$$W_0 \cap (\mathrm{h} W_0)^\perp = \mathbb{C}^2,$$

on which the inner product has signature $(1,1)$. The line corresponding to W_1 is

$$\begin{bmatrix} 1 \\ \delta \end{bmatrix},$$

and

$$\int_{S^1} (1, \delta) \, S \, (1, B(\delta))^t \, \tfrac{d\mathrm{h}}{\mathrm{h}} = \delta + B(\delta) = 2\mathrm{Re}\, \delta.$$

Thus, we recover the conclusion of Example 10.14: in the case of a hypergeometric differential operator of type (∞, ∞, N), the the second bilinear relation is satisfied on the maximal domain.

While this calculation does not give any new information, it has the advantage that all necessary data arises intrinsically from the D-module $\mathcal{M}$. In particular, the somewhat arbitrary choice of real form $\Lambda \mathrm{SU}^D_{1,1}$ in Example 10.14 (and section 10.4) has been avoided. In that case it was necessary to translate the extended solution by the constant loop $\gamma_0 = \gamma^{-1} d\gamma$; this is equivalent to modifying the Hermitian inner product by d, which in turn has the effect of changing the real form from $\Lambda \mathrm{SU}^D_{1,1}$ to $\Lambda \mathrm{SU}^S_{1,1}$. □

So far we have considered only a very restricted family of 'Schwarz disk' examples, that is, where $c = 1$, $a + b = 1$. The case $c = 1$, $a + b \neq 1$ can be treated in exactly the same way, as we have $(\overline{fG})^* = fG$ with $f(z) = \pm(1 - z)^{a+b-1}$, and $fG = \pm(1 - z)^{a+b}((\mathrm{h}\partial_\theta)^2 + \text{lower terms})$. The well known modular equations of type $(\infty, 2, 3)$ and (∞, ∞, ∞) belong to this family. However, the case $c \neq 1$—an example being $(2,3,7)$—does not satisfy our assumptions. Here the indicial equation at $z = 0$ is not $s^2 = 0$ (the matrix $\Omega^{\mathrm{h}}|_{z=0}$ is not nilpotent). We do not have a Frobenius basis of solutions of the form $u_0 = f_0$, $u_1 = f_0 \log z + f_1$ (the extended solution L_- cannot be normalized in the usual way). We have $(\overline{fG})^* = fG$ but the function $f(z) = z^{c-1}(z-1)^{a+b-c}$ is zero at $z = 0$. However, our method can be extended to this case; a change of basis in the D-module corresponds to the conformal transformation of $\mathbb{C}P^1$ which maps the Schwarz disk to the half-plane.

In this book we have not considered the difficult problem of deciding when an abstract VHS or abstract quantum D-module arises 'from geometry'. This is a longstanding problem in Hodge theory. On the quantum cohomology side, preliminary work has been done in [56] and [57] for Fano 3-folds, while characterizations of quantum differential equations for certain Calabi–Yau manifolds have been studied in [2], [3], [21], [41], and [132].

At the very least, this involves taking into account the arithmetic properties of genuine VHS and quantum cohomology (see [88]–[90]), which arise from the existence of integral cohomology on the VHS side and Gromov–Witten invariants on the quantum cohomology side. All the hypergeometric differential operators of type (∞, ∞, N) with $N = 2, 3, \ldots$ can be considered to satisfy the differential geometric properties of mirror symmetry on their largest natural domain. Let us return to this example one more time in order to examine its arithmetic properties.

Example 10.16 The hypergeometric differential operator of type $(\infty, \infty, 3)$ is

$$(\partial_\theta)^2 - z(\partial_\theta + \tfrac{1}{3})(\partial_\theta + \tfrac{2}{3}).$$

If z is replaced by $3^3 z$ we obtain the Picard–Fuchs operator of the elliptic curve M_3^3, so for $N = 3$ we have an example of genuine mirror symmetry. The abstract quantum product is the quantum product for M_3^3, which is the same as the cup product, so the Gromov–Witten invariants are trivial. The same applies for any $N = 2, 3, \ldots$; they all give the same abstract quantum D-module $D^{\mathrm{h}}/((\mathrm{h}\partial_t)^2)$, so there is no way to distinguish them on these grounds. However, the behaviour of the mirror transformation

$$q = z e^{f_1(z)/f_0(z)}$$

(which converts the original D-module to the abstract quantum D-module; see Example 10.11 and section 6.7) does depend on N, as we shall see.

Explicit formulae for f_0, f_1 are known in this case (see e.g. section 16.4 of [107]). First, we have the well known Gauss hypergeometric function

$$f_0(z) = \sum_{d \geq 0} \frac{(a)_d (b)_d}{(c)_d\, d!} z^d = F(a, b, c \mid z)$$

which satisfies $f_0(0) = 1$. Here $(x)_d = x(x+1)\cdots(x+d-1)$. When $c = 1$ we have $(c)_d = d!$, and there is a unique solution $f_0 \log z + f_1$ with $f_1(0) = 0$. The function f_1 is

$$f_1(z) = \sum_{d \geq 1} \frac{(a)_d (b)_d}{(d!)^2} \left(\sum_{k=1}^{d} \frac{1}{a+k-1} + \sum_{k=1}^{d} \frac{1}{b+k-1} - 2\sum_{k=1}^{d} \frac{1}{k} \right) z^d.$$

Now, the transformation $z \mapsto 3^3 z$ in the case $N = 3$ has the effect of making all the coefficients in the expansion of f_0 integers—and then (remarkably) the mirror transformation $q = ze^{f_1(z)/f_0(z)}$ has the same property.

With this as a guide, let us make a transformation of the form $z \mapsto Mz$ so that f_0 has integral expansion, then compute $q = ze^{f_1(z)/f_0(z)}$. The results for $2 \leq N \leq 9$ are as follows:

N	M	$q(z) = z + \cdots$ ($q(z) - z$ is given below)
2	2^6	$40z^2+1876z^3+95072z^4+5045474z^5+276107408z^6+\cdots$
3	3^3	$15z^2 + 279z^3 + 5729z^4 + 124554z^5 + 2810718z^6 + \cdots$
4	2^8	$136z^2 + 23476z^3 + \frac{13529248}{3}z^4 + \frac{2762468038}{3}z^5 + \frac{2933928596912}{15}z^6 + \cdots$
5	5^3	$65z^2 + 5425z^3 + \frac{1517375}{3}z^4 + \frac{150640000}{3}z^5 + \frac{15573698600}{3}z^6 +$
6	$2^6 3^3$	$888z^2+1019124z^3+1309259552z^4+1792760408706z^5+ \frac{12788878196390256}{5}z^6 + \cdots$
7	7^3	$175z^2 + 39739z^3 + 10114041z^4 + 2745264984z^5 + \frac{3883301787708}{5}z^6 + \cdots$
8	2^{10}	$520z^2 + 351796z^3 + 266969312z^4 + 216146465474z^5 + \frac{912189544795024}{5}z^6 + \cdots$
9	3^5	$123z^2 + 19719z^3 + 3548057z^4 + 681279501z^5 + \frac{681983227467}{5}z^6 + \cdots$

Apparently there is something special about the cases $N = 2, 3$, but what? The case $N = 3$ is genuine mirror symmetry. It turns out that the case $N = 2$

is genuine mirror symmetry as well: the differential operator

$$\partial_\theta^2 - 4z(4\partial_\theta + 3)(4\partial_\theta + 1)$$

is another Picard–Fuchs equation for M_3^3, but this time for an embedding in the weighted projective space $\mathbb{P}(1,1,2)$ rather than $\mathbb{C}P^2 = \mathbb{P}(1,1,1)$. (We have seen a similar phenomenon in Example 6.38, where the quantum differential equation for $\mathbb{C}P^1 = M_3^2 \subseteq \mathbb{C}P^2$ has a different form from the usual quantum differential equation for $\mathbb{C}P^1$.) There is another Picard–Fuchs equation for M_3^3, from an embedding in $\mathbb{P}(1,2,3)$, and this corresponds to the hypergeometric equation for $N = 3/2$. Although N is not an integer here, there is still a Schwarz disk, and the mirror transformation does indeed have an integral expansion. These three examples, and (partial) generalizations for M_4^4 and M_5^5 are discussed in detail in [83].

In fact the cases $N = 2, 3$ are known to be special in other ways which have (a priori) nothing to do with mirror symmetry: they are precisely the cases where the monodromy group (with respect to the singular points $z = 0, 1, \infty$) is a subgroup of the modular group $\mathrm{SL}_2\mathbb{Z}$, and precisely the cases where the monodromy group is arithmetic (see [118]).

It is clear that the monodromy group plays a crucial role here, as it determines the hypergeometric equation, and indeed for any N the monodromy group satisfies the well known requirements for mirror symmetry such as quasiunipotence of local monodromy at each singular point (see the references mentioned before Example 10.16). However, the relation with modularity or arithmeticity does not yet seem to be understood in general. For example, in [135] (where the cases M_3^3, M_4^4, M_5^5 are studied in detail), W. Zudilin proposes mirror symmetry as a generalization of the concept of modularity. □

10.6 Further developments

In principle, the construction in the previous section makes sense for D-modules with more general dependence on the spectral parameter. That is, we could start from a (self-adjoint) D-module of the form $\mathcal{M} = D^A/(T_1, \dots, T_u)$, where D^A (in the notation of section 8.2) is not necessarily D^{h}. Such objects arise naturally, not just in the case of mirror symmetry, but in other integrable systems such as the KdV equation.

However, the beginner who has reached this point will be dismayed (and the expert may be irritated) by the fact that the book is ending while we have barely scratched the surface of 'quantum cohomology and integrable systems'. We have persistently given the impression that quantum cohomology and the KdV equation are somewhat analogous, but the truth is that *they are actually the same thing:* it was conjectured by E. Witten and confirmed by M. Kontsevich that the higher genus Gromov-Witten invariants when M is a point lead to a specific solution[5] of the KdV hierarchy! This higher genus theory (which will need infinite rank D-modules) has been studied for very few concrete manifolds M, but the evidence suggests that, for each M, the higher genus Gromov-Witten potentials should give a distinguished solution of a hierarchy of integrable p.d.e. We conclude with some brief comments on this new direction, in which the Grassmannian seems likely to be a crucial tool.

Extensions and hierarchies

Throughout this book we have seen examples of D-modules which correspond to solutions of integrable systems. In section 4.4 we gave an 'extension principle' for D-modules, which shows explicitly how non-linear p.d.e. govern such extensions. It is is based on the hypothesis that there exist generators of the ideal of relations which commute 'to some extent'. Conversely, as we remarked at the end of that section, a 'generic' D-module can be regarded as an extension of this type—and, ultimately, as a sequence of extensions of a D-module corresponding to an ordinary differential operator.

The KdV hierarchy, and other standard integrable hierarchies, represent a very special case, where the generators actually commute. (These are all infinite hierarchies, but we obtain a D-module of finite rank by taking a finite number of members of the hierarchy.)

The scheme of section 4.4 is therefore broader than the standard construction scheme for integrable hierarchies. However, some aspects of the standard scheme survive: we find situations where the extension is essentially unique; where the (generalized) commutativity hypothesis leads to extensions which are independent—in physics terminology, variables can be 'switched on and off independently'; and where we have propagation of natural properties such as self-adjointness and homogeneity, and so on.

One could describe this general picture in terms of flat connections, rather than D-modules, or more naively by saying that an integrable p.d.e. is the compatibility condition for another system of linear p.d.e., and that the linear system may be reduced step by step to a series of o.d.e.—which is the classical point of view, based on Frobenius' Theorem. However, such

[5] The solution is not obtained by dressing the vacuum, however (see [5]).

descriptions are too general to be useful in pinpointing the special properties of integrable systems. As we have already mentioned in section 7.6, the advantage of the D-module approach is that it allows us to specify more precisely the kind of connection (or linear system) that is allowed—as in Corollary 4.21, for example.

Quantum cohomology (cf. the remarks in section 6.8) is ideally suited to this approach, because the relations of the quantum D-module reflect the relations of the quantum cohomology algebra and the ordinary cohomology algebra. The extent to which the natural relations of the quantum D-module commute, and the extent to which the latter can be regarded as a natural extension (in the sense of section 4.4) of a D-module involving fewer variables, seem likely to become important aspects of quantum cohomology in the future.

Current research has already produced various 'reconstruction theorems' in quantum cohomology, notably that of Kontsevich and Manin (see [85], and also [73]), which produces the large quantum cohomology as an extension of the small quantum cohomology under certain conditions (see section 9.2). For appropriate manifolds, at least, this may be viewed as an example of the above scheme.

With this in mind, the appearance of infinite hierarchies (or D-modules of infinite rank) in quantum cohomology should not be too surprising.

The theorem of Witten and Kontsevich

The generating function for gravitational Gromov-Witten invariants of genus g is the potential function $\mathcal{F}_g^M(\mathbf{t})$, where $\mathbf{t}$ denotes the infinite sequence of variables $t_i^{(j)}$, with $0 \le i \le s$ and $j \ge 0$ (see section 9.1). It is usual to assemble these into a single function,

$$\mathbb{F}^M = \sum_{g=0}^{\infty} \hbar^{g-1} \mathcal{F}_g^M,$$

where $\hbar$ is a new 'genus expansion' parameter, unrelated to the parameter h used earlier, and to introduce the 'free energy' $Z^M = \exp \mathbb{F}^M$. The result of Witten and Kontsevich (see [5] for a detailed introduction, and references to the original articles) is as follows.

Theorem 10.17 *The function $\partial^2 \mathbb{F}^{point} / (\partial t_0^{(0)})^2$ is a solution of the KdV hierarchy.*

Moreover, this particular solution is characterized by the fact that it satisfies another equation, the 'string equation' (a generalization of the condition that quantum cohomology has an identity element—see [94]). This characterization provides a way to compute Gromov-Witten invariants by solving a differential equation, and is therefore very much in the same spirit as what we have seen so far in the genus zero situation.

The KdV equation and the string equation can be reformulated as

$$L_n Z^{\text{point}} = 0, \quad n \geq -1,$$

for certain operators L_n, which represent 'half' of the Virasoro algebra. These equations are known as the 'Virasoro constraints'.

From another point of view, the theorem says that 'Z^{point} is a τ-function for the KdV hierarchy', since τ-functions $f(x, t, \ldots)$ have the property that $\partial_x^2 \log f$ is a solution. We have avoided τ-functions so far because they merely represent the 'bosonic side' of the more geometrical theory of the the Grassmannian-valued map W, which is the 'fermionic side' (see [5]).

It turns out that the Virasoro constraints have a simple interpretation in terms of W: they can be written in the form

$$A_n W \subseteq W,$$

where the A_n are certain linear differential operators (related to the Airy equation) in the loop parameter h. This result can be found in [78]. It is reminiscent of the constraints describing the theta function solutions of the KdV equation in section 8.5. However, the condition here is stronger, as they determine the solution uniquely.

The Lagrangian cone of Givental

Following Barannikov, we have seen that the Grassmannian-valued map W plays an important role in quantum cohomology because it can be interpreted as a variation of Hodge structure. Givental has developed this point of view in a different direction, by using W to construct a symplectic quantization formalism, which gives a geometrical approach to the rather inaccessible function $\mathbb{F}^M$.

In this theory, the pre-quantized (genus zero) situation is based on studying the image of W, or, more precisely, the closely related space

$$\mathcal{L}^M = \bigcup_{t \in H^*M} \mathrm{h} W(t) \subseteq H^{(s+1)}.$$

If $H^{(s+1)} = H_-^{(s+1)} \oplus H_+^{(s+1)}$ is identified with the cotangent bundle of $H_+^{(s+1)}$, it may be regarded as a symplectic manifold, and then $\mathcal{L}^M$ becomes a Lagrangian subvariety (the tangent space at any point of $hW(t)$ may be identified with $W(t)$, and this is isotropic because of the Frobenius property of quantum cohomology). But a Lagrangian subvariety is locally the graph of a 1-form—and this 1-form turns out to be $d\mathcal{F}^M(\mathbf{t})$, where $\mathcal{F}^M = \mathcal{F}_0^M$ is the usual genus zero gravitational potential.

Taking this 'Lagrangian cone' as the fundamental object, Givental (see [51], [54]) observed that the effect of a linear symplectic transformation of $\mathcal{L}^M$ can be translated into the action of a differential operator on $\mathcal{F}^M(\mathbf{t})$, and, furthermore, in the quantized version, to the action of a differential operator on $\mathbb{F}^M(\mathbf{t})$. The latter applies when the quantum cohomology is semisimple, and involves a reduction to the case where M is a disjoint union of points. This has already yielded new insights (see [22] and [23]), and gives an approach to the integrable hierarchies associated to higher genus quantum cohomology beyond the case of a point. Before discussing these matters further, however, we make some additional remarks about the role of the cone $\mathcal{L}^M$ in ordinary genus zero (unquantized) quantum cohomology.

In the case of a Calabi-Yau manifold M, we have seen that W is just a map into a finite-dimensional Grassmannian. The cone $\mathcal{L}$ reduces to a Lagrangian cone in a finite-dimensional vector space, and it turns out to be special Lagrangian (see [59]). One could say that 'special geometry' is another aspect of differential geometric mirror symmetry, and it is an aspect that is revealed by introducing $\mathcal{L}^M$.

For Calabi-Yau or Fano complete intersections in toric manifolds, the 'Mirror Theorem' (of [49] and [91]–[93]; see the comments in section 6.7) may be expressed in terms of the cone $\mathcal{L}^M$ in the following way. First, the system of GKZ differential equations (see section 5.5 of [26]) associated to the toric data has a natural basis of solutions given by $J = J_{\mathrm{GKZ}}$; this is usually called the 'I-function'. On the other hand, we have the 'J-function' $J = J_M$ associated to the quantum differential equations of M. The Mirror Theorem says that J_{GKZ} and J_M coincide in the Fano case, and coincide up to a coordinate transformation in the Calabi-Yau case. Each of J_{GKZ} and J_M generates a map W (see section 8.2), and hence an associated cone. The cone is a geometrical object unaffected by choice of coordinates, so these cones coincide. Givental regards the I-function and the J-function as sections of this cone. The purpose of the Mirror Theorem is to compute the structure constants of quantum cohomology, and this is achieved by generating the cone from the I-function, then extracting the answer by a procedure which amounts to the Birkhoff factorization of L, where $W = LH_+^{(s+1)}$.

Quantum cohomology and integrable hierarchies

The Virasoro Conjecture predicts the existence of differential operators L_n^M such that $L_n^M Z^M = 0$, thus generalizing one aspect of the theorem of Witten and Kontsevich. This can be formulated for the genus zero potential $\mathcal{F}^M$ alone, where it has been established in some generality, notably by making use of the cone $\mathcal{L}^M$ (see [51] and [52]).

For further information on the Virasoro conjecture and its origins, we refer to [46], [87], and [94]. In addition to the hierarchies introduced by Barannikov in [11], and the approach of Givental (see the comments at the end of [52]), both of which have been mentioned already, a systematic approach to the integrable hierarchy question has been developed by Dubrovin and Zhang, in [38]. This involves the classification of integrable hierarchies with the property that the system may be reconstructed from its 'dispersionless limit'. In the genus zero situation (see [34]), the differential equations governing the gravitational Gromov-Witten invariants are a system of hydrodynamic type which should be the dispersionless limit of the sought-after higher genus system, with the 'genus expansion' playing the role of the 'small dispersion expansion'.

Apart from the case of a point, the case $M = \mathbb{C}P^1$ has been investigated in detail and gives rise to the Toda hierarchy, but in general the higher genus version of the Virasoro Conjecture, and the search for the integrable hierarchy of which $\mathbb{F}^M$ is a distinguished solution, remains elusive. To appreciate the large amount of work taking place in this area, the reader is urged to consult other articles by the authors mentioned here as well as the latest preprints appearing on the internet.

References

[1] M. J. Ablowitz, D. J. Kaup, A. C. Newell, and H. Segur, *The inverse scattering transform—Fourier analysis for nonlinear problems*, Stud. Appl. Math. **53** (1974), 249–315.

[2] G. Almkvist, C. van Enckevort, D. van Straten, and W. Zudilin, *Tables of Calabi–Yau equations*, preprint, math.AG/0507430.

[3] G. Almkvist and W. Zudilin, *Differential equations, mirror maps and zeta values*, Mirror Symmetry V, AMS/IP Stud. Adv. Math. 38, Amer. Math. Soc., 2006, pp. 481–515.

[4] A. Amarzaya and M. A. Guest, *Gromov-Witten invariants of flag manifolds, via D-modules*, J. London Math. Soc. **72** (2005), 121–136.

[5] E. Arbarello, *Sketches of KdV*, Contemp. Math. **312** (2002), 9–69.

[6] M. Audin, *An introduction to Frobenius manifolds, moduli spaces of stable maps and quantum cohomology*, preprint, http://www-irma.u-strasbg.fr/irma/publications/1997/97013.shtml

[7] M. Audin, *Symplectic geometry in Frobenius manifolds and quantum cohomology*, J. Geom. Phys. **25** (1998), 183–204.

[8] V. Balan and J. Dorfmeister, *Birkhoff decompositions and Iwasawa decompositions for loop groups*, Tohoku Math. J. **53** (2001), 593–615.

[9] S. Barannikov, *Quantum periods. I. Semi-infinite variations of Hodge structures*, Internat. Math. Res. Notices **23** (2001), 1243–1264.

[10] S. Barannikov, *Non-commutative periods and mirror symmetry in higher dimensions*, Comm. Math. Phys. **228** (2002), 281–325.

[11] S. Barannikov, *Semi-infinite variations of Hodge structures and integrable hierarchies of KdV type*, Internat. Math. Res. Notices **19** (2002), 973–990.

[12] V. Batyrev, *Quantum cohomology rings of toric manifolds*, Astérisque (Journées de géométrie algébrique D'Orsay) **218** (1993), 9–34.

[13] R. Beals and D. H. Sattinger, *Integrable systems and isomonodromy deformations*, Phys. D **65** (1993), 17–47.

[14] A. Beauville, *Quantum cohomology of complete intersections*, Mat. Fiz. Anal. Geom. **2** (1995), 384–398.

[15] A. Bobenko and A. Its, *The Painlevé III equation and the Iwasawa decomposition*, Manuscripta Math. **87** (1995), 369–377.

[16] F. E. Burstall and M. A. Guest, *Harmonic two-spheres in compact symmetric spaces, revisited*, Math. Ann. **309** (1997), 541–572.

[17] F. E. Burstall, M. A. Guest, Y. Ohnita, and W. Rossman, in preparation.

[18] F. E. Burstall and F. Pedit, *Harmonic maps via Adler-Kostant-Symes theory*, Harmonic Maps and Integrable Systems, eds. A. P. Fordy and J. C. Wood, Aspects of Math. E23, Vieweg, 1994, pp. 221–272.

[19] F. E. Burstall and J. H. Rawnsley, *Twistor Theory for Riemannian Symmetric Spaces*, Lecture Notes in Math. 1424, Springer, 1990.

[20] F. Calogero and A. Degasperis, *Spectral Transform and Solitons: Tools to Solve and Investigate Nonlinear Evolution Equations*, Studies in Math. Applications 13, North-Holland, 1982.

[21] Y. H. Chen, Y. Yang, and N. Yui, *Monodromy of Picard-Fuchs differential equations for Calabi-Yau threefolds*, preprint, math.AG/0605675.

[22] T. Coates, *Givental's Lagrangian cone and S^1-equivariant Gromov-Witten theory*, preprint, math.AG/0607808.

[23] T. Coates and A. Givental, *Quantum Riemann-Roch, Lefschetz and Serre*, preprint, math.AG/0110142.

[24] A. Collino and M. Jinzenji, *On the structure of the small quantum cohomology rings of projective hypersurfaces*, Comm. Math. Phys. **206** (1999), 157–183.

[25] S. C. Coutinho, *A Primer of Algebraic D-Modules*, LMS Student Texts 33, Cambridge Univ. Press, 1995.

[26] D. A. Cox and S. Katz, *Mirror Symmetry and Algebraic Geometry*, Math. Surveys and Monographs 68, Amer. Math. Soc., 1999.

[27] L. A. Dickey, *Soliton Equations and Hamiltonian Systems*, Advanced Series in Math. Physics 12, World Scientific, 1991.

[28] J. Dorfmeister, *Generalized Weierstrass representations of surfaces*, Surveys on Geometry and Integrable Systems, Advanced Studies in Pure Math., Math. Soc. Japan, to appear.

[29] J. Dorfmeister and J.-H. Eschenburg, *Pluriharmonic maps, loop groups and twistor theory*, Ann. Global Anal. Geom. **24** (2003), 301–321.

[30] J. Dorfmeister and G. Haak, *Meromorphic potentials and smooth surfaces of constant mean curvature*, Math. Z. **224** (1997), 603–640.

[31] J. Dorfmeister, F. Pedit, and M. Toda, *Minimal surfaces via loop groups*, Balkan J. Geom. App. **2** (1997), 25–40.

[32] J. Dorfmeister, F. Pedit, and H. Wu, *Weierstrass type representations of harmonic maps into symmetric spaces*, Comm. Anal. Geom. **6** (1998), 633–668.

[33] V. G. Drinfeld and V. V. Sokolov, *Equations of Korteweg-de Vries type and simple Lie algebras*, Soviet Math. Dokl. **23** (1981), 457–462.

[34] B. Dubrovin, *Integrable systems in topological field theory*, Nucl. Phys. B **379** (1992), 627–689.

[35] B. Dubrovin, *Geometry and integrability of topological-antitopological fusion*, Comm. Math. Phys. **152** (1993), 539–564.

[36] B. Dubrovin, *Geometry of 2D topological field theories*, Integrable Systems and Quantum Groups, eds. M. Francaviglia and S. Greco, Lecture Notes in Math. 1620, Springer, 1996, pp. 120–348.

[37] B. A. Dubrovin, I. M. Krichever, and S. P. Novikov, *Integrable systems. I*, Encyclopedia of Mathematical Sciences 4, Dynamical Systems IV, eds. V. I. Arnold and S. P. Novikov, Springer, 2001, pp. 177–280.

[38] B. Dubrovin and Y. Zhang, *Normal forms of hierarchies of integrable PDEs, Frobenius manifolds and Gromov–Witten invariants*, preprint, math.DG/0108160.

[39] J. Eells and L. Lemaire, *A report on harmonic maps*, Bull. London Math. Soc. **10** (1978), 1–68.

[40] J. Eells and L. Lemaire, *Another report on harmonic maps*, Bull. London Math. Soc. **20** (1988), 385–524.

[41] C. van Enckevort and D. van Straten, *Monodromy calculations of fourth order equations of Calabi-Yau type*, Mirror Symmetry V, AMS/IP Stud. Adv. Math. 38, Am. Math. Soc., 2006, pp. 539–559.

[42] L. D. Faddeev and L. A. Takhtajan, *Hamiltonian Methods in the Theory of Solitons*, Springer, 1987.

[43] H. Flaschka and A. C. Newell, *Monodromy- and spectrum-preserving deformations. I*, Comm. Math. Phys. **76** (1980), 65–116.

[44] W. Fulton, *Young Tableaux*, LMS Student Texts 35, Cambridge Univ. Press, 1997.

[45] I. M. Gelfand and L. A. Dickii, *Fractional powers of operators and Hamiltonian systems*, Funct. Anal. App. **10** (1976), 259–273.

[46] E. Getzler, *The Virasoro conjecture for Gromov-Witten invariants*, Contemp. Math. **241** (1998), 147–176.

[47] A. B. Givental, *Homological geometry and mirror symmetry*, Proc. Int. Congress of Math. I, Zürich 1994, ed. S. D. Chatterji, Birkhäuser, 1995, pp. 472–480.

[48] A. B. Givental, *Homological geometry I. Projective hypersurfaces*, Selecta Math. **1** (1995), 325–345.

[49] A. B. Givental, *Equivariant Gromov–Witten invariants*, Internat. Math. Res. Notices **13** (1996), 1–63.

[50] A. B. Givental, *Stationary phase integrals, quantum Toda lattices, flag manifolds and the mirror conjecture*, Topics in Singularity Theory, Amer. Math. Soc. Transl. Ser. 2, 180, eds. A. Khovanski, A. Varchenko, and V. Vassiliev, Amer. Math. Soc., 1997, pp. 103–115.

[51] A. B. Givental, *Gromov-Witten invariants and quantization of quadratic Hamiltonians*, Mosc. Math. J. **1** (2001), 551–568.

[52] A. B. Givental, *Semisimple Frobenius structures at higher genus*, Internat. Math. Res. Notices **23** (2001), 1265–1286.

[53] A. Givental, *A_{n-1} singularities and nKdV hierarchies*, Mosc. Math. J. **3** (2003), 475–505.

[54] A. B. Givental, *Symplectic geometry of Frobenius structures*, Frobenius Manifolds, Aspects Math., E36, eds. C. Hertling and M. Marcolli, Vieweg, 2004, pp. 91–112.

[55] A. Givental and B. Kim, *Quantum cohomology of flag manifolds and Toda lattices*, Comm. Math. Phys. **168** (1995), 609–641.

[56] V. V. Golyshev, *Modularity of equations D3 and the Iskovskikh classification*, Doklady Math. **69** (2004), 443–449.

[57] V. Golyshev, *Classification problems and mirror duality*, preprint, math.AG/0510287.

[58] P. A. Griffiths, *Periods of integrals on algebraic manifolds*, Publ. Math. IHES **38** (1970), 125–180.

[59] M. Gross, *Calabi-Yau manifolds and mirror symmetry*, Calabi-Yau Manifolds and Related Geometries, eds. M. Gross, D. Huybrechts, and D. Joyce, Springer, 2003, pp. 69–159, 227–236.

[60] M. A. Guest, *Harmonic Maps, Loop Groups, and Integrable Systems*, LMS Student Texts 38, Cambridge Univ. Press, 1997.

[61] M. A. Guest, *An update on harmonic maps of finite uniton number, via the zero curvature equation*, Contemp. Math. **309** (2002), 85–113.

[62] M. A. Guest, *Quantum cohomology via D-modules*, Topology **44** (2005) 263–281.

[63] M. A. Guest, *Introduction to homological geometry: I,II*, Integrable Systems, Geometry, and Topology, ed. C.-L. Terng, AMS/IP Studies

in Advanced Mathematics 36, Amer. Math. Soc. and International Press, 2006, pp. 83–121, 123-150.

[64] V. Guillemin and S. Sternberg, *The coefficients of the Duistermaat-Heckman polynomial and the cohomology ring of reduced spaces*, Geometry, Topology, and Physics for Raoul Bott, ed. S. T. Yau, Conf. Proc. Lect. Notes Geom. Topol. 4, International Press, 1994, pp. 202–213.

[65] A. Haefliger, *Local theory of meromorphic connections in dimension one (Fuchs theory)*, Algebraic D-modules, eds. A. Borel, P.-P. Grivel, B. Kaup, A. Haefliger, B. Malgrange, and F. Ehlers, Perspectives in Math. 2, Academic Press, 1987, pp. 129–149.

[66] F. Hélein, *Constant Mean Curvature Surfaces, Harmonic Maps and Integrable Systems*, Birkhäuser, 2001.

[67] C. Hertling, *Frobenius Manifolds and Moduli Spaces for Singularities*, Cambridge Tracts in Math. 151, Cambridge Univ. Press, 2002.

[68] C. Hertling, *tt* geometry, Frobenius manifolds, their connections, and the construction for singularities*, J. Reine Angew. Math. **555** (2003), 77–161.

[69] F. Hirzebruch, *Über eine Klasse von einfach-zusammenhängenden komplexen Mannigfaltigkeiten*, Math. Ann. **124** (1951), 77–86.

[70] N. J. Hitchin, *Frobenius manifolds*, Gauge Theory and Symplectic Geometry (Montreal, PQ, 1995), eds. J. Hurtubise, F. Lalonde, and G. Sabidussi, NATO Adv. Sci. Inst. Ser. C Math. Phys. Sci. 488, Kluwer, 1997, pp. 69–112.

[71] E. L. Ince, *Ordinary Differential Equations*, Longmans, Green and Co., 1926 (reprinted by Dover, 1956).

[72] H. Iritani, *Quantum D-modules and equivariant Floer theory for free loop spaces*, Math. Z. **252** (2006), 577–622.

[73] H. Iritani, *Quantum D-modules and their generalized mirror transformations*, preprint.

[74] A. Its and V. Y. Novokshenov, *The Isomonodromic Deformation Method in the Theory of Painlevé Equations*, Lecture Notes in Math. 1191, Springer, 1986.

[75] T. A. Ivey and J. M. Landsberg, *Cartan for Beginners: Differential Geometry via Moving Frames and Exterior Differential Systems*, Graduate Studies in Math. 61, Amer. Math. Soc, 2003.

[76] M. Jinzenji, *Gauss-Manin system and the virtual structure constants*, Internat. J. Math. **13** (2002), 445–477.

[77] D. Joyce, *Riemannian holonomy groups and calibrated geometry*, Calabi-Yau Manifolds and Related Geometries, eds. M. Gross, D. Huybrechts, and D. Joyce, Springer, 2003, pp. 1–68.

[78] V. Kac and A. Schwarz, *Geometric interpretation of the partition function of 2D gravity*, Phys. Lett. B **257** (1991), 329–334.

[79] N. M. Katz, *On the calculation of some differential galois groups*, Invent. math. **87** (1987), 13–61.

[80] P. Kellersch, *Eine Verallgemeinerung der Iwasawa Zerlegung in Loop Gruppen*, Dissertation, Technische Universität München, 1999. (Differential Geometry—Dynamical Systems Monographs 4, Balkan Press, 2004.)

[81] K. Kenmotsu, *Surfaces with Constant Mean Curvature*, Translations of Mathematical Monographs 221, Amer. Math. Soc, 2003.

[82] A. N. Kirillov and T. Maeno, *Quantum double Schubert polynomials, quantum Schubert polynomials and Vafa-Intriligator formula*, Discrete Math. **217** (2000), 191–223.

[83] A. Klemm, B. H. Lian, S. S. Roan, and S. T. Yau, *A note on ODEs from mirror symmetry*, Functional Analysis on the Eve of the 21st Century, Volume 2, eds. S. Gindikin, J. Lepowsky, and R. L. Wilson, Progress in Math. 132, Birkhäuser, 1996, pp. 301–323.

[84] W. Klingenberg, *A Course in Differential Geometry*, Graduate Texts in Mathematics 51, Springer, 1978.

[85] M. Kontsevich and Yu. Manin, *Gromov–Witten classes, quantum cohomology, and enumerative geometry*, Comm. Math. Phys. **164** (1994), 525–562.

[86] I. M. Krichever, *An analogue of d'Alembert's formula for the equations of the principal chiral field and for the sine-Gordon equation*, Soviet Math. Dokl. **22** (1980), 79–84.

[87] Y.-P. Lee, *Witten's conjecture and the Virasoro conjecture for genus up to two*, Contemp. Math. **403** (2006), 31–42.

[88] B. H. Lian and S.-T. Yau, *Arithmetic properties of mirror map and quantum coupling*, Comm. Math. Phys. **176** (1996), 163–192.

[89] B. H. Lian and S.-T. Yau, *Mirror maps, modular relations and hypergeometric series I*, preprint, hep-th/950715.

[90] B. H. Lian and S.-T. Yau, *Mirror maps, modular relations and hypergeometric series II*, Nuclear Phys. B Proc. Suppl. **46** (1996), 248–262.

[91] B. H. Lian, K. Liu, and S.-T. Yau, *Mirror principle I*, Asian J. Math. **1** (1997), 729–763.

[92] B. H. Lian, K. Liu, and S.-T. Yau, *Mirror principle II*, Asian J. Math. **3** (1999), 109–146.

[93] B. H. Lian, K. Liu, and S.-T. Yau, *Mirror principle III*, Asian J. Math. **3** (1999), 771–800.

[94] X. Liu, *Quantum product, topological recursion relations, and the Virasoro conjecture*, Surveys on Geometry and Integrable Systems, Advanced Studies in Pure Math., Math. Soc. Japan, to appear.

[95] Y. I. Manin, *Frobenius Manifolds, Quantum Cohomology, and Moduli Spaces*, Amer. Math. Soc, 1999.

[96] S. Novikov, S. Manakov, L. B. Pitaevskii and V. E. Zakharov, *Theory of Solitons*, Plenum, 1984.

[97] Y. Ohnita and G. Valli, *Pluriharmonic maps into compact Lie groups and factorization into unitons*, Proc. London Math. Soc. **61** (1990), 546–570.

[98] R. S. Palais, *The symmetries of solitons*, Bull. Amer. Math. Soc. **34** (1997), 339–403.

[99] A. M. Perelomov, *Integrable Systems of Classical Mechanics and Lie Algebras*, Birkhäuser, 1990.

[100] F. Pham, *Singularités des systèmes différentiels de Gauss-Manin*, Progr. Math. 2, Birkhäuser, 1979.

[101] K. Pohlmeyer, *Integrable Hamiltonian systems and interactions through constraints*, Comm. Math. Phys. **46** (1976), 207–221.

[102] A. N. Pressley and G. B. Segal, *Loop Groups*, Oxford Univ. Press, 1986.

[103] M. van der Put, and M. F. Singer, *Galois Theory of Linear Differential Equations*, Grundlehren der Mathematischen Wissenschaften 328, Springer, 2003.

[104] K. Saito, *Period mapping associated to a primitive form*, Publ. Res. Inst. Math. Sci. Kyoto **19** (1983), 1231–1264.

[105] M. Saito, B. Sturmfels, and N. Takayama, *Gröbner Deformations of Hypergeometric Differential Equations*, Algorithms and Computation in Math. 6, Springer, 2000.

[106] H. Sakai, *Gromov-Witten invariants of Fano hypersurfaces, revisited*, preprint, math.DG/0602324.

[107] G. Sansone and J. Gerretsen, *Lectures on the Theory of Functions of a Complex Variable. II. Geometric Theory*, Walters-Noordhoff, 1969.

[108] M. Sato, *Soliton equations as dynamical systems on an infinite dimensional Grassmann manifold*, RIMS Kokyuroku **439** (1981), 30–46.

[109] M. Sato, *The KP hierarchy and infinite-dimensional Grassmann manifolds*, eds. L. Ehrenpreis and R. C. Gunning, Proc. Symp. Pure Math. 49 (1), Amer. Math. Soc., 1989, pp. 51–66.

[110] M. Sato and Y. Sato, *Soliton equations as dynamical systems on infinite dimensional Grassmann manifold*, eds. H. Fujita, P. D. Lax, and G. Strang, Lecture Notes in Num. Appl. Anal. 5, North Holland, 1982, pp. 259–271.

[111] G. B. Segal, *Loop groups and harmonic maps*, Advances in Homotopy Theory, London Math. Soc. Lecture Notes 139, Cambridge Univ. Press, 1989, pp. 153–164.

[112] G. Segal and G. Wilson, *Loop groups and equations of KdV type*, Publ. Math. IHES **61** (1985), 5–65.

[113] R. W. Sharpe, *Differential Geometry*, Graduate Texts in Mathematics 166, Springer, 1997.

[114] B. Siebert and G. Tian, *On quantum cohomology rings of Fano manifolds and a formula of Vafa and Intriligator*, Asian J. Math. **1** (1997), 679–695.

[115] M. Spivak, *A Comprehensive Introduction to Differential Geometry*, Volumes 1–5, Publish or Perish, 1979.

[116] J. Stienstra, *Resonant hypergeometric systems and mirror symmetry*, Integrable Systems and Algebraic Geometry (Kobe/Kyoto, 1997), eds. M.-H. Saito, Y. Shimizu, and K. Ueno, World Scientific, 1998, pp. 412–452.

[117] K. Takasaki, *Integrable systems as deformations of $\mathcal{D}$-modules*, eds. L. Ehrenpreis and R. C. Gunning, Proc. Symp. Pure Math. 49 (1), Amer. Math. Soc., 1989, pp. 143–168.

[118] K. Takeuchi, *Arithmetic triangle groups*, J. Math. Soc. Japan **29** (1977), 91–106.

[119] C.-L. Terng, *Soliton equations and differential geometry*, J. Differential Geom. **45** (1997), 407–445.

[120] C.-L. Terng, *Geometries and symmetries of soliton equations and integrable elliptic equations*, Surveys on Geometry and Integrable Systems, Advanced Studies in Pure Math., Math. Soc. Japan, to appear, math.DG/0212372.

[121] C.-L. Terng, *Applications of loop group factorization to geometric soliton equations*, Proc. Int. Congress of Math. II, Madrid 2006, eds. M. Sanz-Solé, J. Soria, J. L. Varona, and J. Verdera, Eur. Math. Soc., 2006, pp. 927–950.

[122] C.-L. Terng and K. Uhlenbeck, *Poisson actions and scattering theory for integrable systems*, eds. C.-L. Terng and K. Uhlenbeck, Surveys in Differentiable Geometry: Integrable Systems, International Press, 1998, pp. 315–402.

[123] C.-L. Terng and K. Uhlenbeck, *Bäcklund transformations and loop group actions*, Comm. Pure Appl. Math. **53** (2000), 1–75.

[124] M. Toda, *Weierstrass-type representation of weakly regular pseudospherical surfaces in Euclidean space*, Balkan J. Geom. Appl. 7 (2002), 87–136.

[125] G. Tu, *Kac-Moody algebras and integrable systems*, ed. C. Gu, Soliton Theory and its Applications, Springer, 1995, pp. 230–296.

[126] K. K. Uhlenbeck, *Harmonic maps into Lie groups (Classical solutions of the chiral model)*, J. Differential Geom. **30** (1989), 1–50.

[127] C. Voisin, *Variations of Hodge Structure of Calabi-Yau Threefolds*, Scuola Norm. Sup., Pisa, 1996.

[128] G. Wilson, *The modified Lax and two-dimensional Toda lattice equations associated with simple Lie algebras*, Ergod. Th. and Dynam. Sys. **1** (1981), 361–380.

[129] G. Wilson, *Infinite-dimensional Lie groups and algebraic geometry in soliton theory*, Phil. Trans. R. Soc. Lond. **A315** (1985), 393–404.

[130] J. A. Wolf, *The action of a real semisimple Lie group on a complex flag manifold. I: Orbit structure and holomorphic arc components*, Bull. Amer. Math. Soc. **75** (1969), 1121–1237.

[131] J. A. Wolf, *Real groups transitive on complex flag manifolds*, Proc. Amer. Math. Soc. **129** (2001), 2483–2487.

[132] Y. Yang and N. Yui, *Differential equations satisfied by modular forms and K3 surfaces*, preprint, math.NT/0506576.

[133] V. E. Zakharov and A. V. Mikhailov, *Relativistically invariant two-dimensional models of field theory which are integrable by means of the inverse scattering problem method*, Sov. Phys. JETP **47** (1978), 1017–1027.

[134] V. E. Zakharov and A. B. Shabat, *Integration of non-linear equations of mathematical physics by the inverse scattering method II*, Funct. Anal. Appl. **13** (1979), 166–174.

[135] W. Zudilin, *Number theory casting a look at the mirror*, preprint.

Index

A-model 245
abstract quantum cohomology 130
abstract VHS 244
Airy equation 290
arithmetic group 287
arithmetic properties 285

B-model 245
Baker function 197
Barannikov, S. 242, 244, 274
Birkhoff factorization xxv, 88, 90, 103, 116, 131, 138, 166, 173, 204, 208, 218, 219, 244, 252, 267, 277, 291
bosonic 290

Calabi-Yau manifold 138, 243, 255
Cartan, E. 71, 83
Cartan/Klein strategy 71, 83, 84, 174
CMC 66, 76, 84, 173
CMCLab 171
cohomology 3
complex extended solution 208, 269, 273
complex projective space 9, 19, 39, 113, 234, 240
conjugation map 254, 270
connection 71, 78
constant mean curvature: see CMC
convex manifold 13
covariant derivative 72
cup product 3
curvature 75
cyclic element 60, 61

D-module 44, 46, 49, 51, 54, 101, 154, 180, 186, 218, 287
D^h-module 125
Dickii, L. 182
differential geometric mirror symmetry 276, 279
differential geometry 71
Dorfmeister, J. 278
DPW construction 166, 170, 173, 207
DPW potential 208
dressing 199
dual D-module 56, 108
Dubrovin, B. 34, 229, 239, 242
Dubrovin connection 36

elliptic curve 148, 274
Euler field 236
extended harmonic map 165
extended solution 165
extension of D-module 67, 99, 152, 156, 202, 288

Fano manifold 13, 138
fermionic 290
finite uniton number 203, 205, 214, 269
first bilinear relation 246, 271
first fundamental form 74
flag manifold 20, 40, 122, 134
Floer theory 33, 100
Frobenius algebra 229
Frobenius basis 114, 147, 259, 268
Frobenius manifold 116, 223, 229, 230
Frobenius property 106, 111, 135, 272
Frobenius symbolic factors 91, 147
Frobenius' Theorem 79
fundamental construction 186, 202, 243
fundamental solution matrix 47
Fundamental Theorem of Surface Theory 75

Gabber, O. 92
gauge transformation 48, 50, 82
Gauss-Codazzi equations 75, 174
Gauss curvature 73
Gauss factorization: see Birkhoff factorization
Gauss-Manin connection 249
Gelfand, I. 182

generalized flag manifold 211
Givental, A. 29, 32, 33, 100, 112, 242, 290
Givental connection 36
GKZ system 148, 291
grading 109
Gram-Schmidt: see Iwasawa factorization
Grassmannian model xxiv, 118, 182, 197, 265
Griffiths, P. 245
Griffiths transversality 246, 273
Gröbner basis 58, 138
Gromov-Witten invariant 12, 16, 112, 224, 227, 228
Guiding Principle 191

harmonic map 64, 164, 171, 189, 206
Hertling, C. 244
hierarchy 70, 288, 291
Hirzebruch surface 24, 41, 148
Hodge decomposition 245
Hodge diamond 261
holonomy 82
homogeneity 109, 124, 136, 157, 236, 288
homological geometry 29
homology 2
hypergeometric o.d.e. 279, 283, 285
hypersurface 27, 42, 104, 109, 125, 133, 136, 143

I-function 291
infinite-dimensional flag manifold 203, 217, 277
infinitesimal bilinear relations 246
integrable p.d.e. xxii, 66
integrable system xxii, 43, 67, 71, 116, 152, 154, 182, 223, 288
intersection pairing 8, 247
Iritani, H. 135
isomonodromy 239
isospectral deformation 192
Iwasawa factorization 96, 142, 173, 186, 208, 218, 219, 252, 273, 277

J-function 56, 102, 104, 112, 183, 291

Kac-Moody algebra 71, 164
KdV equation 59, 63, 154, 189, 191, 202, 236, 288
Kellersch, P. 221
Kim, B. 32
Kleiman transversality 14
Klein, F. 71, 83
Kontsevich, M. 288, 289
KP equation 159
Krichever, I. 166
Kuramoto-Sivashinsky equation 158

Lagrangian cone 290
large quantum cohomology 152, 224, 289
Lax equation 156, 192, 193
local system 249
loop algebra 164, 184
loop group 118, 165, 184, 270

Manin, Y. 289
MAPLE 138
maximal unipotent monodromy 268
mean curvature 73
minimal surface 77, 85, 176
mirror map 256, 262
mirror symmetry 148, 243, 245, 256, 274
Mirror Theorem 146, 148, 245, 291
mirror transformation 256, 262, 285
mKdV equation 160
modular elliptic function 253, 276
modular group 287
monodromy 52, 82
Morse-Bott decomposition 205

naive quantization 124
nef 138
non-compact real form 249
normalized extended solution 208
Novikov ring 18

o.d.e. 46, 116
opposite Grassmannian 186
orthogonal group 253
overdetermined system 46

p.d.e. 53, 117
Painlevé equation 160
period integral 259
period map 245
periodic flag manifold 217
Picard-Fuchs operator 259

pluriharmonic map 152, 176, 223, 246
polarized Hodge structure 247
potential 224, 229
principal chiral model 165
pseudo-differential operator 195

quantization 124, 153, 290
quantum cohomology 12, 223
quantum D-module 100
quantum differential equations 33, 63, 190
quantum evaluation map 28, 131
quantum product 16
quintic 3-fold 245

real form, non-compact 249
reality condition 66, 165, 255
Reconstruction Theorem 227, 289
regular point 46, 52
regular singular point 52
Riemann surface 250

Saito, K. 229
Sato D-module 197
Sato Grassmannian: see Grassmannian model
scalar Grassmannian 198
scaling 236
Schmitt, N. 171
Schubert decomposition xxv, 107, 205, 214, 215, 219
Schwartz disk 280
Schwartz map 276
Schwartz triangle 280
second bilinear relation 246, 273
second fundamental form 74
Segal, G. 166, 182, 208
Segal-Wilson Grassmannian: see Grassmannian model
self-adjointness 91, 107, 126, 269, 282, 288
semisimple Frobenius manifold 239
Siegel upper half-space 252
sine-Gordon equation 168, 176
sinh-Gordon equation 78
soliton equation 52
special geometry 291
spectral parameter 52, 182, 198, 287
standard monomial 58
stationary solution 206
string equation 289
structure constants 8
symbolic factors of Frobenius 91, 147
symmetry group 71, 83, 89, 179, 190
symplectic group 254
symplectic volume 10

Toda lattice 32, 122, 142
totally umbilic 77, 88
*tt**-structure xxix
twisted Grassmannian 283
twisted loop algebra 164
twisted loop group 164
twistor construction 211, 214, 215, 248, 249, 273

Uhlenbeck, K. 165
unitary group 254
uniton number 214

variation of Hodge structure: see VHS
VHS 243
Virasoro Conjecture 291
Virasoro constraints 290

WDVV equations 152, 223, 225
Weierstrass representation 77, 176
Wilson, G. 182
Witten, E. 288

zero curvature condition 76, 80, 178
Zhang, Y. 242
Zudilin, W. 287

The manufacturer's authorised representative in the EU for product safety is Oxford University Press España S.A. of el Parque Empresarial San Fernando de Henares, Avenida de Castilla, 2 – 28830 Madrid (www.oup.es/en or product.safety@oup.com). OUP España S.A. also acts as importer into Spain of products made by the manufacturer.

www.ingramcontent.com/pod-product-compliance
Lightning Source LLC
LaVergne TN
LVHW011913200926
848484LV00002B/609

9780198565994